SPC and CONTINUOUS IMPROVEMENT

SPC and CONTINUOUS IMPROVEMENT

Mal Owen

IFS Publications, UK
Springer-Verlag
Berlin · Heidelberg · New York · Tokyo
1989

British Library Cataloguing in Publication Data

Owen, M. (Maldwyn), *1936–*
   Statistical process control.
   1. Industries. Quality control.
   Statistical methods
   I. Title
   658.5'62

ISBN 0-98450-795-0 IFS Publications
ISBN 3-540-50481-8 Springer-Verlag Berlin Heidelberg New York Tokyo
ISBN 0-387 50481-8 Springer-Verlag New York Heidelberg Berlin Tokyo

© **1989 IFS Ltd,** Wolseley Business Park, Wolseley Road, Kempston
Bedford MK42 7PW
UK and **Springer-Verlag** Berlin Heidelberg New York Tokyo

Phototypeset by MFK Typesetting Ltd, Herts.

Printed and bound by Short Run Press Ltd, Exeter

Contents

PREFACE

There is no doubt that quality has become a major feature in the survival plan of organisations. With diminishing markets resulting from improved competitive performance and the associated factor of single-sourcing arrangements by the major organisations, it is clear that unless there is a commitment to change, organisations will lose their competitive edge. This will unfortunately mean elimination and the resultant harsh realities that come with it for the employees.

It has been said on many platforms that unemployment is not inevitable. Those organisations which recognise the requirements for survival know that quality, and its association with customer satisfaction, is now a key issue.

Survival programmes based on quality improvement require an unrelenting commitment to include everyone, from the Managing Director down, in an ongoing, never-ending involvement based on monitoring, and improving, all our activities. These Total Quality Management (TQM) programmes, whatever their specific nature, have a common theme of measuring and then improving.

This text describes the philosophy and techniques of one type of involvement programme – Statistical Process Control (SPC). The material to follow suggests that SPC is a major element of any programme and, if properly applied, could be a complete programme in itself.

Measuring and improving means that data must be collected, used, understood, interpreted and analysed, and thereby lies the difficulty.

The West is only too aware of the reputation of the Japanese in terms of quality performance. It may not be aware of the vast training programmes in simple statistical methods which have been ongoing since the early 1950's, and which have proved a key feature in the progress of Japan to world leadership in many areas of activity.

This book covers the essential elements of these techniques. There is a constant theme of relating the techniques to actual practical applications, making use, as far as possible, of company experiences in applying the particular methods.

Genuine company experiences have therefore been heavily leant on in developing the text. The intention has been to provide a book which could be useful to as wide a readership as possible. Senior executives should find the information of value in setting up Quality Improvement programmes and, at the same time, recognise the role they must play in the SPC implementation. Equally, one would like to think that operators would find the material sufficiently well presented, in a basic practical fashion, so that they would be able to glean information and ideas which would assist them in their important task of monitoring, and improving, the processes with which they are involved on a daily basis. At the same time there should be enough material to interest the critical group of middle management/supervision who are caught in the treadmill of production targets, schedules and deliveries.

The book has intentionally not been written for the technical specialists – Quality Managers, mathematicians, statisticians and the like. There are sufficient texts available already to satisfy their requirement. The contents should appeal to those undergoing formal courses in Universities, Polytechnics and Colleges, but this remains to be seen. Educational qualifications are traditionally based on syllabuses measured more in terms of theoretical and academic standards rather than practical relevance to industry and commerce. It will take some time for this to change. Much wider issues are involved here, appertaining to the role of educational institutions to the outside world. What can be said is that in the area of statistical education/training the West is some thirty years behind Japan and it will be left to the reader to judge why this is so.

An attempt has been made in this book to eliminate, as far as possible, any unnecessary mathematics or statistical theory which interferes with understanding of how the techniques operate in practice. Again, it will be left to the judgement of the reader as to how far the book has succeeded in this respect.

Symbols and conventions constantly interfere with the process of technical understanding. No apology is made for the fact that whilst the author is British, and the case studies relate to U.K. companies, there is a heavy emphasis on the use of standard approaches which are N. American in origin. Appropriate references are provided to British Standards, but in view of the current widespread usage, on both sides of the Atlantic, of control chart techniques which stem from major U.S. organisations, it makes sense not to complicate the issue by trying to carry two different approaches.

A conscious effort has been made to widen the rather narrow traditional view of SPC as an activity restricted to machine shop operations. Administrative applications have been introduced, directly or indirectly, so that the message of a company-wide activity can be understood as relating to all areas of the organisation. It is a belief that the book has been considerably enriched by the inclusion of real case study material which has been provided by leading U.K. based organisations. The author owes a great deal to the contacts made over the years with colleagues in a variety of positions in many organisations across the U.K. and elsewhere. They have provided willing assistance and close collaboration and have been a source of encouragement and support. In formally acknowledging the organisations involved in the preparation of the book, the support of many colleagues, too numerous to mention in person, is also acknowledged with thanks.

I am particularly grateful to John Parsloe, Sauer Sundstrand, who has provided invaluable support over many months in reading and commenting on the scripts as they appeared. SPC programmes seem to bring out missionaries for the cause and John is a leading SPC disciple. His co-operation is much appreciated.

I should also like to thank colleagues associated with IFS for their considerable help and encouragement in the writing of this book.

Finally my thanks to my family. My wife has provided invaluable support in converting an untidy script to a neat word-processor output. My family have also allowed me precious time to write a book which one hopes will prove useful to the increasing numbers who are looking for material which may help them on the road to never-ending improvement.

February 1989   MO

ACKNOWLEDGEMENTS

Various diagrams appearing in the book are based on similar ones in an SPC-operator manual. The manual forms part of a training package on SPC which was developed with Further Education Unit/(DES (Pickup) funding in the course of FEU project RP320, "The development of Teaching/Learning materials in Statistical Process Control." The co-operation of FEU in allowing use of this material is gratefully acknowledged. (The package is available from the Training for Quality Unit at Bristol Polytechnic)

Thanks are due to the following organisations who have provided case-study and other material to support the various SPC techniques. Their co-operation and support is gratefully acknowledged.

British Alcan (Rolled Products), Rogerstone, Newport, Gwent
British Steel (Tinplate Group), Swansea
Century Oils, Stoke-on-Trent
Ford of Europe, Statistical Methods Office, Basildon, Essex
Ford New Holland, Basildon, Essex
Goodyear Tyre & Rubber Company, Wolverhampton
Hewlett Packard (Computer Peripherals), Bristol
ITT Cannon, Basingstoke
Jaguar Cars, Radford, Coventry
Lucas Electrical (Starter & Alternator Division), Birmingham
PPG Industries, (U.K.) Ltd, Birmingham
Sauer Sundstrand , Swindon.

Chapter 1   Introduction

1.1 **Introduction**

A change is taking place in an increasing number of organisations in the West. Heavy competition for a reduced market share, coupled with more critical demands from the customer, have resulted in a need to improve efficiency if the organisation is to survive. This change is manifesting itself in various ways. On a technical level, it means that cost is no longer the dominant factor in making purchasing decisions; neither is productivity any longer a guarantee of remaining in business. At the company level, it means the breaking down of barriers between management and operational staff, and between and within departments. For example, dining facilities for management only are gradually being eliminated and reserved car parking is being replaced by equal access areas. The major element in the survival programmes for these organisations is now quality – quality of product, service and working lives. It is evident that organisations must change in order to remain in business, and that those who do not do so will face, sooner rather than later, the harsh realities of redundancies and closures.

Various techniques, management strategies, company policies and government campaigns are involved in this programme of change. Increased employee involvement, improved management, robotics, just-in-time inventories and quality circles are all playing their part in the transformation. However, of all the programmes relating to quality improvement, those which are company-wide are having the biggest impact, and statistical process control (SPC) is amongst the front runners in this respect. For more and more organisations it is becoming a way of life. It is radically affecting the role of purchasing departments, for example. But more than anything else it is providing everyone in the organisation with a common means of communication and a focus for improvement: the control chart. It is also having far-reaching effects in that it is calling into question the part played by educational institutions in preparing employees for work. In doing so, it highlights the rather tenuous relationship between the academic world and that of commercial and industrial activity. As it is so far-reaching, it brings with it a training requirement which has probably been unmatched in recent times. Tens of thousands of people will need training or retraining and there will be a corresponding need to improve the level of other management and supervisory skills.

So what is SPC, and what makes it different from anything else? The title suggest the use of statistics, but it would be a mistake to assume that SPC is simply statistical analysis: it is much more than that. Reference will be made throughout the book to the related management philosophy and the manner in which the programme is being applied in a variety of industries and also in the various departments, manufacturing and non-manufacturing alike, within those industries.

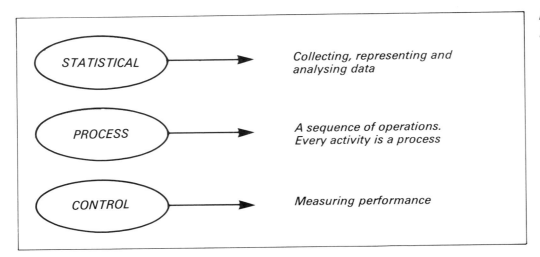

Fig. 1.1 Statistical process control

1.2 **What is SPC?**

It is appropriate to consider the three elements: statistical, process and control (Fig. 1.1).

Statistical. The word 'statistical' implies the collection, representation and interpretation of data. Statistical methods provide a means of assessing risks and predicting results. The word 'statistics' usually conjures up an image of a fearsome topic which can only be studied by those with advanced analytical skills. This misapprehension is due primarily to the effect of our Western educational system, which has over-emphasised the need for academic excellence at the expense of professional capability.

The statistical element of SPC involves the simple handling of data and the understanding of a basic chart. The technique is capable of being understood by anyone, but involves training to various levels of competence so that the techniques are fully appreciated. It appears that the word 'statistical' often evokes fears of academic, theoretical training sessions, and to avoid this, some companies have replaced 'statistical' by 'improvement', 'operator' and even 'manufacturing', with the intention of reducing the fear associated with the word 'statistical'.

However, ultimately, 'statistical process control' is the name that is used, and therefore changing the term 'statistical', for whatever justifiable reasons, brings with it other difficulties. For example, 'manufacturing process control' suggests that the activity cannot be applied outside manufacturing areas. 'Statistical' implies patterns of results, and the interpretation of such patterns, especially by operators, is a key feature of any SPC programme.

Process. Any activity is a process. On the shop floor, it is the combination of methods, people, materials, environment and equipment in the production of a component, or the filling of a container, or the assembly of a unit. However, there are also other processes, perhaps not recognisable as such at first. For example, the recruitment of new personnel, the preparation of the in-house company magazine, the routing of visitors to a certain department and the analysis of accident records are all processes, and they can be measured. The appropriate measure may not be as easily determined as that for turning a bar, for example, but nevertheless, such measures (called process performance measures, performance indicators, or metrics) can be obtained.

Understandably, SPC tends to be associated only with a chart on a machine; however, it must be seen in a wider context as being relevant to any department or any section of an organisation, whether it be manufacturing, service, education or any other.

Control. SPC involves control followed by improvement. Processes are initially brought under control and then improved by reducing the variability about the nominal or, if appropriate, reducing the level of rejects to zero. A controlled situation avoids fire-fighting: the instantaneous reaction to problems as and when they occur. Control means that planning, prediction and improvement can follow.

Control by itself is not enough. Improvement is required: improvement to a new level of performance which leads to opportunities for further improvement, creating a cycle of continuous improvement. In practice there are practical and financial limitations to the extent of the improvements but the aim is perfection.

SPC itself is new, but the message it carries is not. The techniques of SPC date back to the 1920s and were first developed as part of statistical quality control (SQC). The switch from SQC to SPC is recent, and the change of name has had a profound bearing on the sudden resurgence of interest in the statistical aspects. The SQC/SPC relationship will be considered further in Section 1.8.

Some of the background to earlier developments is given below, and this provides the framework around which current interests have been built.

1.3 SQC: from Shewhart to McArthur

1.3.1 Early developments

It is now generally accepted that the first definitive attempts to introduce the philosophy of customer satisfaction took place at the Bell Telephone Laboratories, USA. The manufacturing unit of Bell Laboratories was based at the Hawthorne plant of the Western Electric Company, the scene of early experiments in industrial motivation. In 1924 Dr R.L. Jones was appointed head of a new Inspection Engineering Department for monitoring the quality of the 10 million bakelite telephone receivers produced each year. In 1925 the department was redesignated as the Quality Assurance Department.

Under the auspices of this new department, Jones brought together a wide range of talents. Physicists, engineers, mathematicians and technical experts of all types worked together in introducing systems and techniques which would assist Western Electric to produce better quality telephone receivers, and included among the group's objectives was the development of appropriate statistical methods. Within this team of experts were young engineers and scientists whose names were to become well-known in the field of quality assurance, and one of these in particular was W.A. Shewhart.

Walter Shewhart, a physicist by training, was given the task of studying the potential for industry of applying statistical sampling methods. On 16 May 1924, Shewhart wrote to Jones giving details of the control chart concept, along the lines shown in Fig. 1.2. In 1931 Shewhart published his results in *Economic Control of Quality of Manufactured Product*. In his book he defined statistical quality control as 'the application of statistical principles and techniques in all stages of production, directed towards the manufacture of a product that is maximally useful and has a market'.

Shewhart's concern for the practical applications of his work seems to have been neglected by most of those who read his book. It is unfortunate that the statistical tools were used in isolation by many people, and the publication of his book was seen by academics as a new step forward in mathematical statistics. It was not appreciated that Shewhart was advocating the use of these techniques as part of an overall programme to improve product quality, a philosophy that is as valid now as it was in his time. Another related factor was the influence of the Taylor system of management. Western industry had developed strong lines of demarcation between management and the shop floor, and the philosophy of operator control, whilst not directly referred to by Shewhart, would have been at odds with the thinking of the day. The operator was treated as part of the manufacturing unit and all production decisions were traditionally in the hands of managers and engineers.

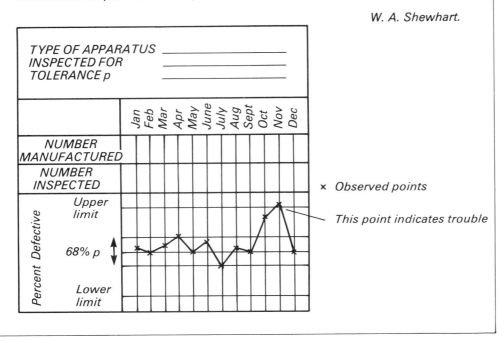

Case 18013

Mr. R. L. JONES:

A few days ago, you mentioned some of the problems connected with the development of an acceptable form of inspection report which might be modified from time to time, in order to give at a glance the greatest amount of accurate information.

The attached form of report is designed to indicate whether or not the observed variations in the percent of defective apparatus of a given type are significant; that is, to indicate whether or not the product is satisfactory. The theory underlying the method of determining the significance of the variations in the value of p is somewhat involved when considered in such a form as to cover practically all types of problems. I have already started the preparation of a series of memoranda covering these points in detail. Should it be found desirable, however, to make use of this form of chart in any of the studies now being conducted within the Inspection Department, it will be possible to indicate, the method to be followed in the particular examples.

W. A. Shewhart.

TYPE OF APPARATUS _____
INSPECTED FOR   _____
TOLERANCE p   _____

| | Jan | Feb | Mar | Apr | May | June | July | Aug | Sept | Oct | Nov | Dec |

NUMBER MANUFACTURED

NUMBER INSPECTED

Percent Defective

Upper limit

68% p

Lower limit

× Observed points

This point indicates trouble

Fig. 1.2 Shewhart's announcement of the control chart

Shewhart's emphasis on monitoring the process involved the use of sampling in a different way to that employed in acceptance sampling. Instead of accepting or rejecting a batch on the basis of a representative sample, Shewhart used smaller samples (usually of size 5 but not necessarily so) which were taken from the process at regular intervals. This resulted in control of the process and early warning of a change from the target level. It implied a move away from the traditional use of large numbers of inspectors employed to inspect bad quality out of production, and suggested a greater efficiency in the employment of staff in the manufacturing areas.

It should not be assumed that work on the practical application of statistics was confined to the USA. In the UK, Dudding and Tippett had been using the technique in the study of electric lamps and textiles respectively. However, Shewhart does merit a major claim to fame and rightly so.

1.3.2 The Second World War

The 1939–1945 war provided the impetus for the extended use of the techniques developed by Jones and his group at the Bell Telephone Laboratories. The need to provide munitions created openings for the application of the new statistical methods. Fewer staff were needed in traditional inspection roles and therefore more staff could be utilised in direct production. The use of Shewhart's ideas also meant that the quality of the product could be better guaranteed.

In the UK, the Ministry of Supply advocated the use of Shewhart's ideas in expanding the war effort. In the USA, intensive training programmes took place, led initially by Shewhart and his younger colleague Dr W. Edwards Deming. Because of the drive of those involved with early developments at the Bell Telephone Laboratories, the effect was more pronounced than in the UK. Quality control here was adopted with less enthusiasm and it was seen by many as a dispensable activity when government controls were removed at the end of the war. There was also a clear misunderstanding of the basic concepts in the UK. Statisticians pushed SQC as a mathematical exercise with no reference to management aspects, and disillusionment was the result. Deming constantly refers to this combination of lack of management involvement and over-emphasis of the academic statistics.

1.3.3 **The early post-war years**

Whatever the reasons, there was an earlier appreciation in the USA of the potential of SQC. This resulted in a rapid growth of professionalism and the formation of the American Society for Quality Control, an organisation with a strong commitment to the application of statistical techniques within the framework of quality assurance. In Europe, a number of SQC programmes were established in particular industries, notably electrical and electronic engineering. However, many of the long-established engineering companies in Britain did not appreciate the relevance of the techniques. With short run/small batch production cycles, they only saw the difficulties of finding a suitable shop-floor charting method, rather than the wider issues of a total company programme.

Meanwhile, on the other side of the world, Japan was recovering from the devastation of war. A large quantity of American capital was poured into the country to re-establish society and regenerate industry, and General MacArthur was put in charge of the project. As industrial production got under way it was not long before the products were seen to be of shoddy quality, and in many instances fit only for the scrap heap. However, Japan was faced with the need to export goods in order to survive; there were few natural resources and if the country was to have a future, it had to be based on producing goods whose quality surpassed that of the competition.

It is not possible to write a book on any aspect of quality without making reference to the resurgence of Japan as a highly successful and much admired trading nation. The story contains a lesson for everyone and it may yet be possible for organisations to adopt ideas and techniques which can be applied as well in the West as they can in Japan.

1.4 **The Japanese phenomenon**

In 1946 the Japanese Union of Scientists and Engineers (JUSE) was set up with the objective of reconstructing Japanese industry. JUSE was aware of the extensive literature on quality control and was particularly interested in Shewhart's work. It was recognised that a new, younger breed of Japanese manager was necessary, as a replacement for the older managers who were prematurely retired. These new managers were sent on missions to the West, where they visited business schools, industrial organisations and professional bodies in order to listen and learn. They ignored the worst practices, picked up the most successful ones and then adapted them for their own benefit at home. It should be noted that the Japanese were quite prepared to learn from others, whereas countries in the West, particularly the British, are seemingly reluctant to accept that other nations may have something to offer. We in turn must now copy the best of Japanese practices and adapt them for our own use.

The Japanese were looking for an external expert to advise them on their approach to improved industrial performance, and Dr Deming was invited to visit Japan shortly after the end of the war. Deming was a statistical consultant employed at the US Bureau of Agriculture. Since about 1939 he had been advocating the use of a statistical approach with a necessarily open form of management. His views were not heeded in the United States but he found a ready audience in Japan, and following a series of visits, in the early 1950s he was invited to address the top executives of Japanese industry. A second presentation followed shortly afterwards and resulted in the introduction of a comprehensive nation-wide training programme in statistical and problem-solving methods. A massive exercise of radio courses, cruises around the seas of Japan and company training courses was undertaken to provide the nation with the tools to understand simple concepts of variation. Deming's involvement was followed by that of Juran, who gave presentations on quality management aspects.

The Japanese recognised the importance of the supervisors in their training programmes by utilising them in quality circles.* The supervisors, having been trained in the techniques themselves, were also trained in leading groups. Quality circles in Japan provided the final element in the programme of company-wide training. It is worth noting that the first quality circle in the USA did not appear until 1974. At the same time the first Japanese car went on show in the USA: it was apparently so heavy and so under-powered that it had difficulty going up the ramp to the showroom. Today, only a few short years later, the Japanese are not only manufacturing high-quality cars in the USA, but first line Japanese supply companies have also moved in. It is also worth noting that the first Japanese quality circle followed some ten years after Deming's visit which had sparked off the statistical training. The importance of this earlier training tends to have been overlooked in the West.

Since the early 1960s the Japanese have gone from strength to strength. They are the market leaders in a wide range of business activities, have a well-trained work force and a proven track record. They have kept things simple in the production sense, and provided the essential management support and commitment. Organisations in the West face an immense challenge, and in order to survive they must adopt (or re-adopt) some of the Japanese approaches to product quality improvement.

In 1981 Deming appeared on an American television documentary entitled 'If Japan can, why can't we?' As a result he was asked by leading American organisations, including General Motors and Ford, to give seminars to company executives on his management philosophies. American management is now learning from the Japanese some 37 years after Deming's presentation to Japanese executives: his views are finally being listened to and gradually implemented by a growing number of organisations. In particular, Ford in the UK is insisting that its supply base must implement SPC programmes and advocating the Deming philosophy as an appropriate framework.

Deming is not the only consultant with programmes for quality improvement but his achievements to date cannot be easily ignored. His fourteen points for management provide an appropriate framework (see Appendix A), and reference will be made to these as the appropriate techniques for SPC are introduced and developed. It is ironic, even depressing, that the Japanese have overtaken the West using management techniques that were available in the USA at the beginning of the last war; it is even more ironic that the statistical basis of their management philosophy

---

* A quality circle is a group of people from the same work area who meet regularly and voluntarily to solve problems of their own choosing and to recommend solutions to management.

was developed in the 1920s by an American and used extensively in North America both during and after the war.

So what went wrong in the West? Why is it only now that organisations are listening to the views of Deming and others and acting on them? What is it about SPC that has suddenly attracted attention? Education, management systems, culture and many other influences have all played a part. Certain specific factors have also been important and some consideration of these is appropriate here.

1.5 **Why go in for SPC?**

Four main factors influence an organisation in its decision to implement SPC. These are listed in Fig. 1.3.

1.5.1 **External pressure**

For an increasing number of companies, there is little choice: they are being told by their customers to implement SPC programmes if they wish to retain custom. Harsh as it may seem, if an organisation has a large proportion of bought-out material, then it is only right that control of quality is implemented at the production stage, which is off-site. The manner in which SPC programmes are being introduced by some organisations is unfortunately still seen by sections of the supply base as reflecting an autocratic approach. Having said that, there can be no doubt about the ultimate effectiveness of SPC. Most customers, however, would surely prefer their suppliers to choose SPC programmes, rather than be coerced into them.

1.5.2 **Internal benefits**

It is far preferable for organisations to opt for a programme of continuous improvement without external pressure. The benefits of SPC are many and varied and will be obvious to most; Fig. 1.4 suggests some of them. There must be feelings of regret in many board rooms that a decision was not taken to implement SPC years ago, in that a chance was lost to gain a competitive edge.

1.5.3 **Survival**

There is fierce competition in the market place; there are a limited number of suppliers of a particular product, and the market share means that eventually, one or more of these suppliers will go out of business. Which company will it be? How will that decision be made? What can be done to influence the situation?

Change is necessary if an organisation is to carry on in business, and recognising

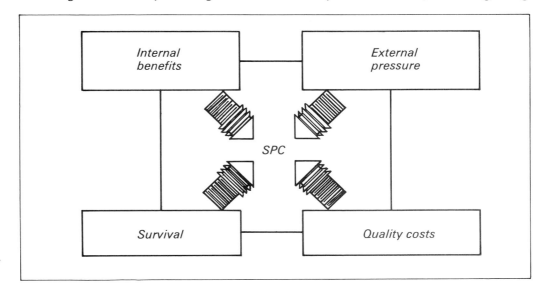

Fig. 1.3 Reasons for implementing SPC

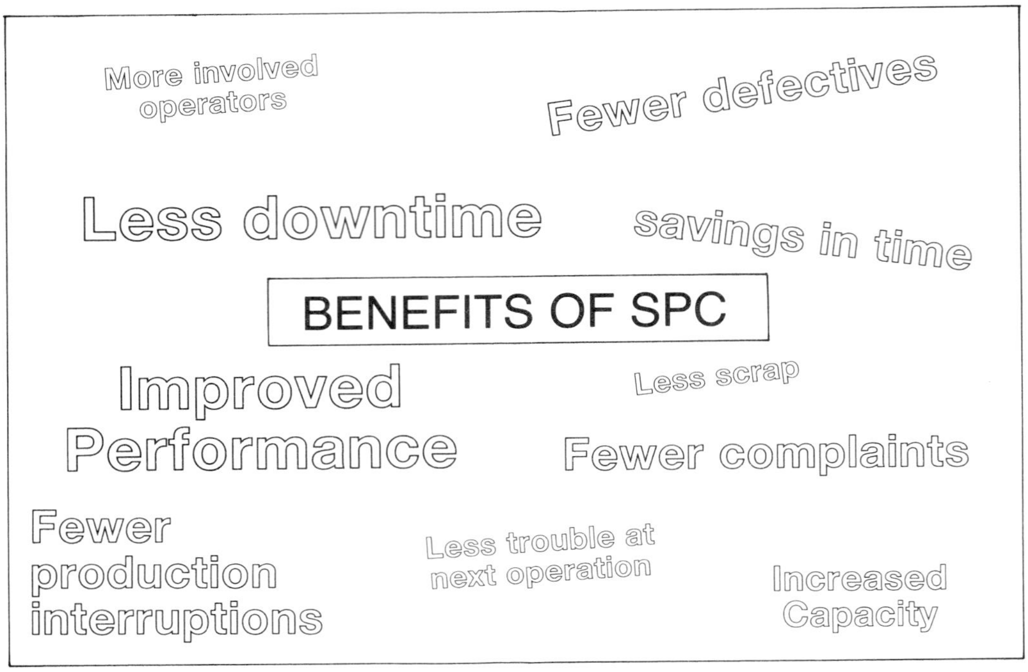

Fig. 1.4
Benefits of SPC

the need to change is the first prerequisite. This may take the form of recognising that scrap levels are far too high, that materials bought in are based on the premise of the cheapest, that company operators have not been provided with the right opportunities to use their skills and knowledge concerning the machines or processes they work with. One clear-cut way of assessing a company's performance is to look at the costs associated with quality.

1.5.4 Quality costs

Far too many companies in the UK operate in ignorance of the quality costs associated with their operation, and this was particularly so before the wave of interest in company-wide improvement programmes. The activities of the Ford Motor Company and others have meant that other organisations have had to focus more on the determination of quality costs and the subsequent necessary actions.

An initial study of a company's quality costs will show a breakdown which can be typically represented by Fig. 1.5. These results may come as a shock. External failure, such as the costs associated with warranty, complaints and returns, provides the major contribution. When these figures are related to the percentage of sales, as in Fig. 1.6, the result is even more dramatic.

The first step is for the organisation to recognise that it must cut into the external failure costs. This is done by realising that an inferior quality product must not reach the customer in the first place, and that there must be an increased emphasis on appraisal. Internal failure costs therefore increase in the short term as a result of preventing (by inspection) an unacceptable product from leaving the supplier's premises in the first place. Gradually, it is being recognised that an emphasis on prevention costs, i.e. those associated with quality planning, design and training, will result in an overall reduction in the total cost of quality.

The diagram shown in Fig. 1.7 is schematic and does not attempt to indicate either accurate figures or an indication of time scale; but sufficient evidence exists from companies who have made inroads into quality cost improvements that this generalised diagram is quite representative of typical Western organisations. Crosby, an American consultant, has used the phrases 'quality is free' and 'the gold in the mine'. It is suggested that, on the basis of Fig. 1.7, approximately l0% of sales costs can be saved by increasing prevention costs. But why stop at l0%? A pro-

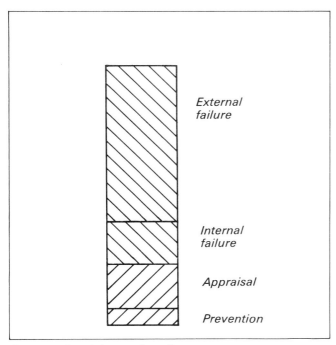

Fig. 1.5 Breakdown of quality costs

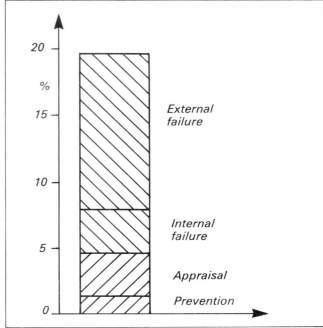

Fig. 1.6 Quality costs as a percentage of sales

gramme of continuous improvement implies a reduction of this figure to its lowest possible level.

It is a matter of some concern that for an unacceptably high number of Western organisations, the true level of quality costs is not known. Companies often continue in operation on the basis that they have made a profit, but the waste entailed in providing this profit is not known. In the same vein, many organisations have stated that they do not intend to implement programmes such as SPC because they cannot afford to buy new measuring equipment. This is a false economy. The money that has to be spent up front will result in substantial savings later. In any

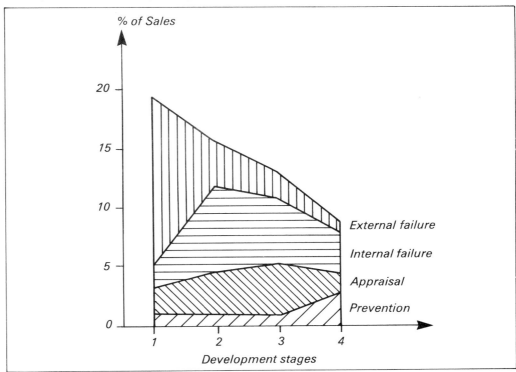

Fig. 1.7 Quality cost patterns over time

case, the option to spend or otherwise will not be there shortly because of decisions taken by the customer. SPC programmes do cost money but they pay for themselves many times over as the programmes begin to yield results.

One of the current problems in industry is the lack of expertise in determining quality costs. If higher education provided the right training, accountancy courses would be placing a far greater emphasis on this area of work. This would mean that as graduate accountants moved upwards to positions of executive authority within industry and commerce, they would be aware of the need for effective quality costing; however, this is not happening. As a result, the quality department has been attempting to calculate costs with the help of appropriate support documents such as various national standards on the determination of quality costs.

All of the four factors discussed above, and possibly others too, are influencing many organisations to adopt programmes such as SPC. These programmes are based on prevention as opposed to detection. Some further explanation of this is necessary.

1.6 **Prevention versus detection**

Traditionally, most companies have operated on the basis of detection schemes as shown in Fig. 1.8. Using a process which combines methods, people, materials, environment and equipment, action has been concentrated on the output. There has been an emphasis on inspection in order that unacceptable items may be weeded out. One hundred per cent inspection is notoriously fallible and is only roughly 80% efficient. This means that a system based on inspection depends on repeated inspections taking place to raise the good product output as near to 100% as possible. It should be pointed out that operating an inspection-based system is expensive and in many ways demotivating; furthermore, it is based on tolerances. This will be considered further in Chapter 2.

This traditional way of doing things is associated with scrap, rework, recycling, warranty claims and concessions. It is far better to use a prevention system such as that represented in Fig. 1.9. Here, a system is set up which, as far as possible, prevents defects from occurring. One has to go back to the source in order to put the system right, rather than just patch it up. This means ensuring that the design is right, but even before that it means making sure that one's interviewing procedures are correct so that the right design engineer is appointed in the first place. This in turn implies that the design engineer must have been trained correctly. This raises the issue of education and training (a topic that Deming has justifiably strong views on) which will be referred to constantly throughout this book.

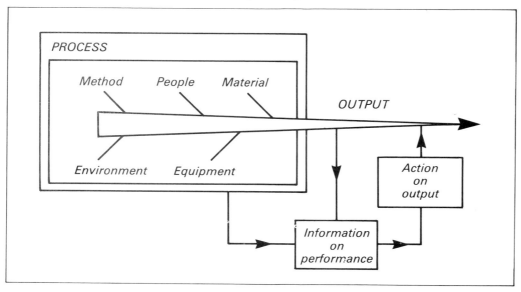

Fig. 1.8
Detection system

Fig. 1.9 Prevention system

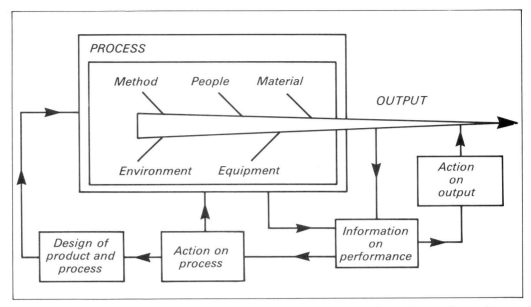

Prevention systems also imply that the operators are given the tools and possibility of access to correct the operations under their direct control. If they are not in a position to directly influence these operations, then SPC, when correctly implemented, means that they will be encouraged and expected to work with supervisors and others in correcting the process as early as possible. The control chart provides the key for instigating action.

The running of organisations using the practices of inspection cannot be justified. Costs are excessive, as shown previously, and it must be remembered that total quality costs are the result of accumulating a series of much smaller costs. Fig. 1.10 emphasises the point. For a typical production operation failure, costs rise as appropriate action is delayed. Nor is it simply a case of going back as early as possible to a particular operation on the machine: one must retrace one's steps back through the process itself, and if necessary, to the supplier of the raw materials. Getting it right the first time produces a knock-on effect: problems which have appeared further down the line disappear as soon as the faults at the earlier stage have been eliminated.

This practice of working back up the production line to remedy problems at source is a common element of SPC programmes. It is the basis on which Ford, and others, are approaching SPC programmes at the supply base. If 60% or so of a

Fig. 1.10 The problems of non-corrective action

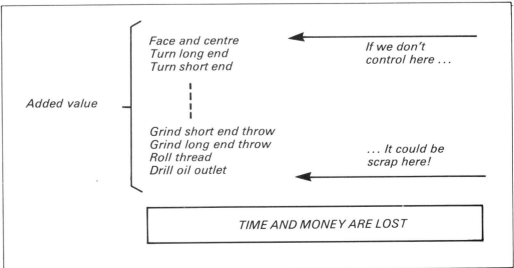

product is affected by off-site operations, it makes sense to insist that any problems are eliminated before the product arrives on the premises. This link with earlier machines is associated with the notion of the internal customer. It is useful to consider this in more depth.

1.7 Quality as customer satisfaction

It has been stated earlier that every activity is a process. Certain activities are more difficult to define as processes than others, but there is a basic theme common to all of them which is summarised in Fig. 1.11. Goods or services are received from the company's internal supplier up-line. The product received is then operated on and handed on to the customer, who in turn repeats the procedure. There is thus a linking of activities as represented in Fig. 1.12.

Process 2 is the customer for the output of process I, process 3 is the customer for the output of process 2 and so on. Hence, in order to satisfy the final (normally external) customer, there is a need to reduce the errors at each stage. 'Quality' then comes to mean 'customer satisfaction'. There are various definitions of what is meant by 'quality', but 'customer satisfaction' is a term which is easy to understand, makes no reference to specification limits and can be used in any area within the organisation.

It has been estimated that approximately 30% of everyone's daily activities are wasteful. Time is spent on checking other people's work, chasing up items which are not where they should be in the system, putting right features which have gone wrong, reworking, apologising to customers, and so on. The list is endless and represents a horrific waste of time, money and effort. If the process can be got right then this waste can be eliminated. SPC is the ideal approach to adopt because it involves controlling the process and improving it. It is in looking at all activities as processes that the key difference between SPC and SQC emerges, and a few words on the difference between the two may be useful.

1.8 From SQC to SPC

The development of the control chart from Shewhart's time heralded in the era of SQC. This approach, despite Shewhart's writings, was product-based as opposed to process-based. Industry took the view that the statistical techniques, and in particular the control chart, would be used to control the quality of the product. The charts were introduced (and very often prepared and produced) by the quality department, and were limited to manufacturing applications. Other tools were available, including sampling schemes, using the notion of acceptable quality level (AQL) as a basis for operation. This led to goods inwards departments using sampling tables to choose representative samples of an appropriate size in order to accept or reject material, as illustrated in Fig. 1.13.

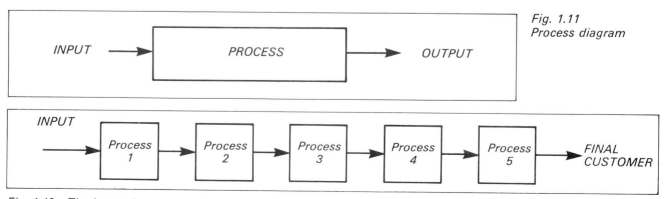

Fig. 1.11
Process diagram

Fig. 1.12   The internal customer chain

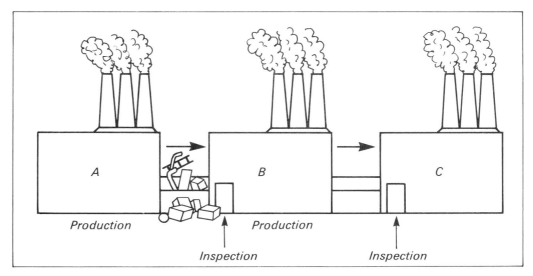

Fig. 1.13 Supplier–customer link: inspection based

The use of sampling schemes as part of control is open to question. The use of an AQL accepts a certain level of defective material as the norm and is the basis of agreement between customer and supplier. Industry has operated in this way far too often. Production at A for company B is followed by inspection at goods inwards so that an acceptable product (cleared through inspection directly and/or re-sorted) can then be produced at B for customer C and so on. This approach is at odds with the concept of continuous improvement. Instead of checking the product after it has been made, efforts should be concentrated on the control of the processes at the point of production. A's product should be monitored at the earliest stage of production so that it can then be shipped to B with a much greater degree of confidence.

There does appear to be confusion concerning the differences or similarities between SQC and SPC. Statements appear and articles are written which suggest that SPC and SQC can be interchanged at will and that it does not really matter which of the two are used; it is also assumed that there is nothing new about SPC and that it has been around for a long time. It is certainly true that the techniques of SPC have been available since Shewhart's day; however, any claim that SPC and SQC are basically the same thing implies a misunderstanding of the true interpretation of SPC. SQC is operated by the quality department with no direct reference to the operator. SPC, on the other hand, implies operator control in a genuine company-wide programme. SPC is therefore a prevention-based system operating at

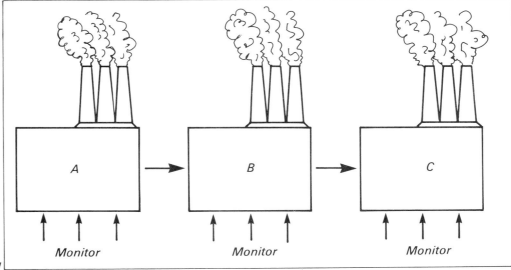

Fig. 1.14 Supplier–customer link:SPC based

an earlier stage than SQC. Whilst some of the techniques are the same, SPC does introduce an extra dimension of defining the process, controlling it and then improving it. In this respect, SPC is relatively new.

SPC and SQC will not be treated as equivalent in this book. It has been written on the basis that the process, and its improvement, is at the heart of quality improvement. Sampling schemes and AQLs will not be considered, and tolerances will only be referred to in the sense of an initial reference point from which to measure improvement. Reference has been made earlier to company-wide quality improvement programmes such as total quality control (TQC) or company-wide quality improvement (CWQI). How do these compare with SPC?

1.9 **TQC and SPC**

Both TQC and SPC are company-wide. It could rightly be argued that staff who receive visitors and ask them to sign in are part of a TQC programme. They should be courteous and welcoming because they are the first point of contact for the visitors; unacceptable behaviour here will, at best, upset the visitor and may, at worst, result in the loss of a vital contract which could have followed. So are reception staff part of an SPC programme or not?

They are certainly operators in the same way as are machinists or invoice clerks. The conventional problem-solving techniques to be discussed in Chapter 3 could apply just as well in the solution of problems in the reception area as in the machine shop. Reception staff have a check sheet (the visitor's book) which when completed gives useful daily, weekly and monthly figures which can lead to appropriate action. Information on peak attendance times, early and late enquiries, telephone calls after official hours, the number of daily visitors and the allocation of parking spaces could all be made use of to improve the service to the customer.

Care is needed in restricting SPC to only those areas in which the control chart has conventionally operated. If one accepts that SPC is company-wide, then all employees are involved and this is consistent with the definition of any activity as a process. If one starts from the TQC position, then the basis of measurement for any activity is the performance indicator, either departmental, group or individual. For example, how many telex messages were not processed at the end of the day? How many invoices were processed error-free? How many letters were successfully written? How many deliveries satisfactory? Data plotted on the performance indicator charts produces a pattern from which appropriate action may be taken, but these plotted points are essentially a form of control chart. It is only one step more to carry out the necessary calculations so that performance-based lines can be put on the chart. One interpretation of SPC, therefore, is that it adds on an extra layer of training and a greater understanding of the chart.

Both SPC and TQC use the notion of the internal customer. Both, therefore, are based on the definition of quality as customer satisfaction. Both require long-term training programmes with top management commitment. Depending on the nature of the organisation, one approach could be preferable to the other. For example, the construction of an offshore oil rig would lend itself more readily to a TQC programme rather than an SPC programme. The reverse would be true in the case of a company which was mass-producing oil pumps. Both programmes, however, are company-wide as opposed to other quality improvement schemes such as quality circles or certification schemes (e.g. BS5750/ISO9000).

Any programme for quality improvement will be a combination of the best of what is available, and adapted to suit the organisation's particular requirements. No one programme contains the complete solution, and whilst it is difficult to reject the logic behind SPC within a Deming-type framework, there will be benefits available by considering other supporting activities.

1.10 Quality, productivity and people

Fig. 1.15 provides a summary of some of the issues covered in this first chapter. It must be added that at the end of the day, a company will only survive by making a profit. More items need to be made, more customers served and more paperwork done, but these must all be of the right quality in order to satisfy the external customer. Whatever the organisation, the common denominator is the workforce. Employees must be involved, encouraged and listened to, because in order to provide a satisfactory product to the customer, the workforce must be provided with the right environment in which to work.

Fig. 1.15 also illustrates the sensitive balance between production figures and products of the right quality, and the input of the employees in producing the goods or services. The relationship between quality, productivity and people is vital to the success of the organisation, and different aspects of this will recur in the following chapters.

Summary

This first chapter has provided the background to the development of SPC and has emphasised the philosophy which allows an SPC programme to develop. A brief summary is given below.

- SPC is the use of statistical methods in monitoring and improving any activity.
- SPC should be promoted because of its benefits rather than because the customer requires it.

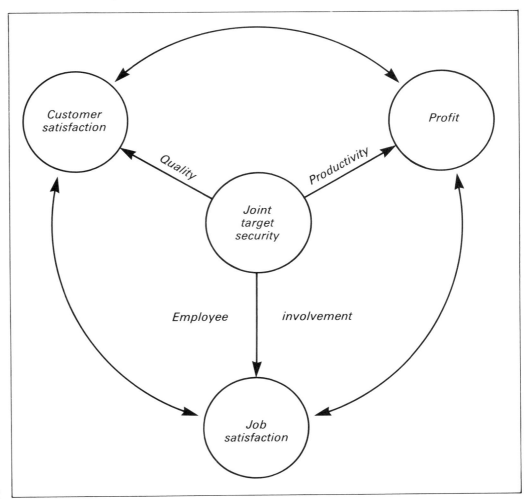

Fig. 1.15 Quality, productivity and employee involvement

- True quality costs should be determined.
- Quality is customer satisfaction.
- SPC and SQC are not the same thing.

It has been stated that a key aspect of SPC is to obtain predictable processes that produce consistent results. In order to decide whether a process is predictable or not, one needs data. A good starting point, therefore, is to look further at the data: what form it takes and how it can be handled and represented. The next chapter covers these aspects, and in doing so provides an opportunity to become familiar with handling numbers and recognising patterns.

Fig 1.2 is based on a diagram in *Managing to Achieve Quality and Reliability* by Frank Nixon (McGraw-Hill, 1971). The author is indebted to the family of the late Frank Nixon for permission to use the diagram.

Chapter 2   Data Collection and Representation

2.1   Introduction

Control of a process is only possible when information is available to indicate the state of that process. In the past, and in too many cases at present also, data has been collected for the purpose of fire-fighting. Data should really be collected so that processes can be monitored as a first step towards improving them. It is no longer adequate to talk in general terms: for example, that machine A is always breaking down, that the coffee from the vending machine seems cold, or that the car park always looks full on Thursdays. Unstatistical comments of this nature are based purely on opinions, and if appropriate action is to be taken in solving these problems, then relevant, truthful data needs to be sought. After collecting the data, it may be that what seemed to be a problem in the first place is no longer one.

In another sense, data is necessary in order to measure variation. Variation rules our lives: no two people are identical, and everyone differs to a greater or lesser extent in height, weight, colour, behaviour and so forth. The difference between one person and another may be small, but it is present and measurable. The same is true in the world of commerce and industry. No two items produced by a machine are the same; assemblies differ, reject levels change, fault figures fluctuate and absentee records alter. Excessive variation is the major cause of many problems in industry, and in order to counteract this, the nature of the variation and how to measure it must first be understood. A significant factor in the success of Japanese industry is an understanding of variability, and this has been a constant theme in Deming's work.

Having established that data is needed from the processes, it has to be said that in many organisations, data already exists. It may come as a sophisticated computer print-out, although this is often in a form which is of no real value in solving the problem in hand. At the other extreme it may only be figures recorded almost at random by an inspector in a notebook; these figures may be mistakenly assumed at some later stage to be a useful guide to what was happening. The intention was well meant but the method and analysis were worthless. Industry must have many examples of inspection-based approaches using similar unscientific attempts to keep a vague check on events. The data to be collected depends on the problem in hand and, more specifically, on what type of data it is.

2.2   Data type

Industrial and commercial data generally comes in two types, which are shown in Fig. 2.1. Continuous, or variable, data relates to quantities which are measured, whereas discrete, or attribute, data relates to counted items. Not all data necessarily takes this form. For example, there may be grading, as in food tasting where products are ranked A, B, C and so forth, depending on the results of the assessment. Data such as this is rather specialised, and therefore the analysis will concentrate on the two previously defined categories.

As the applications of SPC widen, it becomes clear that even at this very early stage there are sometimes problems to be overcome. It is not easy in some organisations to decide on appropriate measures as a basis for collecting data. The various processes may be highly technical, unlike a simple process where a hole is drilled in a plate. The problem-solving techniques which are discussed in Chapter 3 will be of assistance here, both in terms of how to define a process and how to measure it.

A key element of SPC is that any feature that is capable of being measured (as opposed to being categorised on a 'go/no-go' basis) should in fact be measured. This means that 'go/no-go' gauging as an integral part of production should now become past history. Component dimensions must now be measured so that process performance can be determined. However, this still leaves many other characteristics to be analysed as attributes: the number of dented panels, the proportion of unsatisfactory deliveries, the number of misdirected telephone calls, or the number of scratched windscreens. It may well be technically possible to devise a piece of equipment which will be capable of actually measuring the length of a scratch on the windscreen, but this is a wasteful activity because the scratch should not be there in the first place. The aim in these situations is to apply SPC in a way which reduces the number of scratches to zero.

2.3 Data collection

As a first step, thought must be given to various preparatory issues. Does the data exist? If so, is it readily available in the correct format? Where is the data to be collected from and how? How many staff are available to collect the data and who is going to do it? When will it be carried out and how? Questions such as these need resolving early on.

Specifying the time period

Decisions based on some of the points mentioned previously will affect the time period involved in collecting the data. SPC programmes and their knock-on effects to suppliers (and their suppliers in turn) are providing a very wide range of possible applications of data collection and subsequent charting. In some instances, it takes

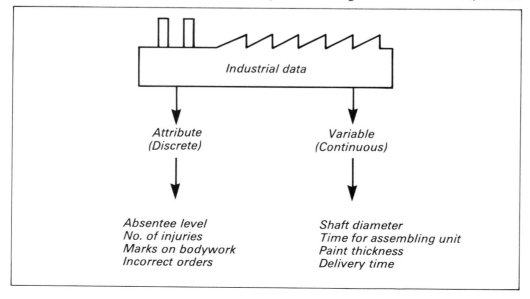

Fig. 2.1 Industrial data categories

months to collect sufficient information; for example, in the production of large assemblies. Other industries produce vast numbers in a short time; for example, millions of electrical connectors can be produced in a week.

It is clear that this extreme of time scales means attention has to be given to the way in which the data is to be recorded. The design of an appropriate check sheet thus becomes an important factor.

2.4 Designing a check sheet

The form of the check sheet will depend basically on whether it is variable or attribute data that is being collected.

For attributes, the chart must indicate the different types of fault allowed for. Fig. 2.2 shows a typical form of check sheet.

It can be seen that the check sheet allows for a number of different types of fault, as well as providing other information such as the model corresponding to the data being collected, the date and so forth. In this case, the features in the diagram relate to a packing audit system, but they could equally well be administrative figures. For

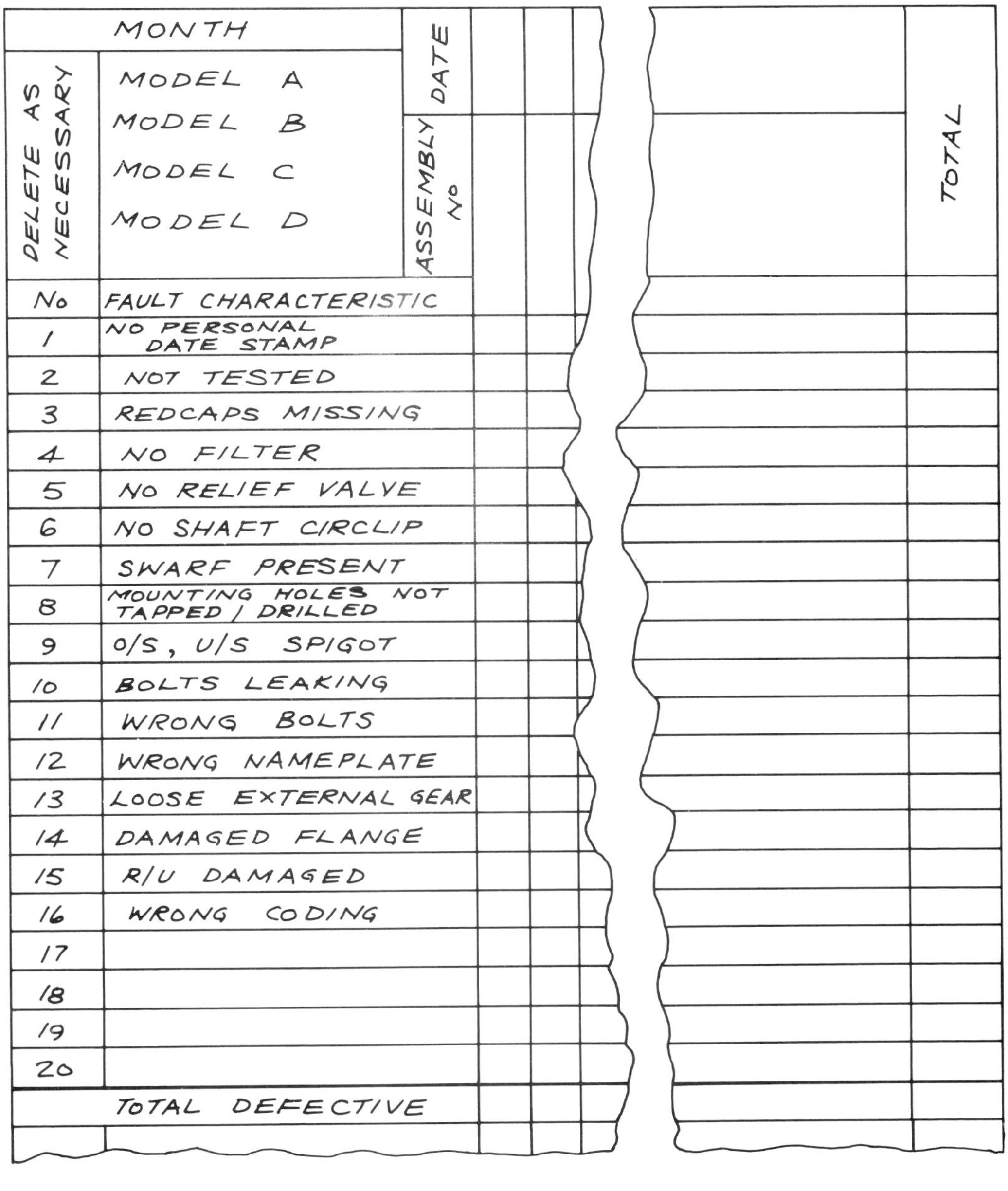

Fig. 2.2 A typical check sheet

example, Fig. 2.3(a), (b) and (c) shows how the fault characteristic column of Fig. 2.2 is amended accordingly. It is easy to overlook the importance of this form of check sheet in providing the basis for ranking priorities for action and generating data for a control chart. Reference will be made to this later, particularly in Chapters 3 and 12.

For variables, the chart should allow for recording how many values of a particular size, weight and so forth occur in the sample chosen. Usually the first step in variable analysis is simply to record the numbers as they become available. Fig. 2.4 shows a typical set of data. A hundred readings are available, but this does not mean that the sample size should always be 100. For example, machine capacity studies are usually carried out with a sample of 50, because anything less than that will not provide enough detail for a pattern to emerge. One hundred is sufficiently large to give the required picture. However, it may not be possible to obtain convenient numbers like 50 or 100, and within these limits, one should concentrate on the information gained from the data rather than worrying about obtaining 100 values.

It is not possible readily to see a pattern in Fig. 2.4. Choosing a value which represents the group in terms of its location on a scale, or its variability, is not easy. Matters can be simplified by making use of tally marks.

2.5 Tally charts

The first step is to determine the least and greatest of the set of readings: in this case, 2.495 and 2.505. The check sheet, or tally chart, allows for each number in the sequence 2.495, 2.496 . . . 2.504, 2.505 to be entered in the first column. A tally mark is then entered opposite each number as appropriate, and the marks accumulated in bundles of five to form groups of completed and part completed five-bar gates. It may be easier to separate each group of five by vertical lines and a typical check sheet for variables will then be as shown Fig. 2.5. Note that often two additional columns are included, one for the total number of readings in each category (known as the frequency) and a second for the percentage of the total frequency.

Some terms may be useful here. Consider, for example, a group extending from 2.4965 to 2.4975. The central value, 2.497, is called the class interval. The value 2.4975 is the upper class limit and 2.4965 the lower class limit. The class width is 0.001 (which is clearly also the difference between each central value).

In generating the tally chart, care should be taken to cater for the presence of wild values. For example, a reading of 2.486, if present, is sufficiently different to the rest of the readings to suggest that it does not really belong to the group. It may have occurred because of a gauge fault, an error in reading, an error in transcription,

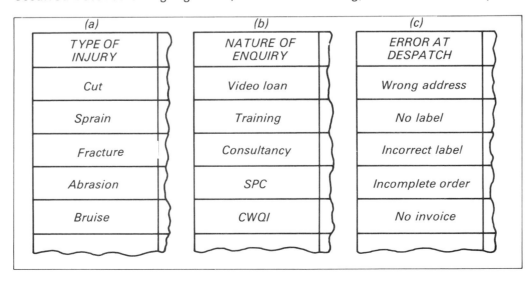

(a)	(b)	(c)
TYPE OF INJURY	NATURE OF ENQUIRY	ERROR AT DESPATCH
Cut	Video loan	Wrong address
Sprain	Training	No label
Fracture	Consultancy	Incorrect label
Abrasion	SPC	Incomplete order
Bruise	CWQI	No invoice

Fig. 2.3 Various fault characteristics

Fig. 2.4
Set of variable data

2·502	2·500	2·498	2·503	2·501	2·500	2·500	2·495	2·504	2·500
2·505	2·498	2·502	2·499	2·498	2·499	2·499	2·498	2·500	2·498
2·498	2·500	2·501	2·501	2·500	2·499	2·500	2·503	2·501	2·497
2·502	2·498	2·502	2·500	2·499	2·497	2·502	2·498	2·500	2·500
2·499	2·499	2·501	2·501	2·501	2·501	2·501	2·498	2504	2·501
2·499	2·500	2·502	2·501	2·499	2·498	2·500	2·496	2500	2·498
2·504	2·497	2·501	2·500	2·500	2·501	2·503	2·500	2·497	2·500
2·499	2·499	2·500	2·502	2·501	2·499	2·500	2·500	2·499	2·501
2·500	2·498	2·500	2·499	2·501	2·501	2·500	2·500	2·496	2·499
2·497	2·500	2·499	2·502	2·500	2·501	2·500	2·498	2·500	2·502

material at the end of the bar, or an impurity in the mixture. Whilst it is valid to exclude these wild values when drawing up the tally chart, the presence of such a value must be investigated and the cause removed.

In the example presented here, the class intervals formed themselves naturally as a result of the actual values being the central values. In some cases the range of values may be excessive and it is then appropriate to group the values in convenient classes. For example, readings at 2.496 and 2.497 would be placed in a single class extending from 2.4955 to 2.4975. The class width would then be 0.002 and the central value 2.4965.

Grouping should be avoided as far as possible because of the lack of accuracy resulting from any further calculations based on grouped figures. Improved access to calculators and computers has meant that it is now possible to carry out calculations based on individual readings very quickly. Grouping of readings should

READING	*TALLY MARKS*								*TOTAL*	*%*
2·505	/								/	
2·504	///								3	
2·503	///								3	
2·502	Ⅲℍ	////							9	
2·501	ℍℍ	ℍℍ	ℍℍ	///					18	
2·500	ℍℍ	ℍℍ	ℍℍ	ℍℍ	ℍℍ	////			29	
2·499	ℍℍ	ℍℍ	ℍℍ	/					16	
2·498	ℍℍ	ℍℍ	///						13	
2·497	ℍℍ								5	
2·496	//								2	
2·495	/								/	
							TOTAL		100	

Fig. 2.5   Check sheet
for variables

therefore be restricted to graphical presentations only. If grouping is necessary, then one must decide how many groups to choose.

As a general guide, one should choose ten intervals when setting up the tally chart. Too many intervals will produce a fussy picture with too much irrelevant detail (see Fig. 2.6). If there are too few intervals, then information is being lost (see Fig. 2.7). It is best to keep things simple. Fortunately, in many practical situations of data gathering of this nature problems of grouping do not arise.

The pattern represented by the tally marks is really a more formal representation of the build-up of the items themselves. The data in Fig. 2.4 could relate any measured quantity: component diameter, delivery time, package weight and so forth. However, the nature of the figures would indicate that component diameter is a typical variable in this case.

2.6   **The histogram**

If the components are stacked according to diameter, then the pattern of variability can be represented as shown in Fig. 2.8. The figure generated is a histogram, usually represented in its more familiar form in Fig. 2.9. For all practical purposes it

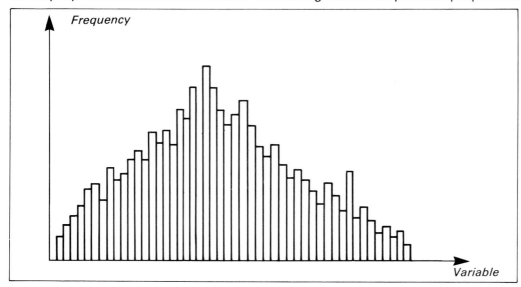

Fig. 2.6   Histogram with too many intervals

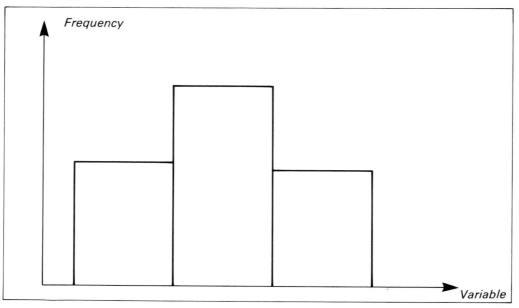

Fig. 2.7   Histogram with too few intervals

is sufficient to accept that the vertical scale is a measure of frequency. However, it will make it easier to link the practice of SPC and the statistical theory on which it is based if an extra detail is explained and this aspect will be covered in Section 2.12 at the end of this chapter.

It is very easy to under-estimate the usefulness of the histogram. It provides a quick, useful picture of what is happening and it has not been valued enough in this respect. It is used particularly in the early stages of carrying out a machine capability study (see Chapter 14), and Fig. 2.10 shows the relevant section of a typical study. Similarly, many company control charts include a section for building up the histogram over time, and Fig. 2.11 shows part of a typical control chart.

At this early stage it must be stressed that the histogram is not used in any sense as a control feature. It is used as part of a control chart only in a secondary sense to provide additional information to support the main picture appearing on the chart itself. More critically, using the histogram alone as a means of control is very misleading, because a time sequence is missing. Chapter 7 will cover this in more depth.

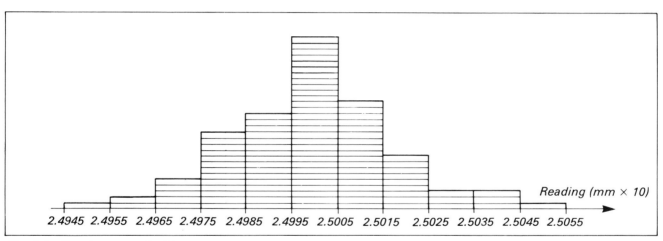

Fig. 2.8   Development of a histogram

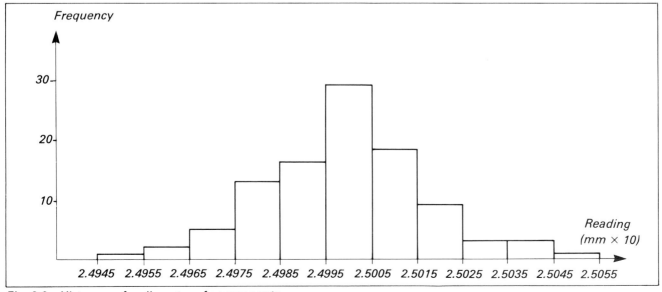

Fig. 2.9   Histogram for diameter of components

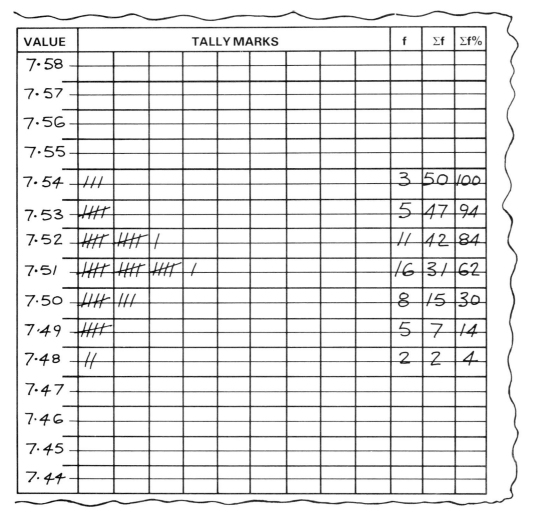

Fig. 2.10   Tally chart used in machine capability study

VALUE	TALLY MARKS									f	Σf	Σf%
7·58												
7·57												
7·56												
7·55												
7·54	///									3	50	100
7·53	ЖГ									5	47	94
7·52	ЖГ ЖГ /									11	42	84
7·51	ЖГ ЖГ ЖГ /									16	31	62
7·50	ЖГ ///									8	15	30
7·49	ЖГ									5	7	14
7·48	//									2	2	4
7·47												
7·46												
7·45												
7·44												

2.7 The histogram and capability

The histogram is concerned with capability, not control. 'Capability' means the ability of the machine or process to generate readings which, over time, build up a pattern which falls within the specification limits quoted. The histogram can be used as an indicator of performance by relating it to the upper specification limit (USL) and lower specification limit (LSL). Since variability can be expressed at this early stage by the base width of the histogram, Fig. 2.12, for example, shows that Process A is incapable, and Process B is capable.

Conventionally, the mid-point between USL and LSL is the target value, or nominal, and an obvious aim of any machine or process is to make sure that the setting is on nominal. A comparison of the position of the peak of the histogram with the nominal will therefore indicate whether there is a need for resetting. Fig. 2.13 shows that Process D is correctly set, whereas Process C requires resetting.

By considering the relationship between the histogram and specification limits, a range of possibilities present themselves. These are shown in Fig. 2.14 and raise the following questions:
- Is the setting correct?
- Has the variability improved?
- Is the process capable?
- Is there an inspection feature operating?

Fig. 2.11 Tally chart
used in process control

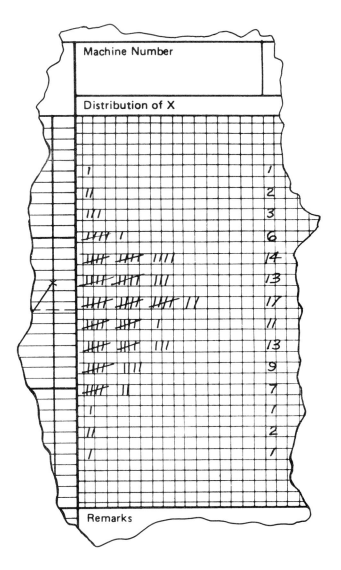

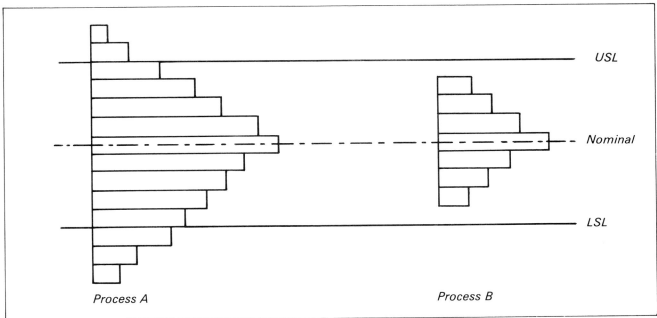

Fig. 2.12 Capability comparisons

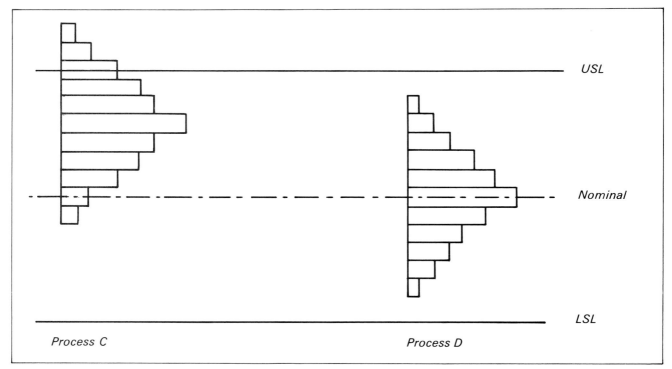

Fig. 2.13   Setting comparisons

- Is it a single head or multi-head machine?
- Have products from two suppliers been mixed?
- Are there wild values present?
- Are false figures being recorded?

The histogram provides an invaluable tool in helping to provide solutions to these and other questions. There is no denying its value as a simple indicator of performance. It is sad that its use in this way has not been emphasised more in the many statistical courses on offer, instead of the more mathematically based techniques which are promoted for academic rather than practical reasons.

2.8   Generating a histogram in practice

In some cases it is possible to produce a histogram by using the physical characteristics of the item concerned. For example, a histogram for hardness can be obtained when tennis balls, golf balls and so forth are bounced off a plate and fall into an appropriate slot, as shown in Fig. 2.15.

Similarly, electrical connectors can be categorised by width by suspending them between pins placed at increasing distances apart.

These practical, visual applications emphasise that the application of SPC must be seen to be relevant to the work area. This is particularly so at the operator level and has important implications when considering the training involved.

2.9   Simplifying the data

SPC techniques are basically simple, and they should not be complicated unnecessarily. The data in Fig. 2.4 represents the actual readings of 100 items which were being studied. They could be direct readings from a gauge, which have been recorded manually, or they could be readings from a computer print-out following the entry of each reading into the computer. Alternatively (and this is increasingly common) they could be values displayed on a screen as a result of an electronic interface.

Fig. 2.14
Various histograms

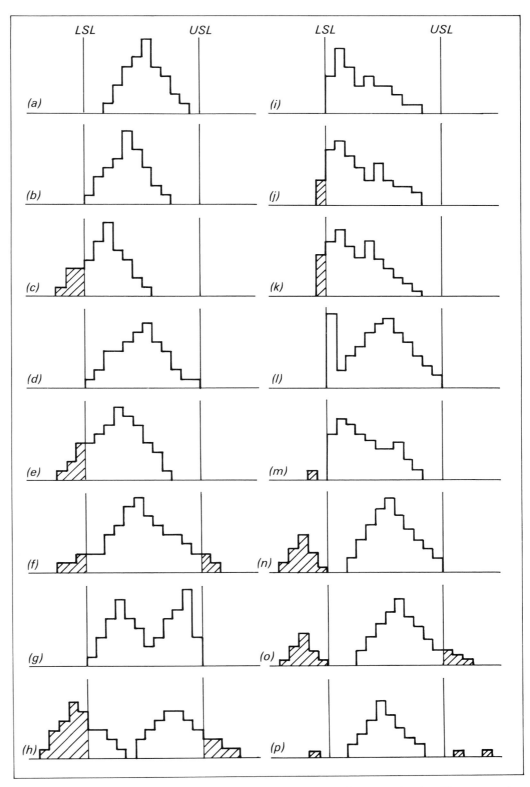

The numbers, however, are more complicated than they need to be. For example, with a gauge zero value at 2.5 units, and one gauge division equal to 0.001 units, a true reading of 2.503 becomes 3 units when coded. The corresponding scale for use in generating the tally chart becomes amended, and the new scale is shown in Fig. 2.16.

This aspect of simplifying data and working with easier numbers should not be

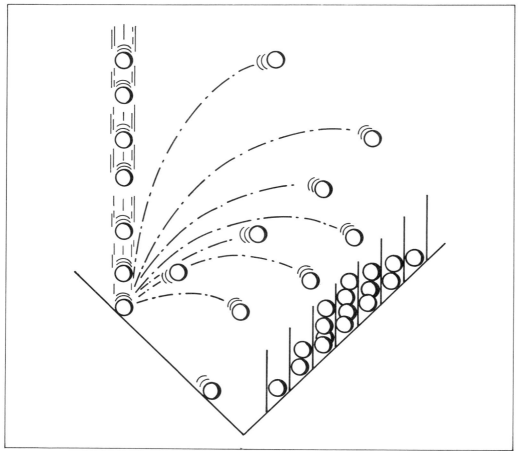

Fig. 2.15   Visual
generation of variability

overlooked, particularly in relation to shop floor training. One of the major barriers
in introducing SPC to operators is their lack of mathematical competence and
confidence, and therefore anything which can make things simpler is to be encouraged. The limited mathematical ability of many operators is not really a reflection
on them but rather a criticism of the education system to which they have been
exposed. This raises training issues, and these will be considered in more detail in
later chapters.

2.10 Limitations of tolerance-based systems

SPC entails the use of performance-based systems, rather than tolerance-based
systems. It will be easier to see why if a brief reference is now made to the rationale
of systems which have depended on the tolerance.

Conventionally, 'go/no-go' gauge limits are set up which correspond to the upper
and lower specification limits for the product. The first one or two items are taken
and measured, and a decision is made regarding the acceptability of the process
setting based on where these two readings fall. If they are near a limit, then the
process is reset; but the decision to reset is based on a minimal amount of
information and there is no real idea as to the amount of variability present.
Resetting may be done in error, which could possibly lead to further resetting, and
so on into a cycle of endless 'knob twiddling' rather than continuous improvement.

There are risks involved in the choice of decision at any stage. Should one adjust
the process or let it run? Should one stop the machine because there may be trouble
when in fact there is not? These risks can be minimised, but only if data is available
which allows sensible decisions to be made. This is not the case with systems based
on tolerances, which are open to question on several counts:

Fig. 2.16   Use of coding

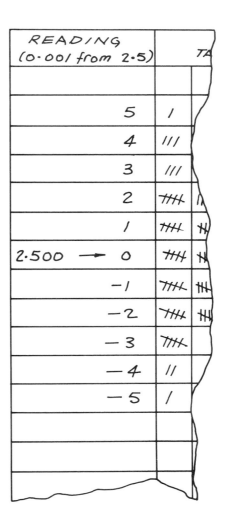

READING (0·001 from 2·5)	TA
5	/
4	///
3	///
2	︴︴
1	︴︴
2·500 ⟶ 0	︴︴
−1	︴︴
−2	︴︴
−3	︴︴
−4	//
−5	/

- Processes never improve using tolerances. The most that can be hoped for is that process performance does not deteriorate, which is indicated by an increased amount of out-of-tolerance product. Any process improvement which may actually be present through reduced variability is not indicated using control systems based on the tolerance. There may have been some justification for using specification limits for control purposes if they are reassessed at regular intervals, but this does not appear to be the case. Once set up the specification limits remain fixed.
- The tolerance value itself is open to question. The choice of USL and LSL at the design stage does not reflect sufficiently the ability of the processes downstream to produce a product which falls within specification. SPC programmes provide a focus to break down these barriers between design and production.
- Processes are made capable by depending on inspection to remove the out-of-tolerance product. The inspection activity is akin to applying a pair of sharp knives, at a constant unchanging distance apart, to cut through blocks of cheese.
- This dependence on tolerance results in problems when items from different groups are put together in an assembly operation. The results are familiar: car doors do not fit the apertures allowed for them, variation in glue and veneer thickness means a lack of fit with layered items, pistons and cylinders have to be graded, and so forth.
- Finally, tolerance-based systems largely ignore the real measurable data that is present.

Tolerances bring with them the familiar scene of inspection, gauging, rework, rejects, warranty, concessions.

The dangers of using tolerances as a means of assessing performance can be seen in Fig. 2.17. No distinction is made between the six cases because all of them produce items within specification limits, and depend on inspection to obtain good products.

The only sensible way to proceed, therefore, is to use the information provided by the process. The process is measured and then improved by constantly reducing the variability about the nominal. Fig. 2.18 illustrates the point. Getting rid of the tolerance and concentrating on the nominal will have a profound effect and require a new way of thinking in British industry, which for years has operated on a principle of producing within specifications. These ideas are certainly not restricted to manufacturing areas; the process industry has operated in a similar way.

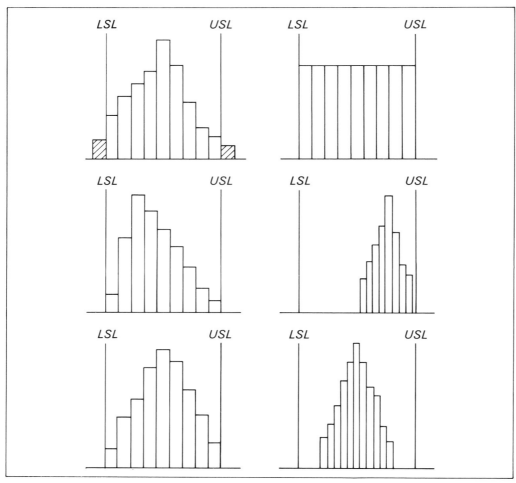

Fig. 2.17 Limitation of the tolerance

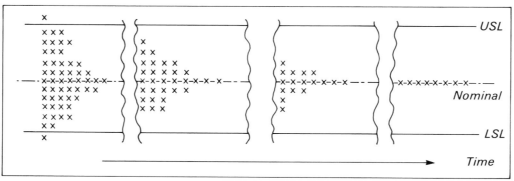

Fig. 2.18 Reduction in variability around the nominal

Plating Shop Process Check Sheet

		Issue		Work Centre		Date		Checked by	

description	check	0630	0730	0830	0930	1030	1130	1230	1330	1430	1530	1630	1730	1830	1930	2030	2130	Specified range	Corrective action
degrease	temp		72°	72°	72°	72°	72°	72°	72°	72°		72°		72°	72°	72°		68-74 C	
	level	R	✓	✓	✓	✓	✓	✓	✓			✓		✓	✓	✓			
electropolish	temp	U	17°	19°	20°	21°	23°	23°	23°	23.5°		24°	B	25°	25.5°	25°		15-25 C	
	level	N	✓	–	–	–	–	–	–	✓		✓	R	✓	✓	✓			
spray	function	N	✓	–	–	–	–	–	–	✓		✓	E	✓	✓	✓			
Cathode clean	level	U	✓	–	–	–	–	–	–	✓		✓	A	✓	✓	✓			
spray	function	R	✓	–	–	–	–	–	–	✓		✓	K	✓	✓	✓			
activate	level		✓	–	–	–	–	–	–	✓		✓		✓	✓	✓			
spray	function	G	✓	–	–	–	–	–	–	✓		✓		✓	✓	✓	G		
tin-nickel	temp	N	71°	72°	72°	71°	71°	71°	72°	71°		70°		70°	71°	71°	N	70-74 C	
	pH	I	2.9	2.9	2.9	2.9	2.9	2.9	2.9	2.9		3.0	B	2.9	2.9	2.9	I	2.5-3.0	
	level	N	✓	–	–	–	–	–	–	✓		✓	R	✓	✓	✓	N		
sump level	function	N	✓	–	–	–	–	–	–	✓		✓	E	✓	✓	✓	N		
sprays	function	U	✓	–	–	–	–	–	–	✓		✓	N	✓	✓	✓	U		
activate	level	R	✓	–	–	–	–	–	–	✓		✓	N	✓	✓	✓	R		
sprays	function		✓	–	–	–	–	–	–	✓		✓	I	✓	✓	✓			
gold strike	temp	T	58°	59°	59°	60°	59°	59°	59°	60°		60°	D	59°	60°	59°	T	25-35 C	
	level	O	✓	–	–	–	–	–	–	✓		✓		✓	✓	✓	O		
sump level		O	✓	–	–	–	–	–	–	✓		✓	D	✓	✓	✓	O		
main gold	temp	N	61°	62°	61°	62°	60°	62°	61°	61°		62°	I	61°	61°	61°	N	55-65 C	
	function		✓	–	–	–	–	–	–	✓		✓	N	✓	✓	✓			
rinses	level		✓	–	–	–	–	–	–	✓		✓	N	✓	✓	✓			
stripper	level		✓	–	–	–	–	–	–	✓		✓	E	✓	✓	✓			
rinses	function		✓	–	–	–	–	–	–	✓		✓	R	✓	✓	✓			

Fig. 2.19 Check sheet as used in a plating process

Fig. 2.19 illustrates a typical check sheet used by inspection staff in a plating process. The procedure is to take a reading of a particular parameter, and to simply tick if the reading is within the limits specified. Continuous improvement requires control charts to be set up for the parameters (though further investigation may well show that some parameters are more critical than others and this will determine the priority areas).

2.11 **Non-symmetrical patterns**

It is a mistake to assume that all patterns are symmetrical. Some characteristics, by definition, cannot produce figures which approach a symmetrical form. Fig. 2.20(a) shows the pattern produced when figures are taken from a process for which negative readings are not possible. Eccentricity of a tube, or parallelism between two surfaces, will provide a histogram for which the peak is displaced to the left of the position of symmetry. This is known as 'positive skewness'. The influence of check weighers in the weight control of prepackaged goods may also give rise to positive skewness.

Fig. 2.20(b), on the other hand, shows a negatively skewed distribution, one where the peak is displaced towards the right from a position of symmetry. The figures could correspond, for example, to lengths of a cut bar, with an end stop present. Generally skewness of this type occurs in practice when there is some physical or other upper constraint on the process.

There are techniques available for analysing non-symmetrical patterns and reference will be made to these later. It is important to recognise that non-symmetrical patterns can exist in their own right, and that such patterns may also occur at times when a symmetrical pattern is expected. The control chart may provide an indication of the presence of the latter, and investigation will be required to determine the cause.

2.12 **Frequency and area**

Fig. 2.9 represents a histogram where the vertical scale corresponded to the frequency. It has been stated that from a practical viewpoint this is quite adequate. However, confusion may arise later if further explanation is not provided at this early stage. Strictly speaking, the vertical scale is chosen so that the area of each rectangle of the histogram represents the frequency. This is achieved by using a vertical scale of frequency/class width, known as the frequency density. Fig. 2.21 illustrates the point. This provides a stepping stone between practice and theory when considering the properties of the normal distribution in Chapter 5.

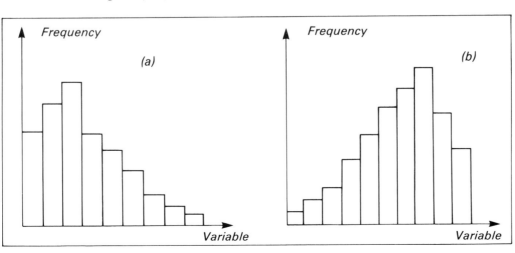

Fig. 2.20 Positive (a) and negative (b) skewness

Fig. 2.21
Frequency as area

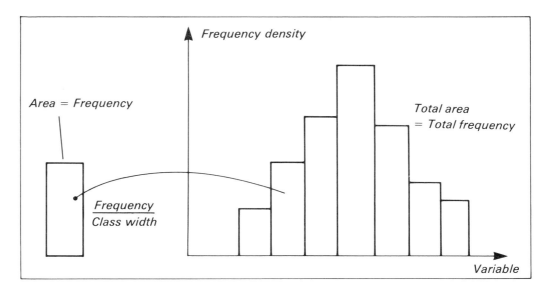

Summary

- Collect data for control and improvement not as part of fire-fighting.
- Variation cannot be measured without data.
- Data may be in attribute or variable form.
- If the feature can be measured (usefully), it should be measured.
- Data collection requires organisation.
- Check sheets simplify the collection of data.
- Avoid grouping on tally charts if possible.
- Choose about ten class intervals on the tally chart.
- Use the histogram as a useful form of graphical presentation.
- The histogram indicates capability, not control.
- Some characteristics generate a visual histogram.
- Simplify the data if possible.
- Discard the tolerance as a means of control.
- Emphasise the nominal as a target for never-ending improvement.
- Not all processes give a symmetrical pattern of readings.
- Take care when using frequency as a scale on the histogram.

This chapter has looked at data collection and some ways of presenting the information. It is not an exhaustive account, and there are many types of graphical presentation available to supplement the material covered here. No attempt has been made to cover these but that does not mean that bar charts, pie graphs and similar methods are not important Use should be made of these to simplify figures and enhance presentation.

Within the last few pages, reference has also been made to setting and variability of processes and in many ways it is now logical to consider how these can be measured. Before doing so, in Chapter 4, it is appropriate to look at some less statistically based techniques which can be of immense value in the SPC programme. There is, therefore, justification in introducing these techniques at this stage. They follow naturally from the check sheet and, in addition, they provide a break before going into the statistical SPC tools in some depth.

Chapter 3   Problem-solving techniques

3.1   Introduction

The Japanese have been making use of simple problem-solving techniques since the early 1950s and are certainly responsible for the introduction of one technique to be considered shortly: the fishbone diagram. Perhaps more than the statistically based techniques, these problem-solving methods fitted readily into the Japanese industrial environment. With less barriers operating between departments and sections, and management and workforce, methods of problem solution which, to some extent, involved group activities came naturally.

The three techniques discussed here have been associated with quality circles, but it would be a mistake to believe that this should always be so. The techniques are just as powerful, and probably more powerful, as part of an SPC programme than they are when limited to quality circle activities. This is not meant in any way as a criticism of quality circles. Indeed, those organisations which have effectively introduced quality circles with the necessary management commitment have seen them provide an appropriate environment for introducing SPC. Several of Deming's points, such as those relating to breaking down barriers, removing obstacles and providing training, will therefore have been worked on, making it that much easier to then introduce and practise the philosophies of SPC. In addition, shop floor training will already have been carried out, some of it highly relevant to SPC programmes. Goodyear Tyre and Rubber Co. at Wolverhampton is typical of an organisation which has made a ready transfer from quality circles to SPC.

Problem-solving tools are different, exciting and effective. They are basically very simple but at the same time relate directly to the everyday activities of the group concerned – and hence their attraction.

3.2   The Pareto principle

An analysis of faults, errors and defects always shows a pattern, as represented in Fig. 3.1. The pattern becomes apparent when relating the number of different types of errors and their contribution to the problem as a whole. A ranking is evident in that the first two categories of fault, A and B, provide the bulk of the total number of faults. This means that if the total number of errors is to be reduced, it makes sense to tackle those categories making the greatest contribution to the problem.

The Japanese have made use of this approach but the principle was originated by Vilfredo Pareto, an Italian economist who studied, amongst other things, the distribution of income and wealth in Italy at the turn of the century. He found that the

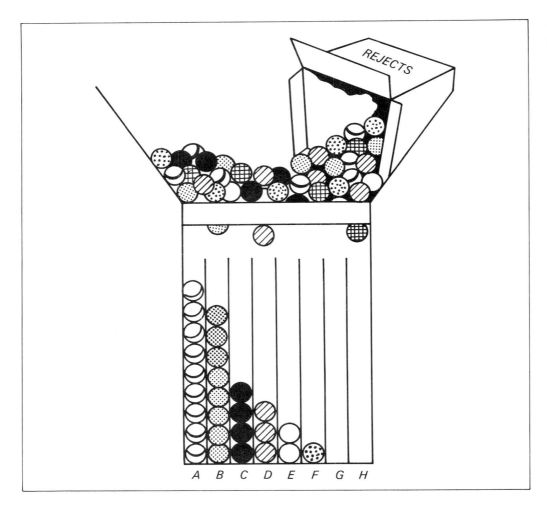

Fig. 3.1   Typical pattern of faults

vast majority of the wealth in society was held by a small percentage of the population.

Fig. 3.1 showed that at the time of analysis only six of the categories of fault were relevant. As time goes on and more data becomes available other categories will begin to register, and the pattern tends towards that shown in Fig. 3.2. It can be seen that with eight categories of fault, the first two contribute by far the greatest number to the total. A useful guide is to make use of an 80/20 rule, i.e. 80% of the problems come from 20% of the fault categories. In other words, 80% of the area enclosed within the Pareto diagram is approximately covered by A and B. This variation on the Pareto principle is attributed to Alan Lakelin, an authority on time management. Lakelin argued , for example, that 80% of sales originated from 20% of customers and that this 80/20 rule provided a simple working tool for deciding priorities.

There is danger, however, in placing too much emphasis on an 80/20 relationship. There is no reason why the figures should not read 70/30, or 60/40, or even 70/40. (Fig. 3.2, for example, represents a 75/25 division.) An exact numerical relationship is not as important as recognising that the diagram shows the 'vital few and the trivial many', as proposed initially by Juran. He quotes the fact that in personnel statistics a small percentage of employees account for most of the absenteeism. Similarly in budget analysis a small number of items account for the bulk of the expenditure. On a personal note, the majority of any one person's telephone calls are to a relatively small number of the contacts in their address book.

In an industrial context the majority of failures, rework, downtime and other quality features are attributable to a relatively few failure categories, scrap types,

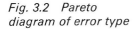

Fig. 3.2 Pareto diagram of error type

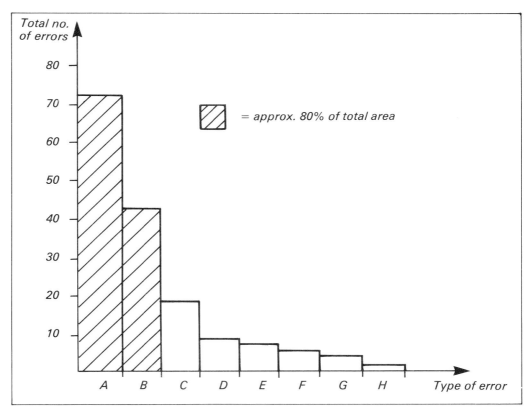

rework classifications, reasons for downtime and concessions. Credits to customers tend to be associated with one or two major features. A relatively few keyboard operators are responsible for most of the keyboard errors. A breakdown of a range of industrial problems shows the same common theme.

The following sequence is suggested when carrying out a Pareto analysis.

3.2.1 Steps to be followed

(a) *List all features*

A comprehensive list should be generated of all the possible features under consideration. Items should not be excluded because it is felt that they may be insignificant. The list may be extended, or even shortened, at a later stage once the data starts to build up, but at this first stage of the analysis an open mind is required when generating the list of items.

(b) *Collect the data*

Using a check sheet based on the list of features, the data should then be collected.

As an example, the check sheet shown in Fig. 2.2 will be used. Data was collected for Model A over a fixed time interval and the results were as shown in Fig. 3.3. It can be seen that following data collection, two extra categories of fault (17 and 18) are now required. In addition, for many of the fault categories no rejects were found. Care must be taken not to categorise too early. Possible fault categories would have been omitted from the list if it had been felt that they were not relevant or appropriate.

(c) *Rank the priorities*

The information from Fig. 3.3 can now be condensed and rearranged in order of frequency of occurrence, as shown in Fig. 3.4.

DELETE AS NECESSARY	MONTH							TOTAL
	MODEL A ✓							
	MODEL B	ASSEMBLY DATE						
	MODEL C							
	MODEL D		B/324	D/517				
No	FAULT CHARACTERISTIC							
1	NO PERSONAL DATE STAMP							1
2	NOT TESTED							
3	REDCAPS MISSING							
4	NO FILTER							
5	NO RELIEF VALVE							
6	NO SHAFT CIRCLIP							
7	SWARF PRESENT							
8	MOUNTING HOLES NOT TAPPED / DRILLED							
9	O/S, U/S SPIGOT							
10	BOLTS LEAKING							
11	WRONG BOLTS							
12	WRONG NAMEPLATE		11	11				35
13	LOOSE EXTERNAL GEAR							
14	DAMAGED FLANGE							3
15	R/U DAMAGED							
16	WRONG CODING							
17	WRONG FLANGE			1				15
18	BODY SCORED							1
19								
20								
	TOTAL DEFECTIVE		2	3				55

Fig. 3.3 Completed check sheet

FAULT	FREQUENCY
Wrong nameplate	35
Wrong flange	15
Damaged flange	3
No personal/date stamp	1
Body scored	1
Total	55

Fig. 3.4 Ranking of faults

(d) *Construct the diagram*

A Pareto diagram can now be drawn. It consists of a series of rectangles the heights (and areas) of which in sequence correspond to the frequencies.

The diagram in Fig. 3.5 is typical of those obtained in a variety of situations from personnel figures to scrap values, assembly faults to sales figures. In this case 'wrong nameplate' is the main problem and the one to be concentrated on. What is not so obvious is the relative contribution made by each category, and for this reason the Pareto diagram can be enhanced as follows.

(e) *The cumulative line*

An added refinement, which brings out the 80/20 approach, is obtained by first calculating the cumulative percentage frequencies as in Fig. 3.6. A new Pareto diagram can now be drawn where the vertical scale is a percentage of the total frequency (Fig. 3.7).

In addition, cumulative percentage figures can be plotted as shown in Fig. 3.8 and the cumulative line drawn. The common vertical axis (in %) is now interpreted as a linear figure for the Pareto diagram and a cumulative figure for the cumulative line. The cumulative line makes it easier to see the build up of the contributions made by the different features.

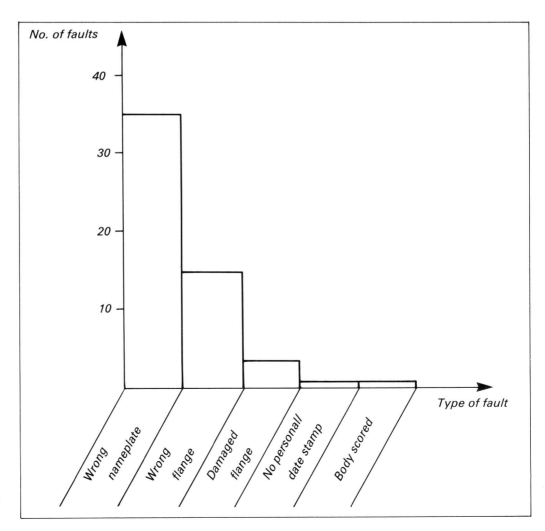

Fig. 3.5 Pareto diagram of frequency of fault occurrences

FAULT	FREQUENCY	% OF TOTAL	CUMULATIVE %
Wrong nameplate	35	63.6	63.6
Wrong flange	15	27.3	90.9
Damaged flange	3	5.5	96.4
No personal/ date stamp	1	1.8	98.2
Body scored	1	1.8	100.0
Total	55	100.0	

Fig. 3.6   Cumulative percentages

In some cases the cumulative line may be drawn in conjunction with a Pareto diagram based on actual values, rather than cumulative percentages. Fig. 3.9 shows such a combination. Care must be taken in such cases to interpret the common vertical scale correctly.

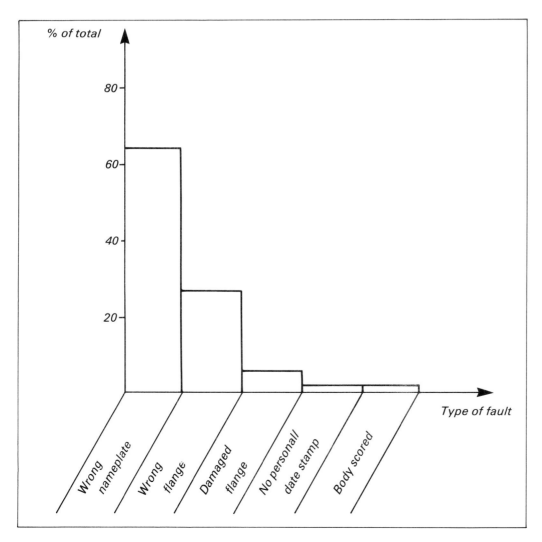

Fig. 3.7   Pareto diagram of percentage of fault occurrences

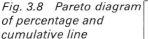

Fig. 3.8   Pareto diagram of percentage and cumulative line

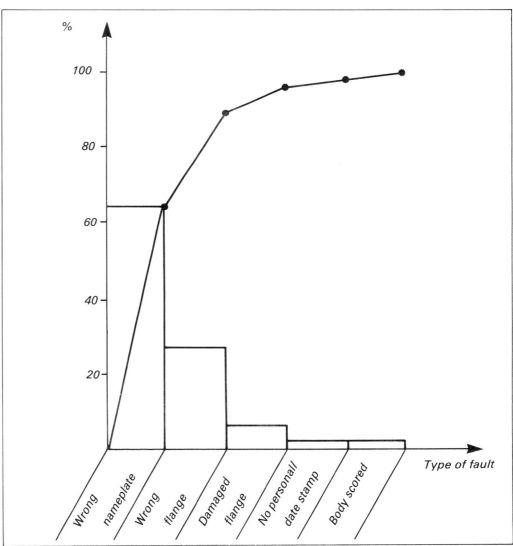

(f) *Interpret the results*

Fig. 3.7 will provide information for progress. It is evident that there is a need to concentrate on wrong nameplate as the major problem, but before taking action it would be advisable to consider other aspects relating to the Pareto diagram.

3.2.2   Other features

(a) *Choice of correct data*

The Pareto diagram in Fig. 3.7 has been based on frequency of occurrence. 'Wrong nameplate' occurred more times than any other fault and therefore is the biggest problem. However, other options for setting up the Pareto diagrams are available, as shown in Fig. 3.10. Cost is very often chosen as the appropriate factor in carrying out a Pareto analysis. Corresponding rankings will almost certainly be different. Fig. 3.11 compares Pareto diagrams based on frequency and cost.

A Pareto diagram based on cost is extremely useful in that it can direct attention to the areas where cost-effective improvements could be made by introducing control charts. There are various reasons for implementing a programme such as SPC, and all of these have a common theme of money. Whilst accepting that employee morale will improve, communications be enhanced, barriers diminished, and so forth, the organisation must nevertheless make a profit to remain in

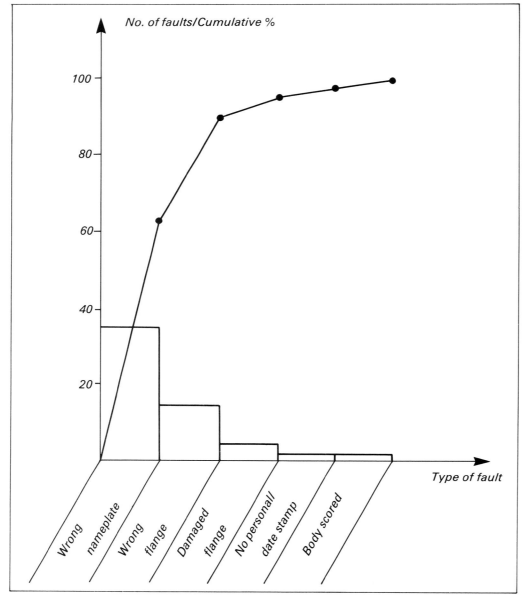

Fig. 3.9   Pareto diagram of number and cumulative line

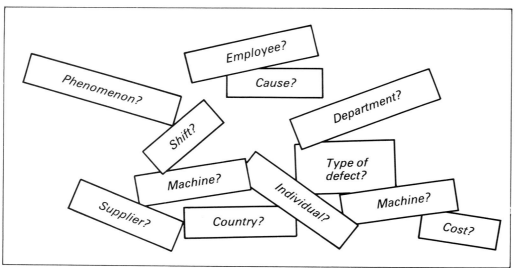

Fig. 3.10   Various features for possible analysis

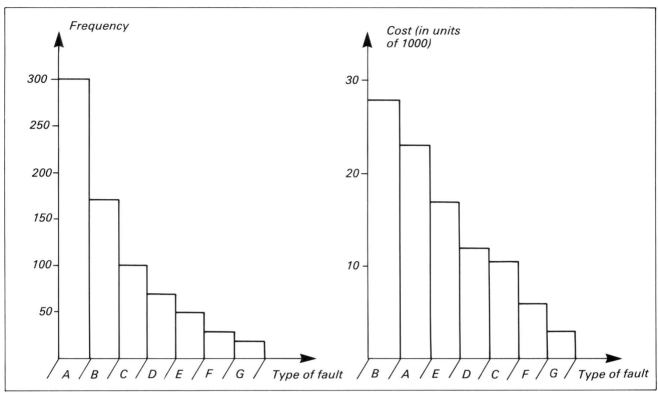

Fig. 3.11   Pareto diagrams comparing frequency and cost

business. Hence the usefulness of the Pareto diagram in this respect.

The other options for the Pareto analysis should not be overlooked, however. Fig. 3.12 shows the same check sheet as before but this time completed for Model B, and over the same time period. Fig. 3.13 shows how the frequencies, and therefore the rankings, can change by altering the basis on which the Pareto analysis was carried out. Priorities for action would now change.

In the same way as the feature to be analysed can vary, then similarly there may be cases where weighting is necessary. There could be a requirement, for legal purposes for example, for certain frequency values to be adjusted according to a weighting factor. Safety requirements might similarly demand a weighting. This will obviously affect the format of the Pareto diagram.

(b) *Showing improvement*

As a graphical technique the Pareto diagram is effective in showing process improvement. For example, Fig. 3.14 shows the effect of a quality improvement programme on the process, using as a base the cost figures in Fig. 3.11.

(c) *Number of fault categories*

Industrial data, particularly in computer format, will often present the information so that perhaps 20, 30 or even more categories of fault or error are recorded. In such cases a common practice is to concentrate on the top ten features and graph the remainder under 'others'. This will avoid a Pareto diagram with an unnecessarily long tail. If, at a later stage, the 'others' column on the diagram becomes larger than the previous column when ranked, then some readjustment of the classification is necessary.

This 'top ten' approach can also be used when considering the introduction of SPC within the supply base. Ford, for example, has insisted on all its suppliers introducing the programme. Smaller companies may feel that they have not the

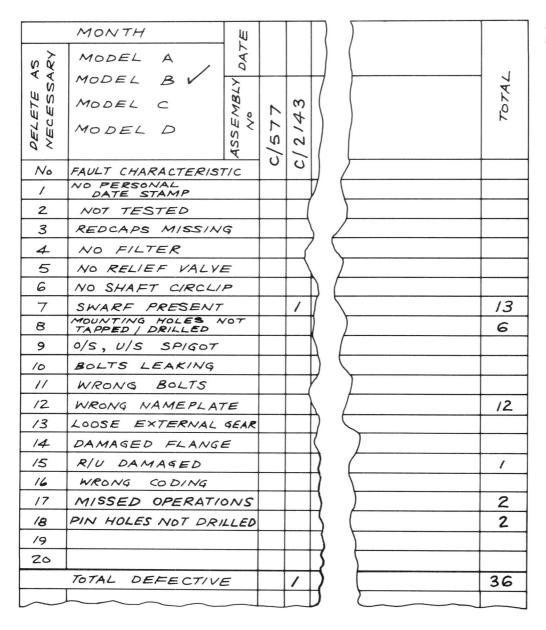

Fig. 3.12   Completed check sheet

resources, let alone the authority, to lay down such requirements. In such cases it may be appropriate to concentrate on the top ten suppliers (by percentage of expenditure) when SPC implementation is being considered.

(d) *Subclassifications*

In some instances the information provided by a Pareto analysis may not be detailed enough. The detail required may then be obtained by carrying out a further Pareto analysis on each major problem in turn. Fig. 3.15 shows how this approach assists in analysing figures for minor injuries occurring in a plant over a 12-month period.

Extracting the information from a problem by using a Pareto analysis in this way is a useful technique. It also assists in resolving a common difficulty which is that of defining the problem correctly in the first place. The problem is not always clear and it is worth making sure at an early stage that the problem as generally understood is the one which is being specified and, hopefully, solved.

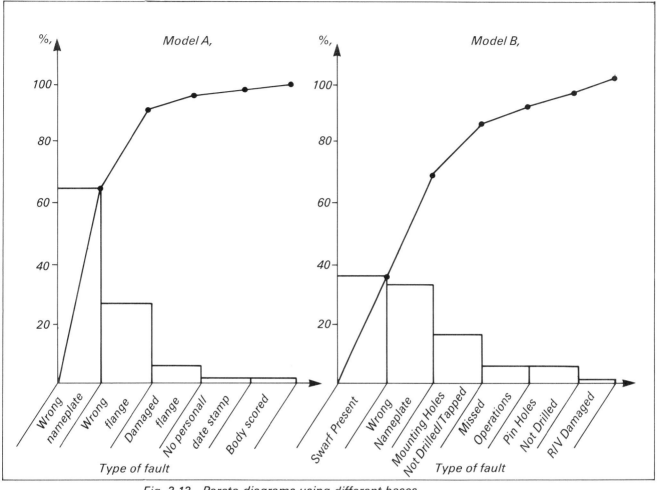

Fig. 3.13   Pareto diagrams using different bases

(e) *Simplifying things*

The various steps in listing features, drawing the Pareto diagram and constructing the cumulative line can be neatly considered in one diagram. Fig. 3.16 is based on an approach used by a Japanese organisation. As with most things they turn their hand to, the Japanese seem to have found a neat, simple way of achieving the end result.

The usefulness of the Pareto principle cannot be emphasised enough. It often provides surprising results in that what were felt to be major problems prior to a Pareto analysis are then found to be minor ones. In so doing it directs attention to the major problem to be tackled. However, there is a difference between concentrating on trivial problems at the expense of major ones, and solving the trivial ones, if convenient, to get them out of the way. It makes sense to eliminate really minor irritating items whenever possible.

Once the key problem has been highlighted it needs to be tackled. Reference to Fig. 3.4, for example, suggests concentrating on the problem of 'wrong nameplate'. What is the best way to proceed? One way is to give free rein to ideas as to how the high level of 'wrong nameplate' items can be reduced. Ideas as to the cause of the problem need to be generated then handled in a structured way. Generating ideas is the basis of brainstorming.

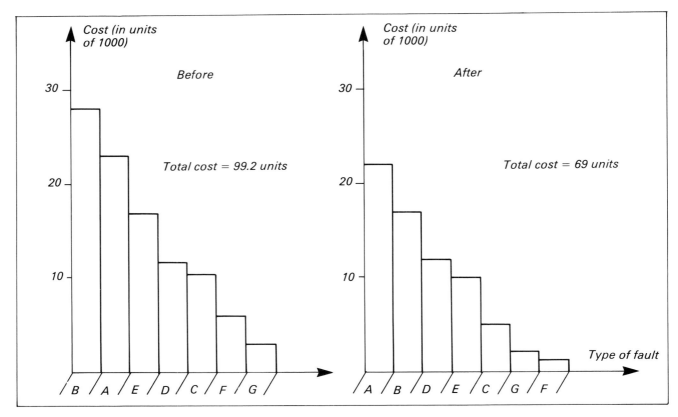

Fig. 3.14   Pareto diagrams showing improvement

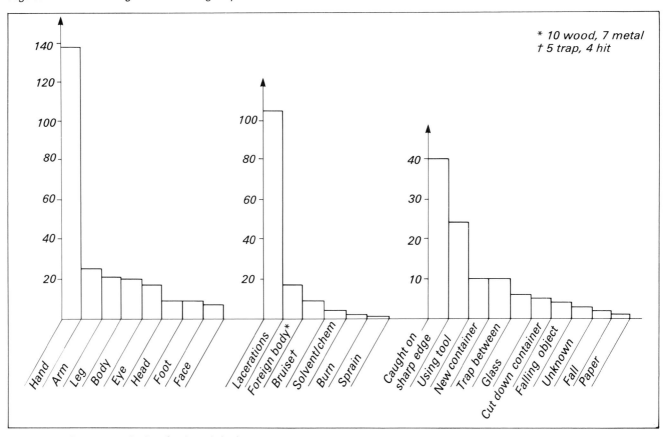

Fig. 3.15   Pareto analysis of minor injuries

Fig. 3.16 Condensed Pareto analysis format

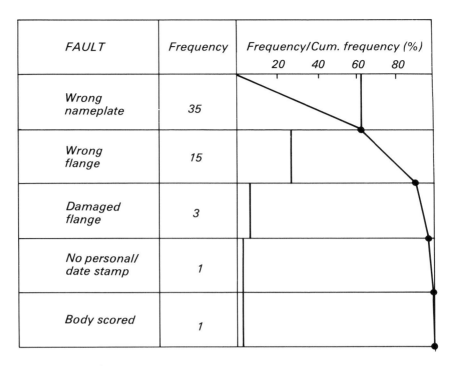

FAULT	Frequency	Frequency/Cum. frequency (%)			
		20	40	60	80
Wrong nameplate	35				
Wrong flange	15				
Damaged flange	3				
No personal/ date stamp	1				
Body scored	1				

3.3 Brainstorming

A traditional feature of British industry, and possibly Western industry in general, is that decisions tend to be taken autocratically. The scene is very familiar. A problem arises and a meeting is called by the manager responsible. Opinions are sought and a decision taken to proceed along a certain line of action. Democracy will be seen to operate in that all present have a chance to comment, but personalities and status interfere with the ability of the group to use all the expertise that is present. Some members feel diffident about expressing views, particularly ones which may reflect on their colleagues, or else fear their ideas are insignificant, even stupid. No one wants to feel foolish in these circumstances and hence the natural inclination is to hold back – and the opportunity is gone.

Brainstorming, a technique developed by Osborn in 1963, avoids these difficulties. Osborn advised that certain rules should be followed for the technique to be effective, and any deviation from these rules results in an activity which is not true brainstorming.

The four basic rules are:

- No criticism is allowed.
- Wild ideas are welcome.
- A large number of ideas are sought.
- A combination and improvement of the ideas is looked for.

Brainstorming works best with a group of six to twelve people, although these numbers are not rigid. With too small a group the brainstorming sequence will not operate effectively. With too large a group it becomes difficult to handle. Before starting the brainstorming session it is vital to identify the topic to be brainstormed. The topic must be clear and well understood by everybody taking part.

The session leader may be chosen from within the group, although often the role falls naturally on, for example, a supervisor. The choice of a leader other than the supervisor could cause conflict, and it is here that further training is often required as the SPC programme is introduced.

The task of the session leader is to bring out ideas from the group, who operate within the rules given below. The ideas are written down on a flip chart or an OHP transparency – not on a board because they could be erased. The flip chart has an advantage over a transparency in that as each sheet is completed it can be displayed for reference during the brainstorming session and at a later analysis stage.

For maximum effect the following rules must be followed:

- Each group member in rotation is asked for one idea per turn.
- If a member has no idea to offer, he or she just says 'pass'.
- No idea should be treated as too obvious or too stupid.
- Ideas should be provided as briefly as possible and it may be necessary for the leader to abbreviate them when recording them.
- No questions are allowed during the session.
- No comments on or criticism of ideas should be made by either the group members or the leader.
- The rotation is continued until all ideas have been exhausted and everyone passes.

After the brainstorming has been completed the large number of ideas must be critically examined and narrowed down.

In many ways the most important aspect of a brainstorming session is the elimination of criticism. Any evaluation of the ideas has to be postponed until a later stage so that all ideas can be brought into the open. If this is not allowed for, valuable ideas could be dismissed because of the inefficient use of time and misplaced criticism.

Whilst brainstorming it is worth while making use of the following verse, attributed to Jerome K. Jerome:

I have six serving men and true
When, what, where, why, how, who

These 5Ws and 1H assist in enabling a full range of ideas to come out.

A properly prepared brainstorming session can be very constructive; a great deal hinges on the effectiveness of the group leader and his or her knowledge and experience of both group leadership and the particular topic for the session. Skilled leaders can bring out the best in the group.

Conversely a lack of success in brainstorming is associated with a lack of leadership skills. In badly led sessions there may be evidence of interpersonal conflicts, and status differences may still influence the activity. A recent example illustrates the point. At an in-house training course, the group of which the Managing Director was a member found it difficult to operate a problem-solving session with an operator as group leader. The Managing Director had insisted that he act as a conventional group member, but it was clear that his position within the company was creating subconscious influences on the group. This was not helped by the lack of leadership and presentation skills demonstrated by the operator chosen to lead the group. Despite the determination of a very open-minded Managing Director to minimise his role, his influence was still operating within the group.

In another sense, the usefulness of brainstorming has not been fully recognised by those organisations which may face problems of implementation. For example, the traditional metal-processing industries have little difficulty in applying control charts to long-running, machine-dominated processes. Control charts can readily be used to monitor shaft dimensions, component thickness or product weight. Applying charts in areas other than these does cause problems, and it is here that the brainstorming technique comes into its own. Two examples may help here.

In an organisation with 200 employees, training for SPC had been carried out at a senior executive level. Project work for the 17 staff involved was an essential element of the course and was backed up by the SPC facilitator with the support of the General Manager. The senior staff were able to choose their projects without too much technical difficulty, although there were some uncertainties regarding the correct type of chart to use. For the next level of training, i.e. supervisory/technical, it was felt that the decision on type of chart, and what to monitor, would not be easily resolved. The General Manager proposed that brainstorming be used to choose process measures. As a result some 100 potential measures of the process were generated. These provided the basis for charting.

A training course in a process industry brought forward a similar proposal from an operative: why not brainstorm what we can measure? Again, a long list of items was generated which provided the SPC facilitator with a large data bank of projects. These were used to assist in the training and to suggest areas where the positive introduction of charting techniques could have most effect.

In conclusion, then, brainstorming provides a new way of involving personnel in looking at problems which affect them. However, it must be carried out properly, according to the rules described above, if the full benefit is to be achieved.

There are alternatives to the form of brainstorming described, and experience with the basic method could suggest greater potential for success with other methods. There can be no dispute regarding the response of the workforce to these approaches. Shop floor operators react positively to becoming involved in brainstorming sessions. They enjoy the experience because it recognises their own knowledge of the process and because the brainstorming sessions are so different to the normal training courses they have previously undertaken. But however interesting the session may be, the shop floor staff will become disillusioned unless there is some follow-up from management.

A brainstorming session will result in a list of ideas relating to the problem in hand. How can this information be used to help solve the problem? A further technique is available. This was devised by Kaoru Ishikawa, who first applied it in the Japanese steel industry. It is associated with cause and effect analysis.

3.4 Cause and effect analysis

In industrial and commercial processes the effect is known but the cause is not clear. Why is it that there are problems with the packaging? Why do companies have difficulties with introducing SPC programmes? Why is there a problem with late deliveries? Brainstorming will have provided a list of possible causes of the problem, but that in itself is not sufficient. To proceed further this list must be analysed, condensed if need be, and rationalised certainly, so that the likely cause is brought out. This is where cause and effect analysis comes in.

3.4.1 Developing the fishbone diagram

The following sequence is adopted when carrying out cause and effect analysis.

(a) *Specify the effect*

The first stage is to write out the problem as the effect, in a box at the end of a main spine.

(b) *Classify the possible causes*

Branches are now added to the main spine, each branch corresponding to a particular category of cause. The conventional categories of cause, based on the early use of the technique in quality circles in the USA, are manpower, materials, methods and machines (the 4Ms). The basic form of the cause and effect diagram is then as shown in Fig. 3.17.

It should be stressed at this stage that this classification is not definitive. Other categories could be adopted, depending on the situation. However, it is worth while using this convention initially because it happens to satisfy many cases and if necessary provides the basis for further adaption.

(c) *Transfer the ideas from a previous brainstorming session*

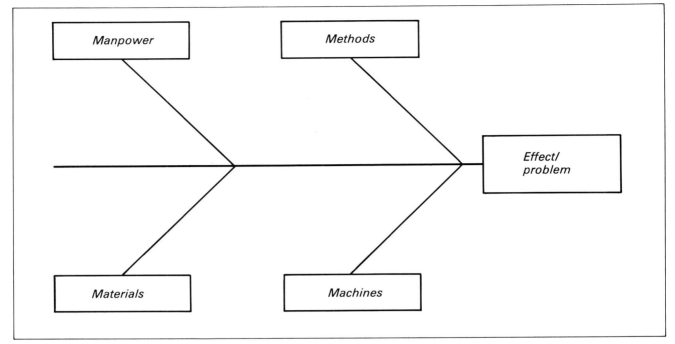

Fig. 3.17   4M fishbone diagram structure

The ideas from a previous brainstorming session are then transferred onto the 4M diagram (or other variation as appropriate). Before doing so it is useful to run through the list of ideas with the group so that they can be narrowed down. Any idea which, outside the brainstorming session, is now seen as inappropriate is eliminated after discussion. Those ideas remaining are allocated to one of the categories. This may be most easily carried out if a code is used. For example:

        *   = Manpower
        ∧   = Methods
        ○   = Machines
        >   = Materials

The relevant symbol is placed against the idea under discussion.

There is no restriction on the number of symbols corresponding to any idea. Here again, the training and experience of the leader is important. Rather than directing the thoughts of the group towards a category the leader should act in a more passive way as a recording agent and simply register the views of the group as they are put forward.

Fig. 3.18 shows a completed 4M fishbone diagram for the problem of an unsatisfactory thread on a component. The shape indicates why it is called a fishbone diagram. It portrays a nice balance of potential causes corresponding to the four categories and is typical of the type of fishbone diagram relating to problem-solving in a manufacturing environment. These four categories are not appropriate for non-manufacturing applications. This point will be dealt with shortly.

(d) *Decide on the likely solution*

The fishbone diagram represents in a more logical form the mass of ideas resulting from a brainstorming session. It therefore helps concentration on those ideas which are more likely than others to produce the solution to the problem. It is at this stage that decision-making procedures and leadership skills are required more than

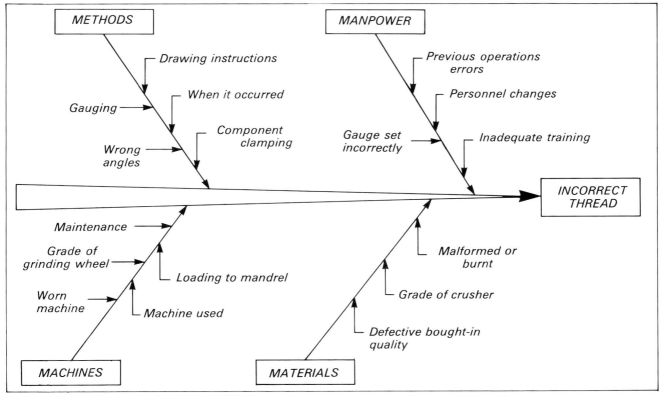

Fig. 3.18   4M fishbone diagram for analysing 'incorrect thread'

ever. The fishbone diagram will not in itself suggest which likely solution to implement or how to arrive at a consensus on that solution. Voting procedures and other methods need to be utilised, and the reader is advised to refer to other sources for details.

3.4.2 Other features of the fishbone diagram

(a) *Choice of structure*

A variation of the 4M theme is to add a further category: maintenance. Sometimes this is environment (or milieu).

A 4M (or 5M) fishbone is a good starting point but it may not always be appropriate. For example, it may result in many ideas being concentrated under just one or two of the main classifications, and therefore an unbalanced diagram. In these cases it is best to choose classifications that provide a more equally balanced fishbone diagram. Fig. 3.19 shows a fishbone diagram for the problem of the delay a customer suffers when trying to pay a bill at a Post Office.

Some thought needs to be given to the choice of category. The headings do not come easily and tend to be specific to the industry concerned. There may well be some overlap in that materials, for example, could remain unchanged when setting up a new classification. The fishbone diagram based on new categories is called a dispersion analysis diagram.

One organisation finds the 7M format always appropriate, i.e. the 4Ms plus marketing, money and management.

(b) *Twig building*

In Fig. 3.19 twig building occurs. To the branch marked 'Training' have been added further branches 'None' and 'Inadequate'. This is one variation of twig building, where all ideas which can be related to a branch are written alongside that branch.

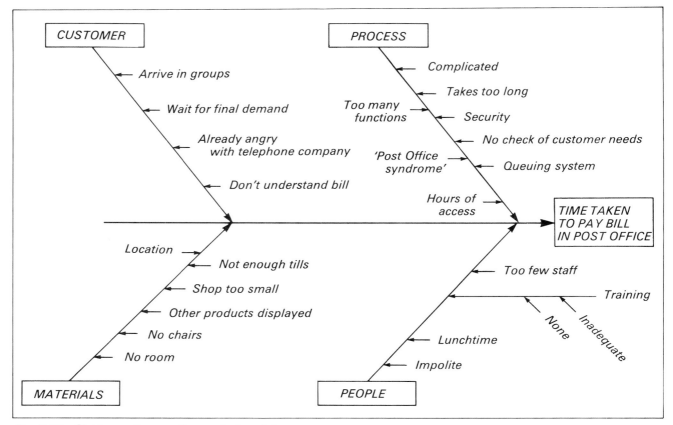

Fig. 3.19   Fishbone diagram for analysing delays

Twig building can also be effective in another sense, in that questions can be raised which seek answers further back in the relevant process. Fig. 3.20 shows typical twig building of this type. At each stage the factors which influence the main cause are considered and then each of these sub-factors in turn analysed, and so forth. Discussion on these sequential sub causes is likely to take place outside a brainstorming session. Here again, leader experience is required to enable the group to identify the sub-cause.

(c) *Process cause and effect diagrams*

A variation on the fishbone diagram is the process diagram. It has the advantage of being readily applied in any organisation because it basically requires a series of block diagrams representing the process.

A typical process cause and effect diagram is shown in Fig. 3.21, which considers

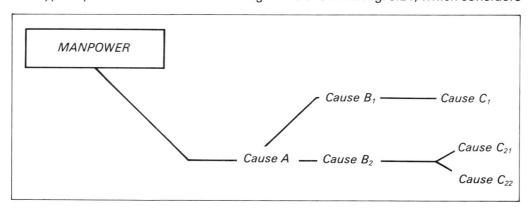

Fig. 3.20
Twig building

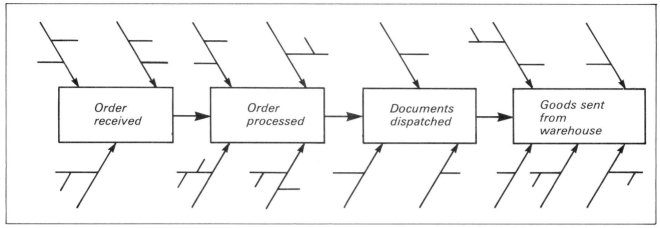

Fig. 3.21   Process analysis

the sequence between receiving an order for an item and dispatching it from the warehouse. This diagram was constructed by carrying out a brainstorming session when all the likely causes of the problem , for each stage of the process, were considered. A cause and effect diagram of this type could then provide useful information on causes which occur repeatedly. For example, training may be a factor which repeats itself.

In this simple form the diagram could be very useful yet at the same time it could be confusing because of the difficulty of handling all the information portrayed. In such cases it is advisable to break the process down into successive stages and consider each stage separately. When doing so, there is no reason why the 4M, 5M or any other variety of classification could not be used at each stage. Such a flow diagram is important also in that constructing it, and then comparing it with the ideal case, provides a catalyst for looking at ways of improvement. Section 3.5 discusses process flow diagrams in more detail.

(d) *Problem or opportunity?*

It is understandable why fishbone diagrams tend generally to consider the reasons for problems, or difficulties, occurring. Late deliveries, excessive scrap, dirty containers and damaged surfaces are all problems which need solving. Sometimes, however, there may suddenly be reduced scrap, clean containers or greatly reduced surface damage. The introduction of control charts at appropriate points is bound to be a major influence in causing improvement in itself. Following that, the chart may then show a special cause to be present which is indicating improvement rather than deterioration. Why not, then, use the fishbone technique to itemise the likely cause of the good feature? Finding the reason for improvement, and integrating it permanently into the process, is more helpful than discovering the reason for deterioration and eliminating it so that it does not occur again.

(e) *Relationship with the Pareto diagram*

A common approach to problem solving is to start with the Pareto diagram and then work through each feature in turn. Fig. 3.22 illustrates the point. 'A' is tackled initially and becomes the focus of a fishbone analysis and subsequent action. 'B' then follows, and so forth. There is no definitive rule in these cases and the sequence for action depends on the time scales involved in introducing appropriate solutions at each stage.

Many problem-solving case studies are available. The one described here illustrates the use of the techniques in tackling an interesting problem.

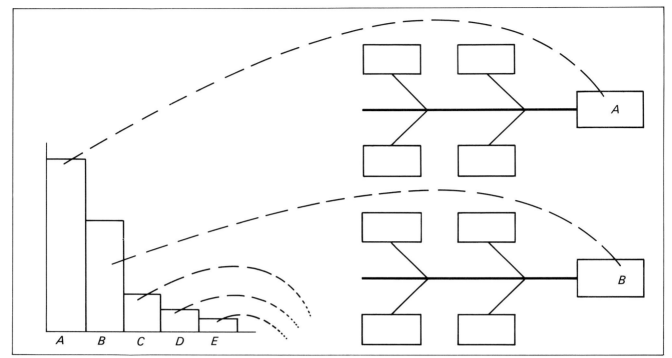

Fig. 3.22   Pareto diagram/fishbone diagram

Case study 1:   The Goodyear Tyre and Rubber Company Ltd

The Wolverhampton plant of this world-wide organisation has introduced SPC as part of its total approach to quality improvement. There has been a long-term training programme covering all operations and all areas of the organisation, both administrative and technical. The company's approach to quality is summed up in the following statement from a publicity document: 'Our basic approach in being the world leader is to constantly achieve higher plateaus of quality, and this is behind our slogan "Out front ... pulling away".'

Control charts are used throughout the organisation and provide a common language at all levels. In particular they are used extensively in clerical, marketing and personnel functions. In addition, shop floor operators are very familiar with them, to the extent of leading business team meetings using control charts on accident levels, or absentee figures, as the basis of decision-making.

Charts in themselves are not enough. The company has recognised the essential need for problem-solving skills. In some cases, as in this case study, the problem-solving techniques were of more direct relevance than charts.

The company has made use of quality circles since they were first introduced in British industry, and their success with quality circle programmes proved a good foundation for the introduction of SPC. Problem-solving skills are therefore known and have been applied over many years. The case study in question emphasises the value of these simple techniques.

The example relates to malfunctioning of reversing bleepers fitted to the company's fork lift trucks. The bleepers give warning to all employees in the vicinity of a reversing truck and play an important part in the prevention of accidents. A system to identify the reasons for failure was therefore set up. Data was collected over a period of 5–6 weeks which enabled a Pareto analysis to be carried out. Far and away the major reason for failure was 'wires pulled out'.

The supervisor and mechanics of the internal transport department held a brainstorming session to examine the various possible causes. A fishbone diagram was subsequently drawn up as shown in Fig. 3.23. From this analysis a solution emerged whereby bleepers would be positioned underneath the body of the truck with the wires trained through internally. Apart from preventing drivers from tampering with the wires, this also had the effect of lowering the noise level – the real cause of the problem. These modifications were put into effect and as a result no instance of bleeper failure due to driver abuse then occurred.

The actual problem was identified and solved here without recourse to charts, although charts were used subsequently to monitor the situation. The case study also illustrates how a relatively minor problem, which could have major safety implications, was solved by using simple techniques.

3.5 Other techniques

(a) *Process flow diagram*

An alternative starting point in problem solving may be to draw a process flow diagram. It could be in a form similar to a conventional computer flow diagram, or it could be kept much simpler, using blocks.

Fig. 3.24 shows a simple flow diagram for invoicing a customer following a training course. There are plenty of opportunities for error, and after considering these, using cause and effect analysis, a new flow diagram could be generated, showing a reduction in the number of stages and hence of the risk of error. If necessary brainstorming could now be used to assess any possible difficulties resulting from introducing the new system.

(b) *Force field analysis*

If the problem to be resolved can be given a numerical value based on intuition, then force field analysis could provide an alternative approach to brainstorming, particularly when the group is large.

A vertical scale, ranging from 0 to 100%, is drawn and each member of the group is asked for an estimation of the level of the problem. For example, for a problem

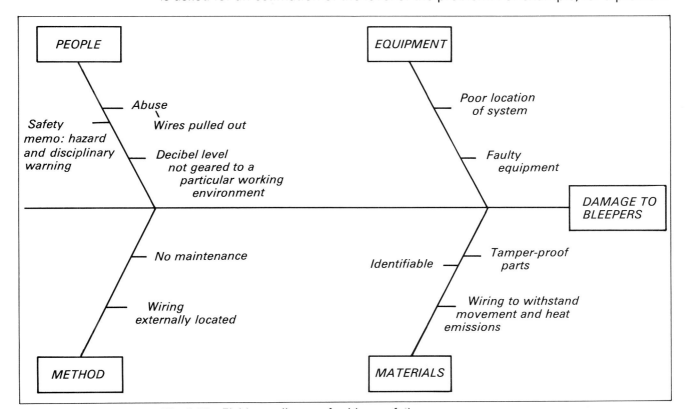

Fig. 3.23 Fishbone diagram for bleeper failure

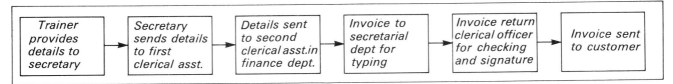

Fig. 3.24   Process analysis of invoicing

defined as 'the chances of becoming a supplier to a leading company' the individual estimates may be 55, 75, 70, 85, 60, etc. These results are plotted on a tally chart and the mean line determined. Each member of the group in turn is then asked to identify factors which will assist in raising the level of the line. These are known as driving forces. Equally, those factors which work against improvement, the restraining forces, are also identified. Both sets are recorded on the force field analysis diagram as shown in Fig. 3.25. The graph then determines priorities for action so that the restraining forces can be minimised and the driving forces strengthened.

(c) *Failure mode and effect analysis (FMEA)*

At the design stage of a product, or equally in the initial stage of introducing a new feature in a service industry, there is a need to consider all the factors which can influence performance at a later stage. In addition it is possible to consider in an analytical way the reasons why the product may not be produced correctly to the design specification, or why there may be problems when the new feature comes into operation in the service industry.

FMEAs provide a useful technique in these cases and there is now increased interest in this method. Several worthwhile sources of reference have been developed by sections of the automotive industry.

Other tools of this type are available. For example, scatter diagrams are often useful in determining relationships. In one organisation it was recognised that there was a very close relationship between the process and the product and that this relationship could be determined numerically. It was easy to monitor the product, so this was done, and therefore in monitoring the product the process was also being monitored. In concentrating on Pareto analysis, brainstorming and the fishbone diagram it would be a mistake to overlook other techniques which may be of more use in certain circumstances. However, at this stage there could be a

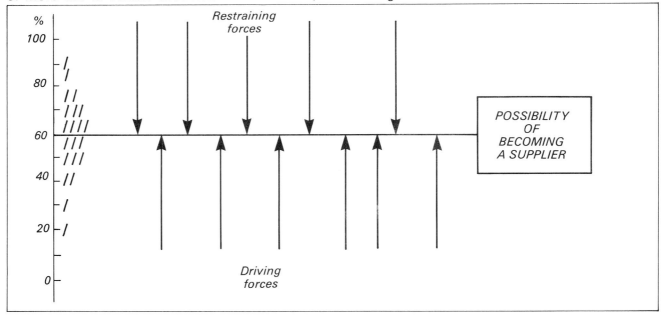

Fig. 3.25   Force field analysis

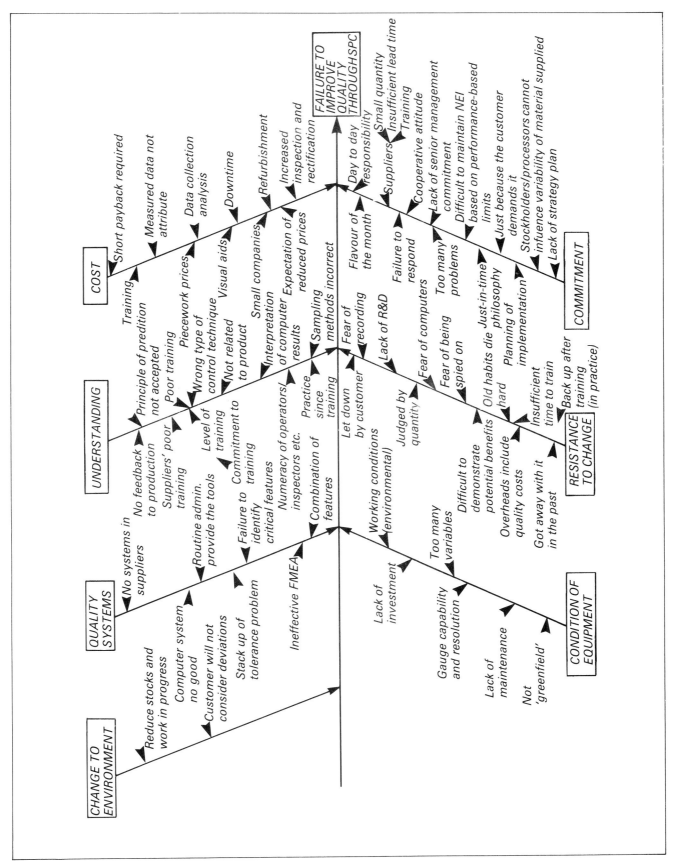

Fig. 3.26   Fishbone diagram for SPC implementation

danger of providing a recipe book of problem-solving techniques rather than indicating the most commonly used methods which are really necessary as part of any SPC programme.

It may be appropriate to conclude this chapter by referring to Fig. 3.26. This fishbone diagram summarises many of the issues relating to SPC. It shows how a problem-solving technique can be used to tackle the introduction of SPC itself, and within the results of that technique are displayed factors familiar to everyone: lack of management support, lack of training, etc. The problems of implementation will be dealt with fully in Chapter 15, as it is only by introducing SPC in a logical programmed manner that the issues listed in Fig. 3.26 will be adequately dealt with.

Summary

- A Pareto analysis highlights the most important problem.
- A Pareto analysis can apply in many situations, not just in a manufacturing area.
- Cumulative lines help to identify the major problem.
- A Pareto diagram on cost is often advisable.
- Sub-Pareto diagrams provide added information for getting to the heart of the problem.
- Try to keep things simple.
- Use brainstorming to generate as many ideas as possible.
- A cause and effect analysis will help to identify the real causes.
- Use 5Ws and 1H.
- Do not be restricted to a standard classification. The 4Ms are often chosen but other groupings may be preferable.
- Use twig building to assist in discovering the likely cause of a problem.
- Use a fishbone diagram on successive stages of a process to pick out common themes.
- Use a fishbone diagram to concentrate on reasons for improvements as well as causes of problems.
- Other techniques are also available to be used as appropriate.

The problem-solving techniques discussed here will no doubt be familiar to many. The growth of SPC, however, is reinforcing their value, and in a much wider context than as part of a quality circle programme. With the commitment of senior management to a programme of continuous improvement these techniques are playing their part in improving processes by involving operators and managers in problem solving groups. As a result barriers are being broken down and progress made in unifying the organisation.

The main technique in SPC is nevertheless the control chart. This involves calculating and plotting values which will affect the process under review. This allows future results to be predicted and therefore very early action taken to correct the process before unsatisfactory material is produced. It means that familiarity with measuring location and variation is necessary, and Chapter 4 covers the essential points required. Without an understanding of these critical ideas there is no value in proceeding with the control chart.

Chapter 4 Measuring location and variability

4.1 **Introduction**

In Chapter 2 it was seen how histograms could be generated from a set of readings. The histograms varied in position, base width and even shape and it follows that if we are to proceed further then some measures of these features are required.

Look again at the set of data represented by the histogram in Fig. 2.9 and reproduced in Fig. 4.1. It is difficult to be specific but the histogram suggests that the process is running at a level of 2.5. At a later stage it may be that a new set of readings could give rise to a situation as shown in Fig. 4.2, where the process appears to be running at a new level of 2.505. This change in process position needs to be measured by choosing values which adequately represent the two sets of readings.

In the same way the variability in a set of readings could change, and Fig. 4.3 indicates how the pattern is amended depending on whether the variability increases or decreases, i.e. whether the values are becoming more or less consistent. In this case the variability has changed but the location has remained constant. This may not always be so.

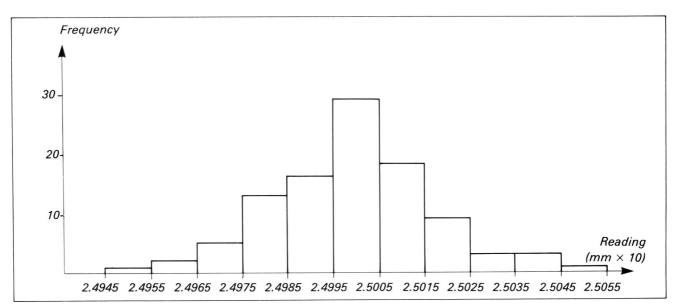

Fig. 4.1 Histogram for diameter of components

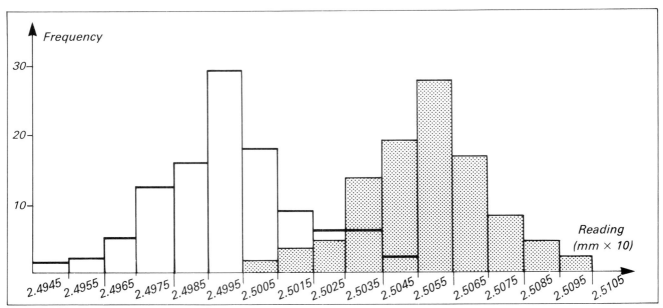

Fig. 4.2   Histograms for different settings

Location and variability are related to accuracy and precision.

4.2 **Accuracy and precision**

Fig. 4.4(a) represents a set of tally marks corresponding to a process which is running at a level above the target, or nominal, of 50. The grouping of the values may be seen to correspond to a record of pistol shots on a target board (Fig. 4.4(b)). The analogy is not perfect because values in a single dimension are being repre-

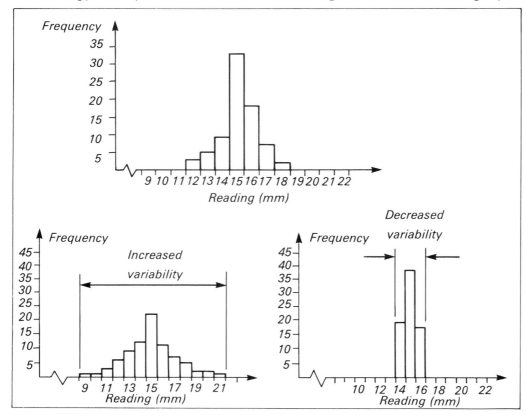

Fig. 4.3   Increased and decreased variability

Fig. 4.4 Precision and accuracy

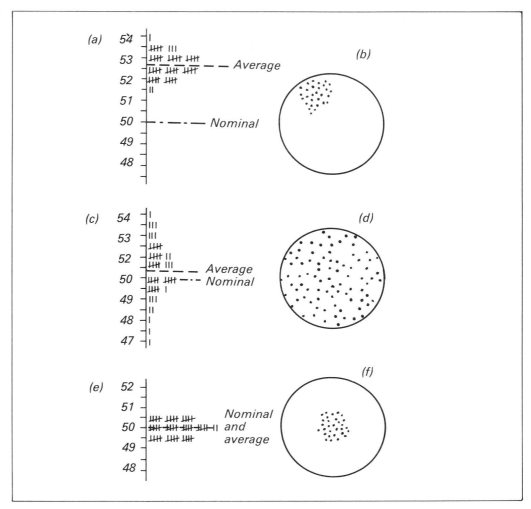

sented by a pattern in two dimensions. However, there is enough of a similarity to show that in the same way as the pistol shots are nicely clustered then equally variability in the readings is not excessive. The process is said to be precise. It is not accurate, though, because whilst the shots are consistent, they are consistently off target. In the same way, the set of readings, as represented by a single value (the average), represents a process running at a level which is off target. The average value as such has not been defined yet. All that is required at this stage is an acceptance that the average is a measure of location of the process.

In Fig. 4.4(c) the reverse is happening. There is a wider spread of readings than in Fig. 4.4(a) and hence the process is less precise. The analogy is appropriate, as in Fig. 4.4(d), where the shots are much more widely spread around the board. However, on average there is a greater chance of hitting the target in this case than there ever was before. In other words, the average value of the set of readings is very close to the nominal.

Anyone doing the shooting in this particular case would wish to have the accuracy of (d) and the precision of (b), and then improve on these as represented by Fig. 4.4(f). Similarly, in the process being considered the aim would be to bring the average value on to the target of 50 and reduce the variability about this figure (Fig. 4.4(e)).

The responsibilities of management and operators are well illustrated in this example. Getting the setting right is very much an operational issue. Reducing the variability, however, is a management responsibility. Deming emphasises that 85% of an organisation's problems can only be resolved by management action.

Improving the variability, therefore, is a direct reflection of the 85% element which is outside the direct control of operators. That is not to say that operators should not have a real input in doing something about improving variability – hence the importance of problem-solving groups. It must be recognised that accuracy and operators go together in the same way as does precision and management.

Measures of accuracy and precision are available, known more commonly as measures of location (setting) and variability (scatter).

4.3 Measures of location

A single representative value of a group of readings is known as the average. The average value locates the distribution at a particular point on a scale and hence the alternative term 'measure of location'.

There are three types of average value: the mean, the median and the mode. These are discussed in turn below.

4.3.1 The mean

The words 'mean' and 'average' often tend to be used as alternatives. For all practical purposes this is acceptable. Purists could rightly point out that the word 'average' is the family name and the mean is just one of three members of that group. The practicalities are that many companies use average charts as mean charts, and provided the general distinction is understood there should be no confusion.

The mean value, sometimes known as the arithmetic mean, is obtained by adding together all the readings in the group and then dividing the total by the number of readings.

For example, Fig. 4.5 shows a set of 20 readings. The readings could correspond to any feature, such as length, weight or number of rejects. A common notation is to use the symbol X to denote the variable (or attribute as the case may be) and to use a subscript to indicate which value of X is being referred to. It is appropriate to use X, not least because most hand-held calculators use the symbol. However, other symbols are sometimes used.

The mean value is then obtained as

$$(7 + 3 + \ldots + 1 + 0)/20 = 75/20 = 3.75$$

It will be seen in Chapter 9 that calculations similar to that above are required when determining the central line on an individual reading control chart.

More generally

$$\text{Mean} = (X_1 + X_2 + X_3 + \ldots + X_n)/n$$

where n is the number of readings available. The mean is donated by $\bar{X}$ (X bar).

This general expression for the mean is cumbersome. It is far neater to write it in

READING	X_1	X_2	X_3	X_4	X_5	X_6	X_7	X_8	X_9	X_{10}
NUMERICAL VALUE	7	3	1	6	2	5	1	3	4	9
READING	X_{11}	X_{12}	X_{13}	X_{14}	X_{15}	X_{16}	X_{17}	X_{18}	X_{19}	X_{20}
NUMERICAL VALUE	4	1	2	9	3	4	4	6	1	0

Fig. 4.5 Table of readings

the form

$$\Sigma X/n$$

where the Greek letter Σ (sigma) represents 'the sum of'.

There would appear to be three situations in particular where a mean is required within an SPC programme:

(a) *Sample mean*

In a great majority of control charts, samples of size 5 are taken. A more detailed discussion on sample size is given in Chapter 6, but reference to calculating the sample mean is appropriate here.

Fig. 4.6 shows a section of a control chart for variables. The data block allows for the individual readings and the total of the five readings in each sample, as well as the sample mean itself. If a calculator is being used then the ΣX space is not required and $\bar{X}$ can be entered directly. Calculating the mean mentally is often preferable, however, and the process can be simplified if the numbers are reduced to whole number form (by coding), as in this example. Alternatively, provided the figure before the decimal point is the same for all values in the sample, the figures after the decimal point in $\bar{X}$ can readily be determined by doubling the total of the five decimal quantities and then dividing by 10 (1/5 = 2/10).

Operators of an older age group feel more comfortable with traditional mental arithmetic routines than they do with the calculator. The reverse may be true for younger employees, who are more used to keyboards and visual displays. It is important that considerations of this nature are taken into account when designing SPC training programmes – and not just necessarily for the operators. Even some managers do not find the calculator straightforward.

(b) *Grand mean*

Fig. 4.7 shows the section of a control chart including 20 sample means which have been determined as part of an initial analysis of a process. In order to determine the control limit for monitoring $\bar{X}$, the mean of these twenty sample means is required. This, from the definition of a mean, will be given by

$$\Sigma \bar{X}/20$$

This result is know as the grand mean and is denoted by $\bar{\bar{X}}$ (X double bar).

Hence $\bar{\bar{X}} = \Sigma \bar{X}/20$

Fig. 4.6 Data block from (X̄, R) chart

x_1	3	0	-6	3	4	-1	3	7	6	-4	-1	4	-1
x_2	-3	-2	4	1	1	-2	-1	3	0	-1	-4	-6	-4
x_3	2	0	-2	1	-1	-4	1	-2	-1	1	2	5	0
x_4	1	2	-2	-1	-5	-7	4	-3	-2	0	3	-2	-1
x_5	-1	-4	0	-2	2	-2	5	-2	2	2	1	-1	-5
ΣX	2	-4	-6	2	1	-16	12	3	5	-2	1	0	-11
$\bar{X}$	0.4	-0.8	-1.2	0.4	0.2	-3.2	2.4	0.6	1.0	-0.4	0.2	0	-2.2

ΣX	2	-4	-6	2	1	-16	12	3	5	-2	1	0	-11	-3	-5	-12	-7	7	-14	-10
$\bar{X}$	0.4	-0.8	-1.2	0.4	0.2	-3.2	2.4	0.6	1.0	-0.4	0.2	0	-2.2	-0.6	-1	-2.4	-1.4	1.4	-2.8	-2

Fig. 4.7 Table of sample means from (X̄, R) chart

or more generally

$$\bar{\bar{X}} = \Sigma \bar{X}/k$$

where k is the number of samples (which is not the same as the sample size).

(c) *Moving mean*

In Chapter 9 a chart will be considered for which a moving mean is required.

The moving mean is based on successive mean values, which are grouped in twos, threes, fours, etc., depending on earlier decisions relating to the sensitivity of the chart.

Fig. 4.8 shows a section from a data block of sample means for a moving mean chart.

If a moving mean based on three samples is required, then the first moving mean is given by

$$(48.2 + 49.1 + 51.3)/3 = 49.53$$

To determine the second moving mean the first mean is dropped to be replaced by the fourth, giving a second moving mean of

$$(49.1 + 51.3 + 47.2)/3 = 49.20$$

Similarly for the other readings

The mean, however it is used, does require some mathematical dexterity. This is not true of the second average: the median.

4.3.2 **The median**

To obtain the median of a set of readings the readings must first be ranked according to size (either increasing or decreasing).

For example, for a given set of 7 numbers the sequences would be

Before ranking: 17, 19, 2, 6, 11, 23, 4
After ranking: 2, 4, 6, 11, 17, 19, 23

The median is then the middle item in the ranked set, i.e. 11. The median of a set of readings with X as the variable is donated by $\tilde{X}$ (curly X).

If a further figure is added to the ranked sequence, i.e.

2, 4, 6, 11, 17, 19, 23, 26

then the median is actually not present because there are an even number of readings. In such a case the median would be taken as the mean of the two most central readings, i.e. $(11 + 17)/2 = 14$. It therefore makes sense to ensure that as far as possible in SPC applications there is an odd number of readings. Since most control charts operate on a sample of 5, this causes no problem.

Fig. 4.9 shows a section from a control chart for medians. The data block allows for the individual readings and the median of each set. Note that there is no need to

Sample no.	1	2	3	4	5	6	7	8	9	10	11	12
Batch no.	270	271	272	273	274	275	276	277	278	279	280	281
x	48·2	49·1	51·3	47·2	48·3	47·1	46·5	49·1	50·2	49·1	49·8	47·2
$\bar{x}$			49·53	49·20	48·93	47·53	47·30	47·57	48·60	49·47	49·70	48·70

Fig. 4.8   Data block from ($\bar{X}$, moving R) chart

x_1	−1	0	−1	1	−5	1	2	−1	2	0	−1	−1	−3	−1	2	−4	−3	−2
x_2	3	1	0	0	1	−3	−3	−3	−2	0	−2	2	−1	−2	2	1	−1	−3
x_3	−4	1	−3	2	2	1	1	1	−2	2	0	1	2	−2	3	−3	0	1
x_4	2	3	1	−2	1	3	2	1	−1	3	−3	2	−1	−3	1	1	2	2
x_5	1	−2	4	0	3	−1	3	7	3	4	1	4	3	1	1	2	1	1
$\tilde{x}$	1	1	0	0	1	1	2	1	−1	2	−1	2	−1	−2	2	1	0	1

Fig. 4.9   Data block from ($\tilde{X}$, R) chart

record the total of the individual readings for each sample as this is not required for a median chart.

The advantages and disadvantages of the mean and median will be considered in Section 4.3.4.

4.3.3 **The mode**

The mode is the most frequently occurring value. Hence, for a histogram of readings as represented by Fig. 4.10 the mode corresponds to the peak of the distribution, i.e. 127g.

Similarly, for a sample, the mode corresponds to the value which occurs more times than any other. Thus, for a sample of size 5 with readings of 2, 3, 2, 4 and 1, the mode is 2.

The mode is rarely used in SPC, and has been included here principally for completeness. That does not mean that as a statistical measure in a more general

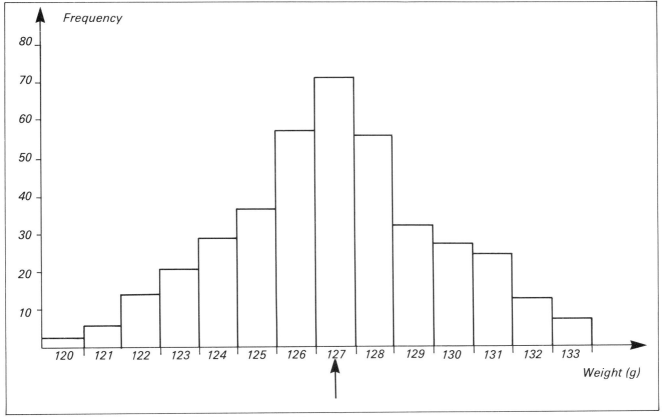

Fig. 4.10   Histogram and the mode

sense the mode does not have advantages. For example, consider an analysis of returns on errors in completing a works maintenance form. The form has 14 questions which require a response. It will follow that some information blocks will provide more errors than others, as shown in Fig. 4.11. In this case section 9 will correspond to the mode because it contains more errors than any other section. In an SPC context, however, there appears to be little evidence of any practical value of this measure.

In a sense, the question of whether or not the mode is relevant summarises the relationship between statistics as a mathematical science and statistical tools as applied in SPC. There is no point in promoting the use of a type of average which plays a very minor part in analysing processes and assisting in improvement – hence the emphasis on the mean and the median. Both have their virtues, and some consideration of their relative advantages and disadvantages would be useful at this point.

4.3.4 **Mean or median?**

As will be seen in Chapters 6 and 9, organisations have an option as to whether they use the mean or the median as the appropriate measure of process performance. Which should they choose?

The median has the distinct advantage of ease of calculation. Basically all that is required is for the operators to plot the values on a chart, count up to 3 and ring the median. Fig. 4.12 shows a typical section from a median control chart. The specific details regarding plotting will be covered in Chapter 9. At this stage, a recognition of the limitations of the use of the median is what is important.

Whilst the median is very easy to calculate, it does have the disadvantage that it remains unaffected if values above (or below) the median are changed. Fig. 4.13 illustrates the point. The mean, on the other hand, is affected if one (or more) of the individual sample values is altered. Hence it is a better representative value than the median.

There is therefore a need to balance the advantage of ease of operation against the disadvantage of lack of precision. It seems the majority of organisations have opted for the use of the mean. Some have chosen the median, but it would be wise to do so only as an intermediate step in moving to a mean chart, or possibly straight

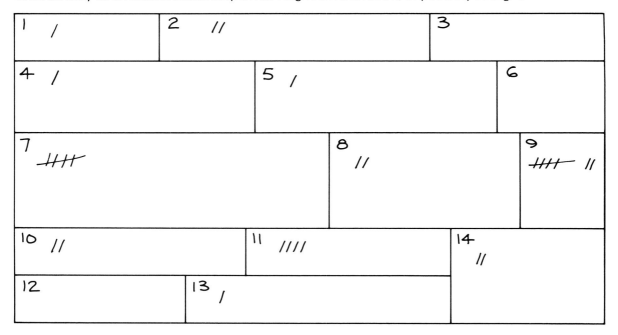

Fig. 4.11   Analysis of errors in form

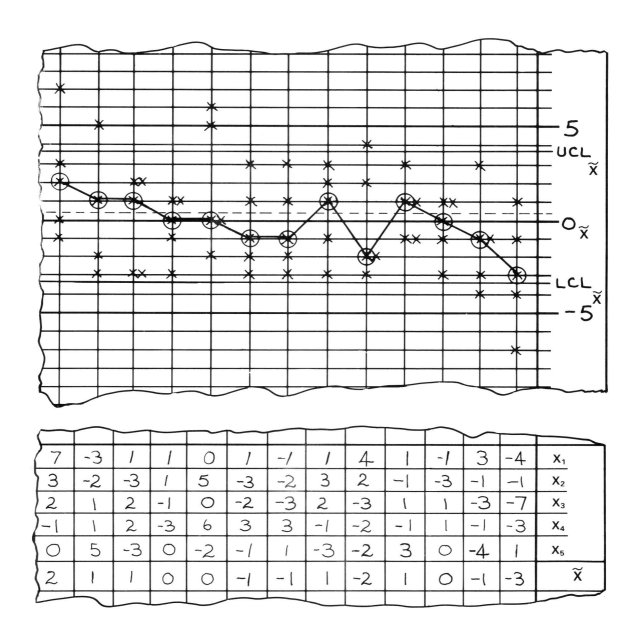

7	-3	1	1	0	1	-1	1	4	1	-1	3	-4	x_1
3	-2	-3	1	5	-3	-2	3	2	-1	-3	-1	-1	x_2
2	1	2	-1	0	-2	-3	2	-3	1	1	-3	-7	x_3
-1	1	2	-3	6	3	3	-1	-2	-1	1	-1	-3	x_4
0	5	-3	0	-2	-1	1	-3	-2	3	0	-4	1	x_5
2	1	1	0	0	-1	-1	1	-2	1	0	-1	-3	$\tilde{x}$

Fig. 4.12 Section from median chart

READINGS					MEDIAN	MEAN
-4	-1	1	3	4	1	0·6
-4	-1	1	5	8	1	1·8
-8	-3	1	3	4	1	-0·6
-8	-3	1	6	12	1	1·6

Fig. 4.13 Table of mean and median values

on to computerised monitoring. The advantage of the mean is worth the extra numerical difficulty involved, and a common feature of successful SPC programmes is the heavy emphasis that has been placed on manual charting using the mean. Handling the problem of the mathematics should not be a major issue, but it does mean that simplification of the numbers and probably an extra element of training are necessary.

As a result of the growth of electronic gauging, means (and other quantities) can be generated and displayed at will. Care is needed here, however. Organisations looking at the resources available should recognise that electronic gauging and computer software is only one small element of the total SPC programme and decisions to purchase should be in line with senior management commitment to the company-wide activity.

4.4   **Measures of variability**

Changing the setting of a process does not normally affect the shape of the distribution, as Fig. 4.14 shows. Equally, calculating the mean or the median is no help in determining the variability, or spread, of a set of readings, as shown in Fig. 4.15. Some other quantities are therefore needed in order to measure the variability. These are the range and the standard deviation.

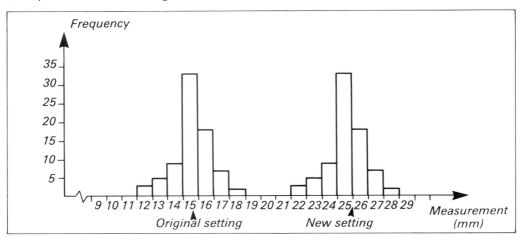

Fig. 4.14   Change of setting

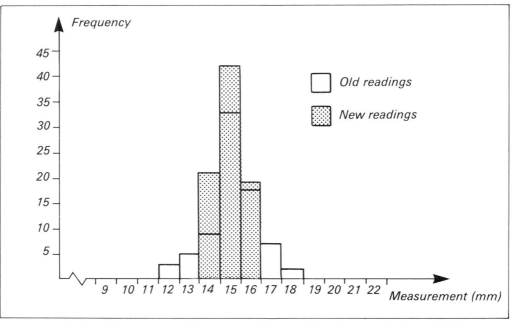

Fig. 4.15   Change of variability

4.4.1 **The range**

The range is the difference between the least and the greatest of a set of readings. For example, for the group of values 6, 4, 7, 1 and 3 the range is 7 − 1 = 6. The range is generally denoted by R, although British Standards have also made use of **w**.

Whilst it can be determined without too much difficulty, the range does have a disadvantage, particularly when used as an indicator of variability in a larger set of readings.

Fig. 4.16 shows a typical histogram. The range by definition is 15 − 6 = 9. The value is determined from the readings of 6 and 15 which correspond to the extremes of the distribution. These extreme values are evidently not representative of the main block of readings and hence the range is rather inappropriate as a measure of the variability within the complete set.

The situation could be even worse, as indicated in Fig. 4.17. Here the presence of wild values distorts the picture even more and the new range of 18 − 3 = 15 is anything but appropriate as a measure.

Fig. 4.16 Histogram and the range

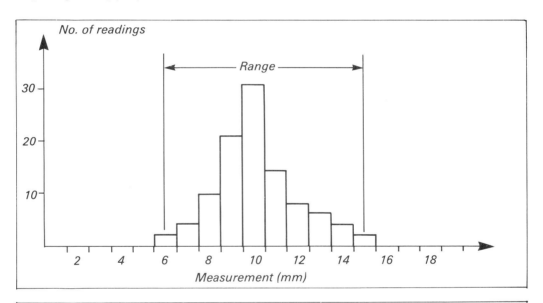

Fig. 4.17 Influence of wild values on the range

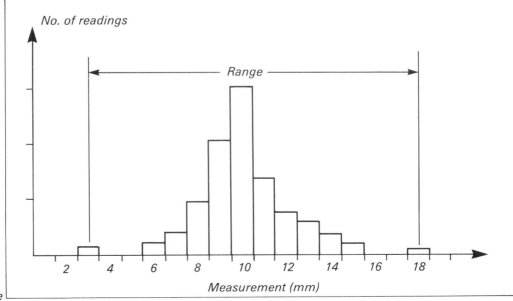

There has to be a better measure and statisticians have provided one: the standard deviation. This is discussed in the following section. However, the range should not be dismissed as being of no value. In fact, when assessing the performance of a process on the basis of a representative sample, the range is a perfectly adequate measure and is widely used in control chart applications. The reasons why the range is used will be considered in Chapter 6. The net result, however, is that a typical control chart for variables has a data block for recording the ranges and a section on the chart for plotting these values. Fig. 4.18 shows a typical data block.

The operators who will be plotting shop floor charts require familiarity with positive and negative numbers. This familiarity is not always present and hence training in basic numerical skills will be needed in such cases. Visual aids are becoming increasingly available to assist trainers in getting over to a group the difference between positive and negative numbers, and how to then calculate a range. All this may seem very basic, but a lack of attention to detail here could render the charts meaningless and SPC operations in general suspect.

The range is a practical measure of variability. It lends itself readily to the shop floor and is a perfectly acceptable indicator of process performance. However, it is still only an indicator, or estimator, and a better measure of variability is required. A quantity is needed which is based on all the numerical values of the group and not just the two extremes. That is where the standard deviation fits in.

4.4.2 The standard deviation

The standard deviation is defined as

$$\sqrt{\frac{\Sigma(X - \bar{X})^2}{n - 1}}$$

Variations of this expression are sometimes used. Equally, different symbols are available. Neither of these factors helps in trying to make the standard deviation clearly understood. The above expression is the one which will be adopted here, in line with the approach of those organisations implementing SPC. The expression will be denoted by s at this stage.

It will be seen that the expression requires the initial calculation of $\bar{X}$ based on a sample of n readings. After $\bar{X}$ has been determined, it is subtracted from each reading in turn (giving a series of $(X - \bar{X})$ readings). These readings are multiplied by each other and then added together (giving $\Sigma(X - \bar{X})^2$). Note that Σ is again used to indicate that the sum of a series of squared differences is required. Finally the expression is divided by 1 less than the number of readings and the square root of the result determined.

All this appears rather complicated and an example will be given to clarify the steps involved. A simple set of numbers will be used such as those provided by the

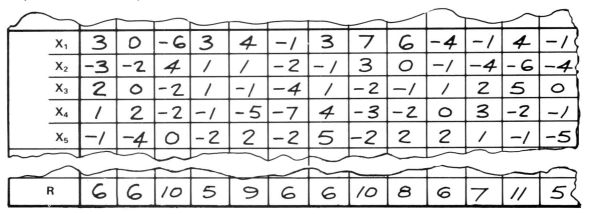

X₁	3	0	-6	3	4	-1	3	7	6	-4	-1	4	-1
X₂	-3	-2	4	1	1	-2	-1	3	0	-1	-4	-6	-4
X₃	2	0	-2	1	-1	-4	1	-2	-1	1	2	5	0
X₄	1	2	-2	-1	-5	-7	4	-3	-2	0	3	-2	-1
X₅	-1	-4	0	-2	2	-2	5	-2	2	2	1	-1	-5
R	6	6	10	5	9	6	6	10	8	6	7	11	5

Fig. 4.18   Data block from ($\bar{X}$, R) chart

Fig. 4.19 Standard deviation calculations

X	$X - \bar{X}$	$(X - \bar{X})^2$
3	2.6	6.76
−3	−3.4	11.56
2	1.6	2.56
1	0.6	0.36
−1	−1.4	1.96
2	0	23.20
ΣX	$\Sigma (X - \bar{X})$	$\Sigma (X - \bar{X})^2$

first sample readings in Fig. 4.6 (or Fig. 4.18). The values are reproduced in the first column of the table shown in Fig. 4.19

In the table $\bar{X}$ is given by 2/5 = 0.4. Subtracting 0.4 from each reading in turn gives the series of values in the second column. Some of these values are positive, some negative. If the readings are totalled, the result is zero (thus verifying a property of the mean). This interesting result is no help, however, in obtaining an appropriate measure of variability. Since using the difference between each reading and the mean does not provide a way forward it is logical to do the next best thing and consider these differences squared. Multiplying each difference by itself makes all the results positive and provides the readings in the third column. A form of average of these squared differences is then obtained by dividing the total, 23.20, by 4, i.e. (n−1). This cannot be a mean value in the true sense because of the use of (n−1) rather than n. However it does represent an averaging out of the readings and the reason for using (n−1) rather than n will be explained shortly. Finally $\sqrt{(23.20/4)}$ is evaluated, giving a value for s of 2.408. What does this actually mean? How can it be interpreted in a practical sense?

In the numerical example just considered a sample of 5 was used. A small sample with simple numbers makes it easier to follow the steps in calculating the standard deviation. In practice it is much more likely that a larger set of numbers will be involved, and in order to provide an interpretation of the standard deviation in such cases Fig. 4.20 will be used. The histogram is symmetrical, or at least reasonably so. In such cases it is known that the width of the base of the histogram is approximately equal to 6s. (An explanation of this is not required at this stage but will be given in Chapter 5 when the normal distribution is considered.) This relationship provides a practical interpretation of s. For any distribution which shows a fair degree of symmetry, i.e. the peak is not displaced too much to one side or another, then the base of the histogram, as a rough measure of variability, is approximately 6 standard deviations (6s).

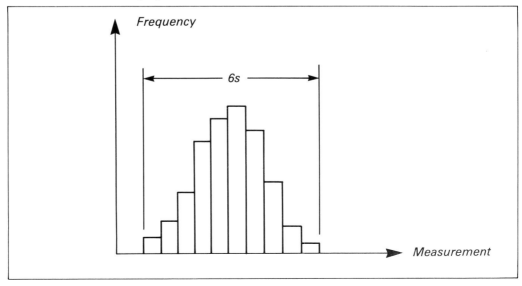

Fig. 4.20 Practical interpretation of the standard deviation

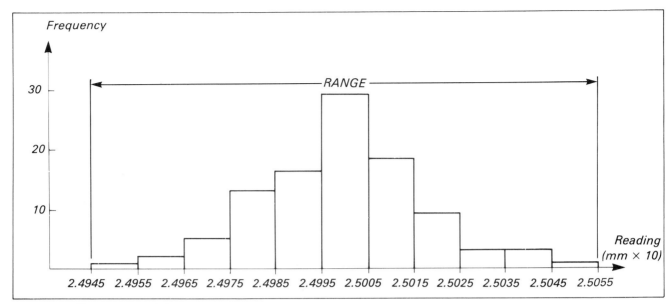

Fig. 4.21   Histogram and the range

This can be verified by considering an actual set of figures. Fig. 2.4 showed a typical set of readings which were represented as a histogram in Fig. 2.9 and displayed here again in Fig. 4.21. The values range from a minimum of 2.495 to a maximum of 2.505, giving a range of 0.010. For the total of 100 readings the standard deviation, s, can be shown to be 0.00181, and hence 6s = 0.011, which can be compared to the range value of 0.010.

Calculating s in practice

In practice the standard deviation can be readily obtained by the use of an appropriate pocket calculator, though the exact routine for determining s (and $\bar{X}$) may well vary depending on the make of calculator used. However, a typical keyboard layout will be as shown in Fig. 4.22. The numerical values are based on the readings analysed earlier and for which the appropriate calculations were shown in Fig. 4.19.

The following points will be helpful:

- ΣX is the sum of the 5 values. This figure of 2 agrees with the value obtained directly by adding together the 5 readings.
- ΣX^2 is the sum of the squared readings, i.e. the sum of X^2 not $(X - \bar{X})^2$. It is not actually required for the standard deviation calculation but there could be a need for recourse to this figure for other reasons. The value was not obtained previously but it would in fact be given by:

$$[3^2 + (-3)^2 + 2^2 + 1^2 + (-1^2)]$$

which is 24.
- n is the number of readings involved, i.e. 5. It is good practice to check this figure after the input of a set of readings. If the figure displayed does not tie in with the number of readings available, then an immediate error is highlighted.
- $\bar{X}$ is the mean value. If $\bar{X}$ and n are correctly displayed, then the $\bar{X}$ value must be that obtained from $\Sigma X/n$, and this should be the same as that determined from the original calculations (0.4).

ΣX^2	ΣX	n	$\bar{X}$	σ_n	σ_{n-1}
24	2	5	0.4	2.154	2.408

Fig. 4.22   Typical calculator display

- σ_n (pronounced sigma n) is one form of the standard deviation sometimes adopted. It is not the same as σ_{n-1}.
- σ_{n-1} (pronounced sigma n − 1) is the same as the measure that we have been calling s. The value of 2.408 is the same as that obtained by direct calculation.

It will be evident that the name sigma is used for both Σ and σ. This is because in the Greek alphabet Σ is the capital (upper case equivalent) version of the letter and σ the small (lower case equivalent) version.

4.5 **Sample and population**

It is somewhat unfortunate that different symbols are often used for the same statistical measures, almost at random with no consistency of approach. There would appear to be no hard and fast rules. The following definitions are based on the most common industrial practice. However, text books and articles in journals do tend to go their own way as regards notation, and so it is not surprising that confusion reigns. The following notation, therefore, is only a guide to the various symbols as related to the population or a sample from the population.

The symbol $\overline{X}$ is used for the mean of a sample, as defined previously. If it was the population from which that sample was taken that was of interest then exactly the same formula would be used to calculate the mean. However, the number of values used to determine the mean would correspond to the number of values in the population – for example the number of components produced over a long period of time by a machine. The Greek letter μ (pronounced mu) then represents the mean of that much greater group.

Similar distinctions are used for the standard deviation. The actual formula used for determining the standard deviation of a group (whether a sample or a population) is

$$\sqrt{\frac{\Sigma(X - \overline{X})^2}{n}}$$

Note that n, the actual group size, is used in the calculation. If the population standard deviation is being determined then the symbol σ is used.

In many practical situations the sample results are used to work back to find out how the process is performing, i.e. the evidence provided by the sample is used to estimate the corresponding values for the population. Given a sample mean of $\overline{X}$ then it can be shown that this result is what is called a best estimate of μ. A convention usually adopted is to indicate such a best estimate by putting a cap ($\wedge$) over the appropriate value. Hence $\overline{X} = \hat{\mu}$ Thus, at any instant, the sample means shown in sequence (see Fig. 4.6) can be taken as estimates of the corresponding population mean.

In the same way, a best estimate of the population standard deviation, $\hat{\sigma}$, is needed. Without going into the statistical reasoning, it can be shown that using

$$\sqrt{\frac{\Sigma(X - \overline{X})^2}{n}}$$

	Mean	Standard deviation
Sample	$\overline{X}$	s
Population	μ	σ
Population based on sample result	$\hat{\mu}$	$\hat{\sigma}$

Fig. 4.23 Sample/population notation

for the best estimate of σ based on a sample, does not give such a good result as using

$$\sqrt{\frac{\Sigma(X - \overline{X})^2}{n - 1}}$$

In other words the best estimate of σ, i.e. $\hat{\sigma}$, is given by

$$\sqrt{\frac{\Sigma(X - \overline{X})^2}{n - 1}}$$

which we have called s.

From now on this convention will be adopted here. In so doing the sample evidence is being used to work back to give the best possible estimate for that population measure. In other words, s and $\hat{\sigma}$ will be treated as one and the same thing.

In many ways reference to σ_n on calculators is a hindrance rather than a help. Sample results are really only there to provide a guide to the true population results. There is a school of thought which suggests eliminating σ_n from a calculator altogether. It would no doubt help the understanding.

There is an alternative way of determining $\hat{\sigma}$ and that is to make use of the mean range, $\overline{R}$. There will be more on this in Chapter 6. What is needed now is some further explanation on the use of 6s as a means of assessing, in a simplistic way, process capability and process improvement.

4.6 **Measuring improvement**

In Chapter 2 the relationship between a histogram of performance and the specification limits with which the performance was to be compared was assessed in a purely visual way. No attempt was made to provide some numerical measure of this relationship. The standard deviation provides a way of expressing in a numerical sense whether the readings represent a capable process.

It is necessary at this point to distinguish between a set of readings produced by a machine and that produced by a process

A machine analysis requires the individual readings to be obtained in sequential order. The subsequent pattern will represent a picture of the performance of the machine using the same material, and unchanged operating conditions, over a short time interval. Hence Fig. 2.10 represented the results of a machine capability study. A process analysis is based on a series of samples taken over a defined period of time. The corresponding pattern of points as shown in Fig. 2.11 is a secondary result of the analysis, but even so the pattern is useful as a guide to process performance

The simple idea of comparing a histogram with specification limits can therefore apply to either a machine or a process. Further detail will be provided in later chapters.

In Fig. 4.24 it will be assumed that it is a process that is being considered. For process A the spread of the readings is such that 6s is greater than the tolerance. Hence Tolerance/6s is a fraction which is less than 1 and corresponds to an incapable process. Process B, on the other hand, shows an improvement. The spread of the readings now matches the tolerance and Tolerance/6s is equal to 1. The process is just capable. Further improvement results in process C, where the standard deviation has been reduced further so that 6s is very much less than the tolerance. This represents a very capable process. The expression Tolerance/6s therefore provides a means of assessing improvement, the aim being to make this value as large as possible over time.

Fig. 4.24 Process improvement

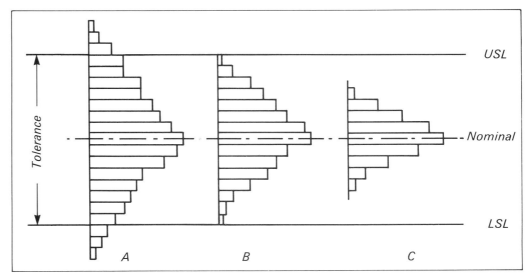

Tolerance/6s is a cumbersome expression, though, and for convenience, this variable is designated C_p. This notation is one which has been adopted by the West from the Japanese, and other indices are also available, in particular C_{pk}, which provides a measure of the setting in relation to the nominal (see Chapter 8). Progress in process improvement can therefore be represented by a corresponding increase in C_p value, and Fig. 4.25 summarises this.

It is probably fair to say that before the emphasis on SPC programmes by companies such as Ford, most Western industry was operating in an environment where C_p was less than 1. In other words, organisations were working with inherently incapable processes and were succeeding in maintaining operations because of a complete dependence on inspection-based systems. Unacceptable product was being inspected out at the end of the line with the associated result that quality appraisal and failure costs were excessively high.

An increasing emphasis on preventive programmes has meant that organisations have been aiming for C_p values in excess of 1 as a direct reflection of capability. However, there is a long way to go. C_p values of anything between 4 and 6 have been the norm for many years in Japan. The Japanese obsession with reducing variability about the nominal has resulted in better products, more consistency and greater reliability. The West has a long way to go, but there are nevertheless heartening signs.

For example, those organisations that have adopted the full company-wide implications of SPC have worked on achieving certain C_p values within a given time scale. The time scales involved are being measured in months if not in years. Catching up with the Japanese will be difficult enough even if the education system were geared to the task and the training departments of Western industry were

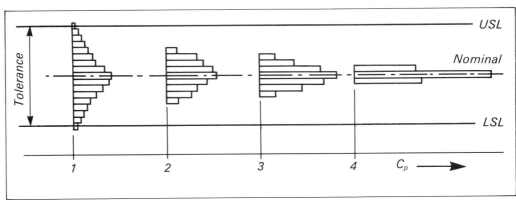

Fig. 4.25 Process improvement and C_p

primed for the job. However, with a training programme in SPC that is considerably adrift of the competition in terms of both relative position and scale, the situation is not as good as one could have wished.

In practice, C_p values can be determined much more quickly by using data from a control chart rather than by evaluating the standard deviation. However, a notion of the interpretation of C_p is not out of place in this chapter.

4.7   **Continuous improvement**

Ultimately, organisations will concentrate on the nominal for any process and reduce the variability about the nominal. A first stage, however, is to get the process setting on target. Fig. 4.26 illustrates the two features involved: making sure that the setting is on nominal and at the same time working on the variability to reduce it about the nominal figure.

Whatever the variability of the readings, an operator's responsibility is to ensure that the setting is right. The responsibility should be delegated to the shop floor by management, not in a dictatorial way but in a way which accepts that only the shop floor personnel have the knowledge and experience to carry out such a task. They should also have had the appropriate training. Reducing the variability about the nominal is a management responsibility. Having said that, management should encourage the shop floor to become involved in working together with them to reduce the variability about the nominal. Problem-solving groups provide the forum for improvement. It may be practical to split responsibilities but it is not advisable. A mutual approach to reducing the variability by looking at changes in the system is a sign of positive management.

The right-hand section of Fig. 4.26 emphasises the new way of thinking. If the variability is such that reject product is produced because of excessive variability, then optimum setting can no longer be considered. For example, it is no longer tenable to think in terms of reducing reject level to a minimum beyond one specification limit so that items beyond the other specification limit can be reworked. Those days are gone. The only way is to concentrate on reduced variability, and Fig. 4.25 summarises the approach.

With the setting on nominal, the variability is worked on so that an increasingly narrow band of readings is obtained. Even when this band comes within the tolerence zone, improvement must continue. How far an organisation proceeds

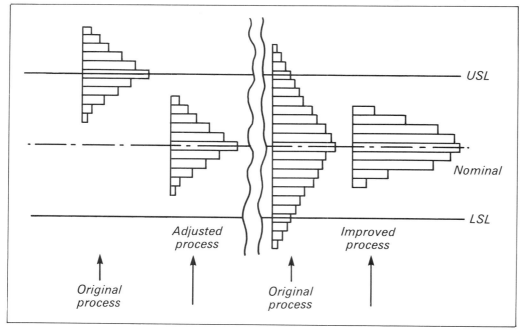

Fig. 4.26 Process adjustment and process improvement

down this path will be determined by commercial considerations. The philosophy of continuous improvement demands company-wide commitment. Concentrating on fine-tuning one process will not provide the overall company improvement which is being sought. Other processes which have not yet been considered will need to be tackled and an SPC programme must be structured to cater for all the processes involved.

Summary

This chapter has covered the basic measures for location (setting) and variability. It has introduced ideas on measuring improvement and emphasised the need to work on processes within a constructive wide-scale programme, so that there is a continuous reduction in process variability. These ideas are summarised here:

- Accuracy corresponds to location and precision to variability.
- Location problems can be resolved by operators. Variability problems require management action with operator involvement.
- The mean and the median are two types of average used in SPC.
- The mean has considerable advantages over the median.
- Σ means 'the sum of'.
- Control charts make use of a grand mean and also of a moving mean.
- Range and standard deviation are used to measure variation.
- Range is usually used in control charts because of its simplicity.
- Standard deviation is a better measure but is more complicated.
- For a reasonably symmetrical distribution the base width is approximately 6 standard deviations (6s).
- Always use the (n−1) form of the standard deviation formula.
- A machine and a process are not the same.
- Relating 6s to the tolerance gives a useful measure of improvement.
- Continuous improvement of variables requires continuous reduction in the variability about the nominal.

This chapter has considered quantities which can be used to represent setting and variability. The histogram has again been referred to a great deal and the notion of a sample as representing the population has been introduced, as has the fact that the base width of the histogram is approximately 6s. All these ideas need expanding and bringing together. They form the statistical basis of charting, and some appreciation of the normal curve and the effects of sampling is necessary before proceeding to the control chart. This will be provided in the following chapter.

Chapter 5 The normal distribution and sampling

5.1 Introduction

The statistical basis of the control chart for variables is the normal distribution. Even for attribute charting the normal distribution can be used as an acceptable practical basis, and therefore some knowledge of the properties of this distribution is necessary.

Before considering the normal curve one or two simple ideas on probability are required.

5.2 Measuring risk

Almost any standard text book on probability will make reference to the tossing of coins, or the throwing of dice. The examples are quite appropriate as an introduction to probability in a practical way. For example, Fig. 5.1 shows the results obtained when dice are thrown three at a time, and a record made of the score on each individual dice.

The results are based on a sample of 20 throws, i.e. 60 records in total. On the assumption that the dice are not biased and that they roll freely onto an even surface, there should be no more chance of one particular side being displayed than any other. This is reflected in the results. There appear to be more fives and less threes, but this is purely due to chance. With 60 throws, in fact, each side would be expected to turn up 10 times. As it happens only one of the sides turned up that number of times, but there is an indication that the top of each column is stabilising on the expected value.

If more results were taken then this trait would become more apparent. For example, with 600 results in total a picture appears like the one shown in Fig. 5.2. Here the top of each bar is quite close to the expected value of 100. If the number of throws was increased yet again to 6000, 60 000 and so forth, then the results based on the actual throws would become closer and closer to those based on the expected straight line (corresponding to 1000, 10 000 and so forth). In other words, what is known as a probability distribution can be fitted to the results of dice being thrown. In this case a rectangular distribution holds true, as shown in Fig. 5.3.

The situation changes if the results obtained for the total score recorded when three dice are thrown are considered. Again the dice are thrown 60 times. In this case, as Fig. 5.4 shows, the pattern is anything but rectangular and reflects the fact that there are far more ways of getting a total score of 9, for example, than there are of obtaining a total score of 4. Once again, if more results are obtained then the pattern becomes clearer. With 600 results in all a typical set of observed readings is as shown in Fig. 5.5. A rectangular distribution would clearly not be valid as an

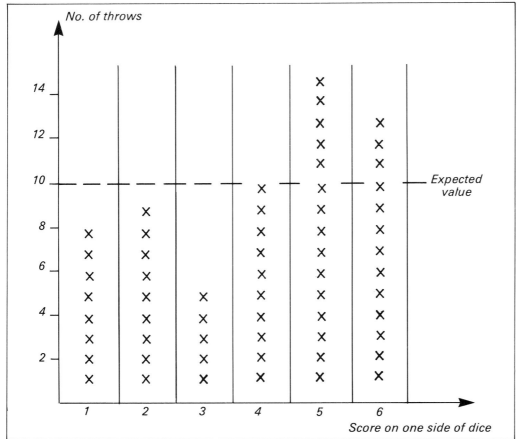

Fig. 5.1   Record of dice scores

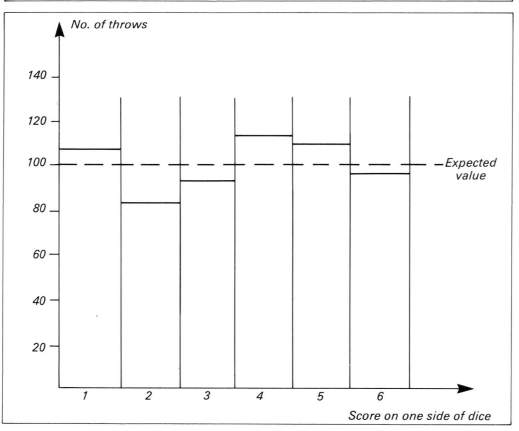

Fig. 5.2   Comparison of actual and expected scores

Fig. 5.3   Rectangular distribution

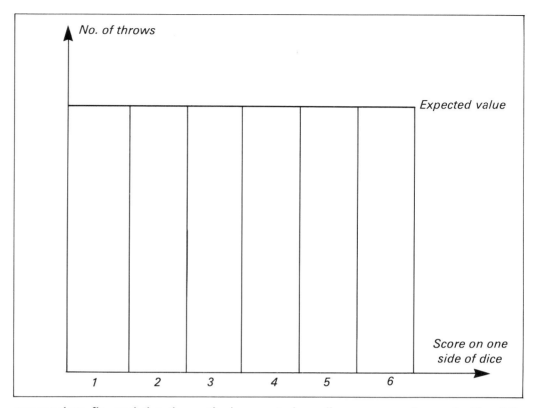

appropriate fit, and the theoretical expected readings are as shown in Fig. 5.6. Nothing would be gained by going into the details of how these expected readings have been generated. As the number of throws increases then the pattern of the observed readings gets closer to the symmetrical model shown in Fig. 5.6.

At this stage it will be useful to do some simple calculations. Out of the 600 throws of the three dice, 53 registered a total score of 8, for example. Thus the experimental probability of obtaining a total score of 8 is 53/600, which can then be compared with the theoretical value of 58/600. It may now be possible to recognise that probability can be expressed as a fraction. The maximum value of this fraction is 1, given by the chance of obtaining a total score which is either 3, or 4, or ..., or 17, or 18. Similarly the minimum value of this fraction is 0, which is the chance of obtaining a total score which is not either 3, or 4, or ..., or 17, or 18.

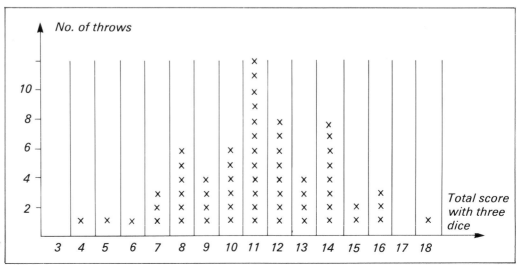

Fig. 5.4   Record of total score on three dice

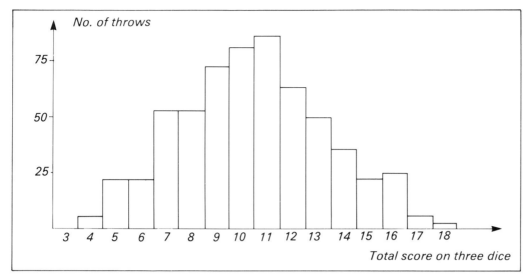

Fig. 5.5   Results of 600 throws

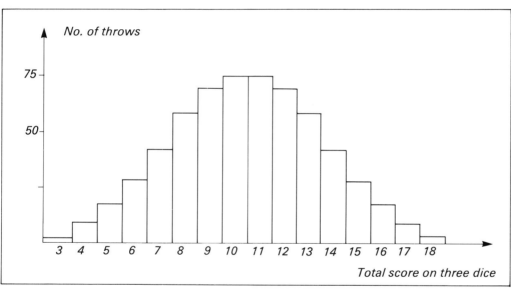

Fig. 5.6   Expected results for 600 throws

The dice example is sufficient to introduce the basic ideas of probability. These ideas now need to be related to the practice of manufacturing industry.

5.3 From theory to practice

In Chapter 2 a histogram was produced from a set of 100 readings (Fig. 2.9). Fig. 5.7 shows the same histogram again, but this time, for guidance, the actual frequencies have been indicated for each class interval.

Since the total number of readings is 100, the number in any given interval can be used to calculate the probability of obtaining a measurement in that interval. Fig. 5.8 illustrates the point. For example, 29 of the readings occur in the interval 2.4995 to 2.5005. There are l00 readings altogether and therefore an experimental value for the probability of obtaining a reading in the interval 2.4995 to 2.5005 is given by 29/100 = 0.29. In a similar way the probability of obtaining a reading in the interval 2.4985 to 2.4995 is 16/100 = 0.16.

Not only can other individual probabilities be evaluated in this way, but an important rule relating to probability also follows. If all the probabilities are added together (i.e. 0.01 + 0.02 + ... + 0.03 + 0.01) the total is equal to 1. This corresponds to the total area under the histogram. (It will only actually equal the area if frequency

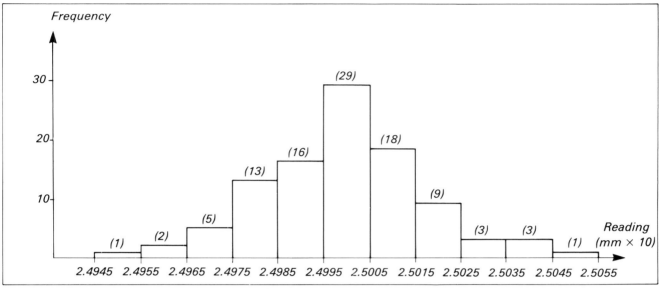

Fig. 5.7 Histogram for diameter of components

density rather than frequency is plotted on the vertical axis.)

There is no requirement to go into any depth regarding probability issues. A few important points only need emphasising.

- Total probability is numerically equal to 1. For example, for the information represented in Fig. 5.7, it is certain (i.e. there is a probability of 1) that a reading somewhere between 2.4945 and 2.5055 will be obtained.
- The symbol p is usually used for the numerical value of the probability. For example, in the case of Fig. 5.7, the probability of obtaining a reading in the interval 2.5015 to 2.5025 is given by p = 9/100 = 0.09. It may be useful to relate some of these probability figures to the last column shown in Fig. 5.9. This tally chart is the same as that shown in Fig. 2.5 but with the percentage column now completed.

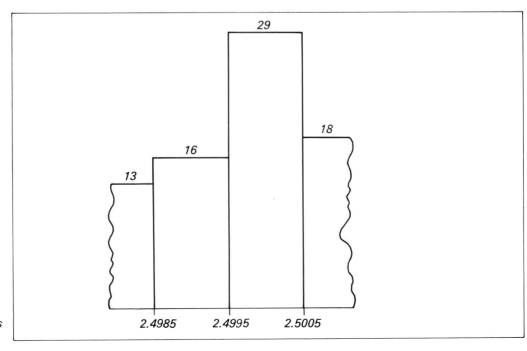

Fig. 5.8 Probability as an area

READING	TALLY		MARKS							TOTAL	%
2·505	/									/	0·01
2·504	///									3	0·03
2·503	///									3	0·03
2·502	ⅢⅢ	////								9	0·09
2·501	ⅢⅢ	ⅢⅢ	ⅢⅢ	///						18	0·18
2·500	ⅢⅢ	ⅢⅢ	ⅢⅢ	ⅢⅢ	ⅢⅢ	////				29	0·29
2·499	ⅢⅢ	ⅢⅢ	ⅢⅢ	/						16	0·16
2·498	ⅢⅢ	ⅢⅢ	///							13	0·13
2·497	ⅢⅢ									5	0·05
2·496	//									2	0·02
2·495	/									/	0·01
								TOTAL		100	/

Fig. 5.9 Check sheet and percentages

- There are two laws of probability. Fig. 5.9 assists in providing an illustration of the addition law. If the probability of obtaining a reading of either 2.502, 2.503 or 2.504 is required, for example, the answer is the sum of the separate probabilities (i.e. 0.09 + 0.03 + 0.03 = 0.015). The second law of probability can best be explained using the rules relating to control chart patterns. This will be done in Section 6.12.

- It is sometimes beneficial to relate probability values on a scale such as that shown in Fig. 5.10. Many familiar probability values relating to dice, coins and cards can be recorded on such a scale. In the context of control charts it will be found that constant reference will be made to a probability value of 0.0013 (which is about 1 in 1000), and lesser reference to the value of 0.025 (which is exactly 1 in 40).

5.4 Histogram and probability distribution

In Section 2.12 the notion of the area under the curve corresponding to the frequency was introduced. Now, instead of defining the vertical scale on the histogram as frequency density, it can be defined as relative frequency density,

which is

$$\frac{\text{Relative frequency}}{\text{Total frequency}}$$

Fig. 5.10   Probability scale

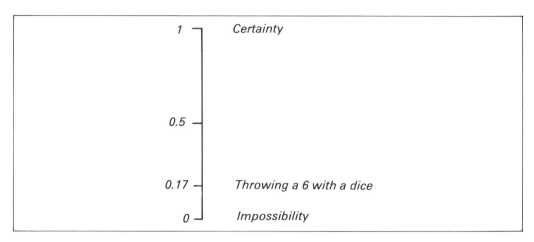

i.e.

Frequency/Class width
Total frequency

Fig. 5.11 illustrates the point. The total area under the histogram is now 1 and this ties up with the fact that total probability is unity. This relationship is necessary when the properties of the normal distribution are considered.

5.5 The normal distribution

In Fig. 5.6 a symmetrical pattern was generated. The normal distribution can be considered as a good fit to this, but a preferable way of introducing the normal curve is to make use of readings such as those represented in the familiar histogram shown in Fig. 2.9.

In practice it is often acceptable to superimpose a normal distribution on to the histogram as shown in Fig. 5.12. It is to be expected that a perfect fit is rare. However, the steps in generating a smooth curve can readily be seen in the

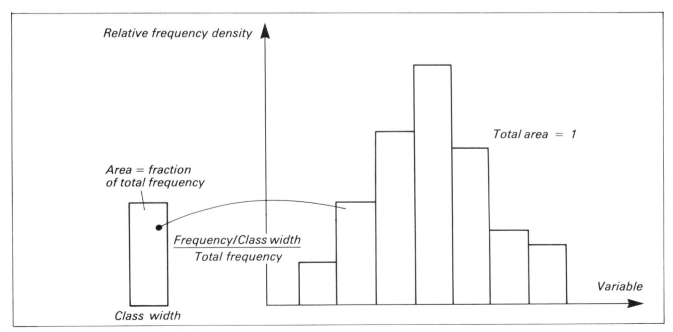

Fig. 5.11   Relative frequency as area

sequence in Fig. 5.13. By successively reducing the class width and at the same time taking more and more readings, it is possible to see the histogram approach the bell-shaped form.

5.5.1 **Properties of the normal distribution**

The normal distribution has important properties:

(a) *The total area under the curve is equal to 1 (or 100%)*

(b) *The area between any two vertical lines represents the probability of obtaining a reading in this interval (Fig. 5.14)*

This means that there is a direct comparison with the procedure whereby the area of a rectangle corresponds to a probability value. The value obtained from the normal curve is the theoretical one. The histogram gives the practical one.

(c) *The curve is perfectly symmetrical about the mean value*

All the curves in Fig. 5.15 are normal, but they have different mean values corresponding to different positions on the horizontal axis. In addition, since the base

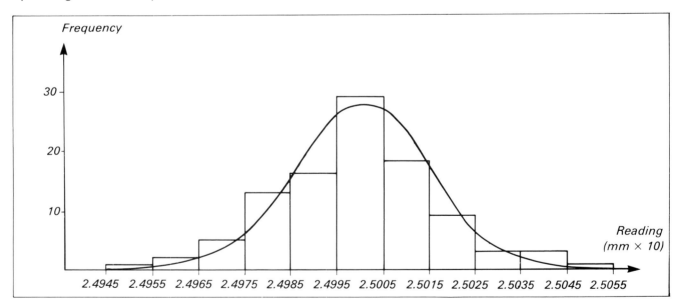

Fig. 5.12   Histogram and the normal curve

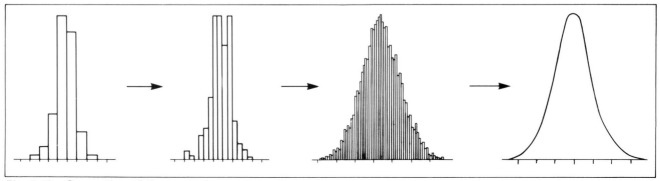

Fig. 5.13   Generation of the normal curve

Fig. 5.14 Probability as an area

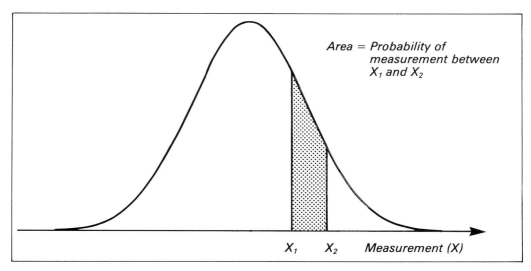

Area = Probability of measurement between X_1 and X_2

X_1 X_2 Measurement (X)

width of each curve is an indication of variability, then the wider the base the larger the standard deviation. All these curves, however, possess common properties.

Instead of dealing with various normal distributions with different means and different standard deviations, they can all be reduced to a common form. Fig. 5.16 illustrates the procedure. In practice the position of any normal distribution with respect to some absolute origin is not important. Of far more interest is the value of the mean, i.e. the position of the line of symmetry of the curve. It is logical, therefore, to reduce all normal distributions to a common reference point which is the position of the mean. This is obtained by subtracting the mean value from each value of X (for each separate distribution), giving $(X - \mu)$. The interim unit of measurement is now not the actual value of X but the distance from the respective mean. As the width of each distribution depends on the standard deviation, so it follows that the value of the particular standard deviation needs to be taken into account. This is done by expressing the quantity $(X - \mu)$ as a multiple of the respective standard deviation, i.e. a further quantity defined as $(X - \mu)/\sigma$ is introduced. This is denoted by Z.

The result of this is a particular normal curve (sometimes referred to as the standard normal curve) where the horizontal scale is defined in units of standard deviation measured outwards from the mean (Fig. 5.17). It is common practice to build up the idea of the standard normal curve using populations and the corre-

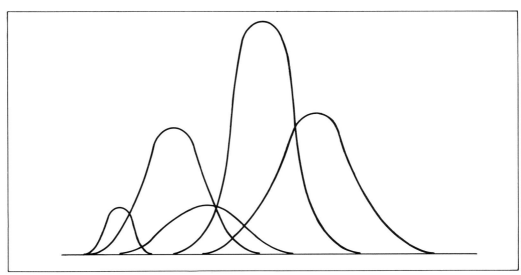

Fig. 5.15 Examples of normal curves

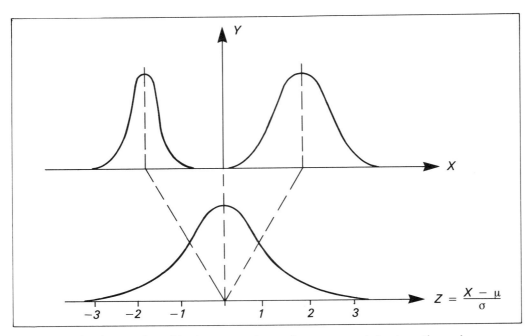

*Fig. 5.16
Standardising normal
curves*

sponding (μ,σ) notation. Having seen how the curve originates, there is now no reason not to revert to the more familiar ($\overline{X}$,s) notation as in Fig. 5.17 onwards.

(d) *The curve provides a practical interpretation of the standard deviation, s*

In Fig. 5.18 imagine a ruler (the tangent) rolling around the normal curve. As it rolls over the top of the curve towards *B* it is rotating in a clockwise direction. After *B* it rotates in an anticlockwise direction. Hence at *B*, where the curve crosses the tangent, the distance *AB* is 1 standard deviation.

It is also of interest (although more mathematical than practical) that the tangent at *B* cuts the base of the curve at 2s away from the mean.

(e) *The properties of the normal distribution make it possible to determine the areas enclosed within given multiples of s*

Fig. 5.19 explains why, as indicated earlier, the base width of a symmetrical histogram is approximately 6s. Practically all measurements relating to a mean value $\overline{X}$ fall within 3s on either side of $\overline{X}$, i.e. the probability of obtaining a reading within $\overline{X} \pm 3s$ is nearly 1 (0.9973 in fact).

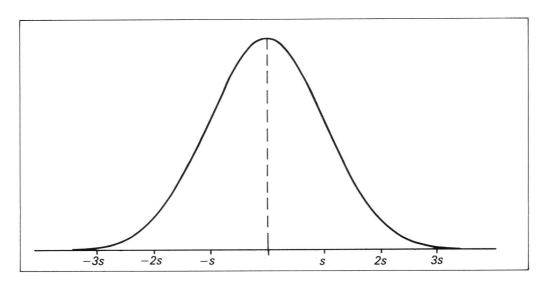

*Fig. 5.17 Standard
normal curve*

Fig. 5.18 Standard deviation and the normal curve

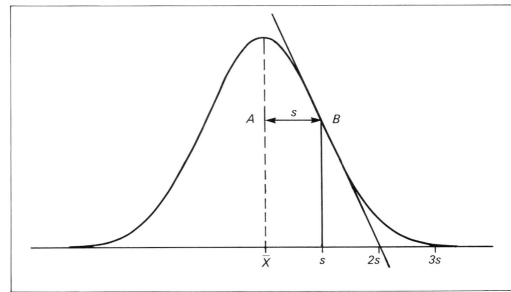

Fig. 5.19 Areas and multiples of s

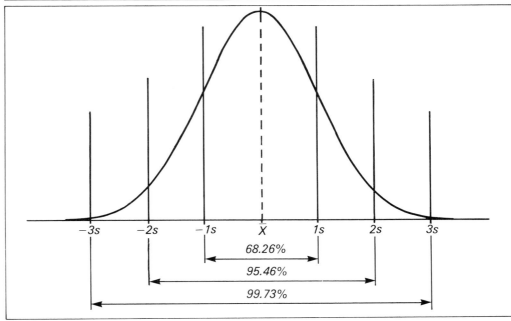

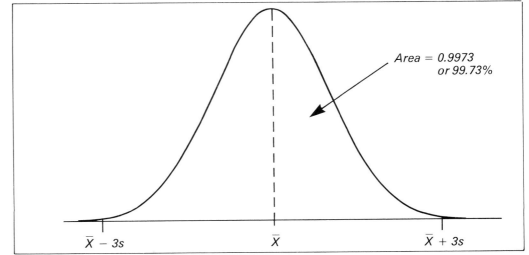

Fig. 5.20 Area and 3s

In terms of the relevance of these areas to the control chart, the curve needs to be looked at in a different way (Fig. 5.20). If 99.73% of readings fall within $\bar{X} \pm 3s$ it is equivalent to saying that $(1 - 0.9973) \times 100$, i.e. 0.27%, of the readings fall outside $\bar{X} \pm 3s$. In other words, the probability of obtaining a reading in one particular tail of the normal curve is 0.135% (Fig. 5.21).

The idea that all readings fall within a band width of 6s and that equally there is only about a 1 in 1000 chance of obtaining a reading beyond $\bar{X} + 3s$ (or similarly beyond $\bar{X} - 3s$) forms the basis of the performance-based control chart. The control lines will correspond to this 1 in 1000 risk and they provide a guide for detecting the presence of what are called special causes which must be sought out in practice.

5.5.2 Other areas under the normal curve

The areas quoted in Fig. 5.19 have been obtained from standard tables. They are based on Z values and the corresponding P_z values. Some explanation is necessary.

The earlier step of converting all readings to units of standard deviation measured from the mean is summarised in the result

$$Z = \frac{X - \bar{X}}{s}$$

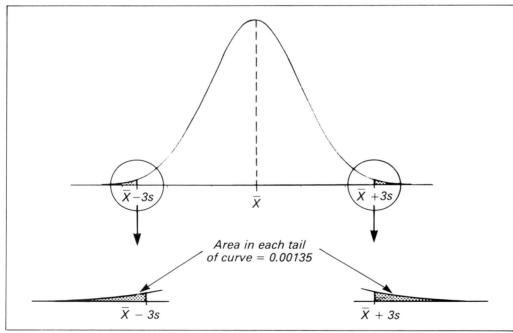

Fig. 5.21   Area in tails of the normal curve

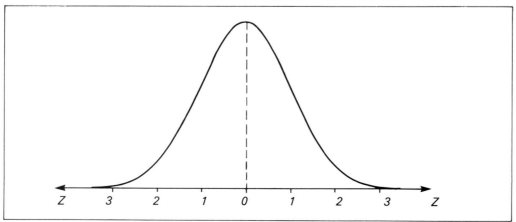

Fig. 5.22   Z values and the normal curve

(If a population is used then the corresponding result is $(X - \mu)/\sigma$.)

The standard normal curve then has Z values plotted horizontally on a scale shown in Fig. 5.22. Z is always taken as positive, whether it is being measured to the left or to the right of the mean. The mathematical properties of the normal curve now enable tables to be generated giving the area in the tail (denoted by P_z) corresponding to given Z values. Appendix B shows a typical table.

The familiar 0.00135 value corresponds to Z = 3. Also, for Z = 4 a P_z value of 0.00003 is obtained. The relevance of this figure will become clearer when machine capability studies are dealt with in Chapter 14. The tables provide the value of P_z (i.e. the probability of obtaining a value greater than Z) corresponding to standard increments of Z from 4.0 down to 0. Fig. 5.23 illustrates the relationship between the Z value (and the corresponding X value) and the area under the curve beyond that Z value (P_z) using Z = 1.33 as an example.

As an example of how to use the tables, the set of 100 readings introduced in Chapter 2 will be used. This is represented in familiar histogram form in Fig. 5.24. Upper and lower specification limits have been introduced at LSL = 2.496 and USL = 2.503. It is also known that the mean of the group, $\overline{X}$, is 2.5 and that the standard deviation, s, is 0.00181. Hence if a normal approximation is used, a curve as in Fig. 5.25 results.

The aim is to ascertain the percentage of product which is outside specification, as represented by the shaded areas.

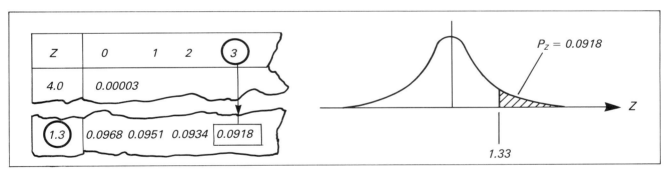

Fig. 5.23   Use of P_z table

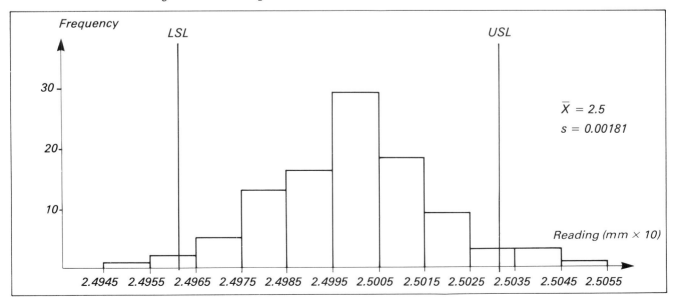

Fig. 5.24   Histogram for diameter of components

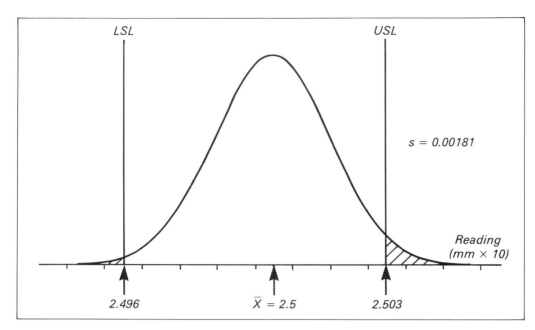

Fig. 5.25 Normal curve for diameter of components

Z_1 corresponding to $X = 2.503$ is given by

$$Z_1 = \frac{X - \bar{X}}{s} = \frac{2.503 - 2.5}{0.00181} = 1.66 \text{ (to 2 decimal places)}$$

Therefore

$$P_z = 0.0485$$

Z_2 corresponding to $X = 2.496$ is given by

$$Z_2 = \frac{X - \bar{X}}{s} = \frac{2.496 - 2.5}{0.00181} = -2.21 \text{ (to 2 decimal places)}$$

The negative sign is always ignored in cases such as this as it is only multiples of the standard deviation that are of interest, whether above or below the mean.

Therefore

$$P_z = 0.0136$$

Hence the percentage of product outside specification is

$$0.0485 + 0.0136 = 0.0621 = 6.2\%$$

It should be noted that there are different forms of P_z tables available. For example, some provide areas from the centre measured outwards. Provided that there is an understanding that the total area under the normal curve is 1 (or 100%) then it does not really matter how the areas are defined initially. Pocket-sized laminated reference cards are commercially available which carry a copy of the P_z table. They provide a quick way of assessing capability, but their use should not be overplayed. As will be seen the control chart is the key technique, and all that the P_z table does is to provide a retrospective guide as to whether the process is capable or not.

5.6 Continuous improvement

Instead of representing processes in the form of histograms as in Fig. 4.25, they can now be represented more appropriately by means of the bell-shaped curve. Continuous improvement can now be better represented as shown in Fig. 5.26.

It is worth repeating the point that the specification limits only provide a reference point from which to measure process improvement. Controlling the process is undertaken by making use of performance-based limits which correspond to the tail points of the normal curve. The actual control limits, however, take into account the fact that in most instances in practice samples are being taken from the populations represented by the normal distributions shown in Fig. 5.26. Some knowledge of the effect of this sampling will be useful in understanding the basics of the control chart for mean and range.

5.7 Sample and process

Chapter 6 will cover in more depth the practical issues relating to sampling, including the sample size to take, frequency of sampling, and numbers of samples to take when first starting up the control charts. Here the pattern built up by sample values and its relationship to the population from which those samples were taken will be considered.

The information partially presented in Fig. 4.6 is shown in full in Fig. 5.27. The 5 readings from the first sample, combined with the readings from the second sample, then the third, then the fourth, etc., start to form a pattern for the individual readings. Provided sufficient readings are obtained – and 20 samples of 5 would be quite acceptable – then the form corresponding to these readings can be taken as a

Fig. 5.26 Normal curve and continuous improvement

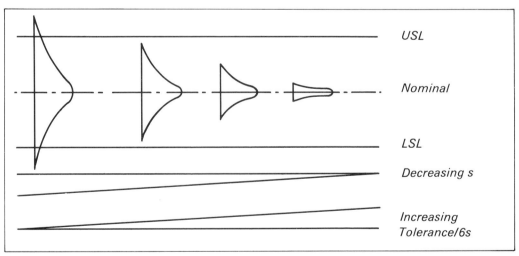

X_1	3	0	-6	3	4	-1	3	7	6	-4	-1	4	-1	-3	-4	-3	-2	5	-3	0
X_2	-3	-2	4	1	1	-2	-1	3	0	-1	-4	-6	-4	0	-5	-4	-3	2	-4	-1
X_3	2	0	-2	1	-1	-4	1	-2	-1	1	2	5	0	-1	4	-1	-1	1	-2	-1
X_4	1	2	-2	-1	-5	-7	4	-3	-2	0	3	-2	-1	0	1	-4	-1	-1	-8	-3
X_5	-1	-4	0	-2	2	-2	5	-2	2	2	1	-1	-5	1	-1	0	0	0	3	-5
ΣX	2	-4	-6	2	1	-16	12	3	5	-2	1	0	-11	-3	-5	-12	-7	7	-14	-10
$\bar{X}$	0.4	-0.8	-1.2	0.4	0.2	-3.2	2.4	0.6	1.0	-0.4	0.2	0	-2.2	-0.6	-1	-2.4	-1.4	1.4	-2.8	-2
R	6	6	10	5	9	6	6	10	8	6	7	11	5	4	9	4	3	6	11	5

Fig. 5.27   Data block from ($\bar{X}$, R) chart

good representation of the process itself, i.e. the population from which these samples were taken.

Fig. 5.28(a) shows a tally diagram for the 100 readings. Fig. 5.28(b) shows the pattern obtained when the sample means are recorded as a tally diagram. It can readily be seen that the individual readings show a pattern typical of many production processes – one which is symmetrical and approaching the familiar bell-shaped curve. Of course there is not perfect symmetry, and a smooth bell curve would not be expected, but allowing for the discrepancy between practice and theory one can be satisfied that a normal distribution is operating.

For the pattern of sample means, the picture is not so clear. Is it normal? Obviously not as it stands. However, there are enough sample means to show the form that is being built up, and what is clear is that there is reduced variation present in comparison with that for the individual readings. This is logical. The mean of any sample set of readings is a balance of the individual readings and it is therefore natural that the distribution of sample means will have a smaller spread. In the same way it would be expected that the pattern of sample means would be located at the same point on the horizontal axis as that of the individual readings, and again this can be seen.

Fig. 5.28 is based on actual readings from a process. It illustrates the relationship, based on sound reasoning, between the population distribution and the distribution of the sample mean.

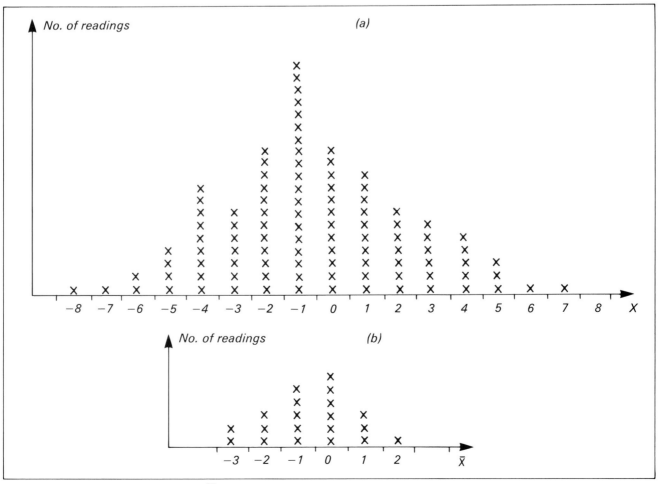

Fig. 5.28   Tally diagrams for X and $\bar{X}$

5.8 Comparing sample and population

Fig. 5.29 shows the normal curves produced when samples of varying size are taken from a normal population. Two results are known to be true:

- The mean of all distributions is the same and is equal to μ.
- The standard deviation of the distribution of sample means is related numerically to the standard deviation of the population (σ). With a sample of size n the former is in fact $\sigma/\sqrt{n}$.

Rather than repeatedly using the cumbersome expression 'the standard deviation of the distribution of sample means', this is referred to as the standard error (S.E.) and denoted by $\sigma_{\bar{x}}$ in this case because sample means are being considered. Hence $\sigma_{\bar{x}} = \sigma/\sqrt{5}$ for sample size 5.

This result can be checked using the set of 100 readings represented in Fig. 5.27. If these are fed into a calculator and the standard deviation calculated, it is found to be 2.945. This is therefore σ, and $\sigma_{\bar{x}}$ will be $2.945/\sqrt{5} = 1.317$. If the 20 values of $\bar{X}$ are used, the standard deviation of these is 1.47. Allowing for the fact that only 20

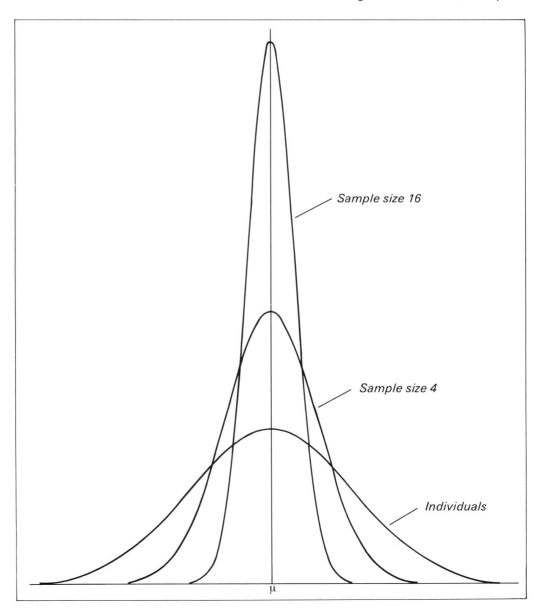

Fig. 5.29 Effect of sample size on variability

sample means are available there is sufficient agreement between the two standard deviation values to show that the theoretical result is likely. Also for the 100 readings the mean is -0.57 and this equals the mean of the 20 sample means as expected.

The effect of increasing the sample size, therefore, is to narrow the spread, as shown in Fig. 5.29. The significance of this result in terms of control charts will be seen in Chapter 6. The effect of changing the sample size is to alter the position of the control limits on charts for variables. Thus, if a sample of size 10 is taken, then the corresponding control limits which reflect the width of the normal curve will be closer together than those which relate to a sample of size 5. This does not mean that the respective processes are different. It has more to do with the sensitivity of the chart and its ability to detect a change in the process setting.

5.9 Sampling from a non-normal distribution

In the analysis thus far the parent population representing the process has been assumed to be normal. This may not always be so. The population could be represented by a skew distribution, for example, in that samples are being taken relating to the ovality of a bar. In fact this does not matter because the relationship between sample and population covered in Section 5.8 still holds true. Irrespective of the form of the parent population it can be shown that the distribution of the sample means approaches a normal distribution. The mean of this normal distribution is the population mean μ and the standard error is $\sigma/\sqrt{n}$. This is known in statistical terms as the central limit theorem

It is understandable that there is some restriction on the validity of this result and it is the sample size n that influences the extent to which the normal distribution can operate. Fig. 5.30 illustrates the logic.

Essentially, as the parent population shows less and less similarity to a normal distribution, then this is reflected in the fact that an increased value of n is required to generate a normal distribution for sample means. If the parent distribution is of the form shown in Fig. 5.30(a), then a sample of size 4 begins to show a trend towards symmetry and agreement with the normal curve. If the parent population is represented by the form shown in Fig. 5.30(b) then 4 is not a sufficien -tly large sample size to smooth out the irregularities.

In practice, in control chart terms, this is no problem. After all, a production operation can hardly be envisaged where there are no items produced on nominal and increasing numbers produced as one moves further away from the nominal — which is the practical equivalent of the top diagram in Fig. 5.30(b). For the great majority of processes the parent population is either sufficiently normal to be considered such or else it is skewed. With a sample of 5, which is typical, it can be taken for granted that the distribution of sample means is as near normal as makes no difference, and the important consequence is that the overall spread of the readings is smaller than that for the individual readings.

Perhaps more of a problem is the range. It must be emphasised that the expression $\sigma/\sqrt{n}$ only holds for the distribution of sample mean. There is nothing to be gained here by providing the equivalent standard error for the distribution of sample range. It is accepted that for a sample of size 5 the range is not normally distributed but skew. This has implications in terms of the relative position of control limits on the range chart, and these will be considered in Section 6.8.3.

Summary

This chapter has introduced simplistic ideas of probability and how they can be related to the normal distribution. A knowledge of the normal curve is essential for an adequate understanding of the control chart, including the reasons for control limits and interpreting the presence of special causes. An associated requirement is

Fig. 5.30 Parent population, sample size and the normal curve (based on a diagram in Statistics for Modern Business Decisions – Lawrence Lapin (1973) Harcourt, Brace Jovanovich)

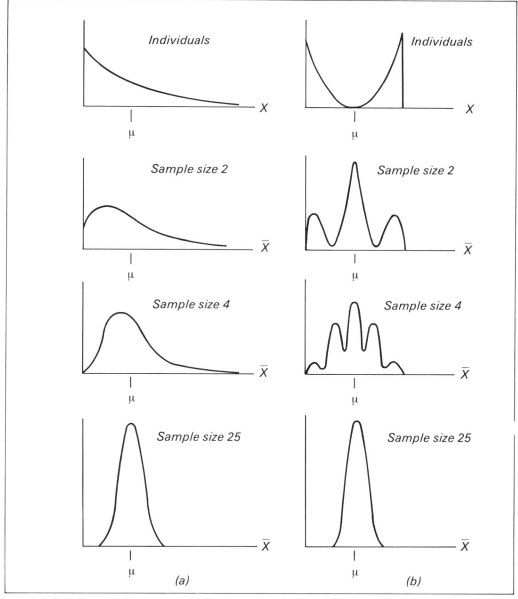

that the relationship between sample distribution and parent population is adequately appreciated. The material has been covered from a statistical viewpoint in this chapter; the practical implications will be dealt with when control charts are considered.

Here is a brief summary of the main points covered in the chapter:

- An understanding of probability is only required in sufficient depth to be able to appreciate the basis on which the control chart operates.
- A value of p = 0.0013 corresponds to the position of a control limit on a chart.
- Standard normal curves are measured in units of standard deviation away from the mean.
- Continuous improvement means gradually reducing the base of the normal curve over time when the curve is set symmetrically on the nominal value for the process.
- The standard error for the distribution of sample means is $\sigma/\sqrt{n}$.
- For any population, the distribution of sample means approaches a normal curve as the sample size increases.

● The normal distribution is essential to developing an understanding of the use of numerical techniques in controlling a process.

The critical technical feature in any SPC programme is the control chart. The basic building blocks which are used to generate a control chart have now been covered and the next few chapters will cover charting techniques in some depth.

Chapter 6 deals with the mean and range control chart for variables. Attributes will be considered later. The interpretation of charts is the same whatever is being plotted and Chapter 8 will focus on this.

Chapter 6   Control charts for variables

6.1   **Introduction**

This chapter will cover the most commonly used control chart for variables: the mean and range chart. Before any control chart is introduced, however, there are certain basic requirements which have to be satisfied. It is too easy to rush into collecting sample readings from a process without giving thought to defining the process, determining the features to be monitored, and specifying the measurement system.

Reference has been made earlier to the process. It may be straightforward to define. It may only involve a single operator on a simple machine. Even so that operator's activity is part of a much larger process with a chain of inputs and outputs, and a resulting need to appreciate the internal customer philosophy. More complex processes require detailed attention and should make full use of the experiences of the personnel associated with the process. Problem-solving techniques are important in this respect.

The Pareto principle is also useful in that it will highlight key problem areas where control charts can be used with maximum benefit. There is little to be gained by introducing charts in areas where the processes are running smoothly and not giving cause for concern. It makes sense to use charts to tackle the big problems and make inroads into solving some of the major quality issues.

What is being monitored? Does everyone understand? How and where is the information to be collected? What type of chart will be used? How acceptable is the measuring equipment? Is unnecessary variability present? These questions must be resolved, and clear guidelines laid down, at a very early stage. This is part of the management of the SPC programme.

6.2   **Charts for variables**

Control charts for variables enable the monitoring of the natural variability occurring in a process where the data is provided in measurable units rather than counted ones. The charts will then be used to reduce this variability around the nominal value. Charts are based on variability due to common causes and are used to determine the presence of special causes. These terms need explaining.

6.3   **Common and special causes**

Previous chapters have introduced the idea of piece to piece variation. If there is one idea more than any other that has permeated Japanese industry it is this recognition that one item will differ from the next one and will be different again to the previous one in a production process. As more and more readings are taken then a pattern of variability develops, as shown in Fig. 6.1.

There are many sources of variability causing this pattern. With engineering components, as those shown, the material used, the performance of the machine, the temperature of the coolant and many other factors will all be operating simultaneously, but not at a uniform level over a time. The net result is that the readings 'jump about' over a limited portion of a linear measurement scale while the natural pattern of variability gradually expresses itself.

These various factors contributing to the final pattern are known as common causes (sometimes also as random causes). They produce a pattern which, under stable conditions, is called a distribution. In the case of variables, the most common form of distribution is the normal distribution. Therefore, if more and more readings are taken the situation shown in Fig. 6.1 can eventually lead to a picture as in Fig. 6.2.

If only common causes of variability are present, then the distribution corresponding to a process remains stable over time. In other words, if the pattern at any one particular time is known, then it can confidently be predicted that at a later time the same pattern will repeat itself. Fig. 6.3 indicates how a knowledge of the behaviour of the process over time provides sufficient information to suggest that at a later stage the distribution will remain unchanged in terms of position, spread and shape. A process which behaves in this way is said to be in statistical control.

The normal distribution has been used as the appropriate model because it satisfies a great many processes. However, even if the distribution is not normal, the same argument can be applied. If the basic distribution is skew, as in measuring parallelism for example, then the skew distribution will remain constant over time and again can be predicted. The statistical reasoning behind the generation of the control limits will differ, but that does not invalidate the principle that a consistent distribution, irrespective of its form, can be used to predict future performance.

A distribution, then, can be defined by its position, spread and shape. The effect of changing any one of these is shown in Fig. 6.4. If the process remains unstable, with changes in position, spread and possibly even shape occurring as time passes, then special causes are said to be present which produce this distortion. The

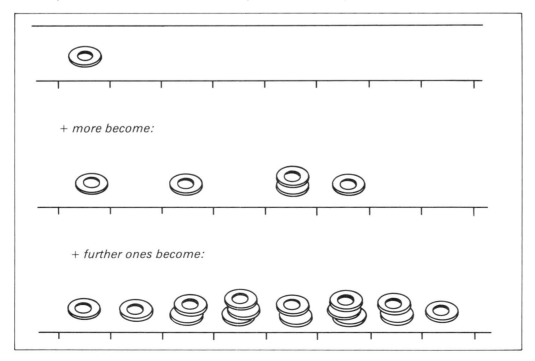

Fig. 6.1 Pattern of variability

Fig. 6.2 From histogram to normal curve

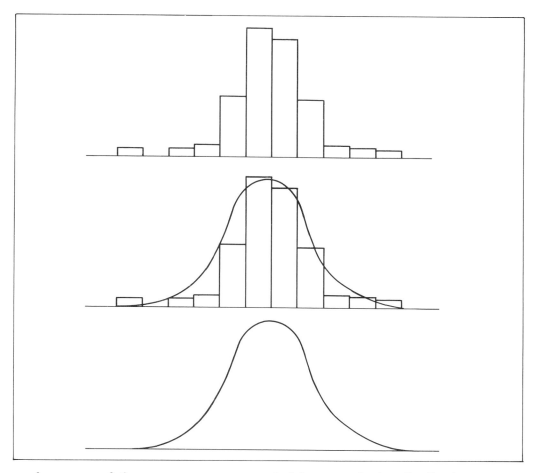

performance of the process as represented by a particular distribution can no longer be predicted, and such a process is said to be out of statistical control, as shown in Fig. 6.5.

If a process is out of statistical control it cannot be known with any degree of certainty how it will perform in the future. Questions relating to delivery times. material usage or production output could not be answered with any degree of accuracy. In other words, there is no consistency of performance. More than that it would not be adequately known how performance could be improved. Again, there would be no information on whether the recurring problems faced daily were the result of management action (or rather lack of it) or operator action (or equally lack of it). Control charts provide the key. They require three stages: measurement, control and improvement.

The first stage in setting up the chart is the collection of data to determine the characteristics of the distribution. The process is then brought under statistical control by eliminating the special causes. Continuous improvement then means reducing the variability about the nominal value. Fig. 6.6. illustrates these stages. Another way of looking at things is represented in Fig. 6.7. It may be too simplistic

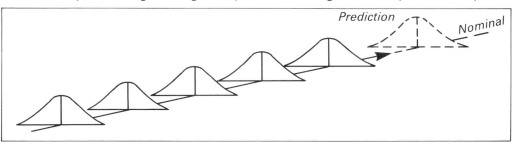

Fig. 6.3 Predictable process

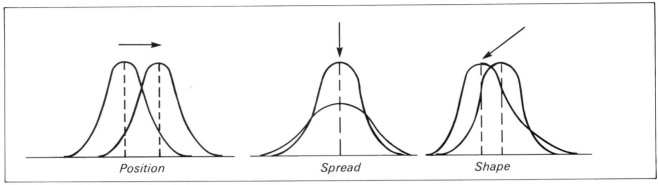

Fig. 6.4 *Position, spread and shape*

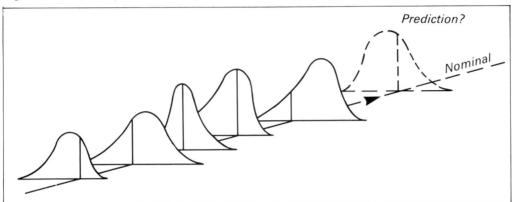

Fig. 6.5
Unpredictable process

for control engineers, but it adequately represents, in block diagram form, the diagrammatic approach used previously.

The first step is to define a measurement of performance for the process. In industrial processes the appropriate measurement of performance is readily obtained: the temperature of the oven, value of chemical constituent, dimension of a machined component, etc. For other processes, particularly administrative ones, the measures cannot be determined as easily. However, they can be found: time taken to ship a product, time to deliver, response time for a particular computer program for example. (It has to be said that administrative applications of charts for variables are not so common as those relating to attribute charts, and Chapters 10, 11 and 12 will provide many applications in attribute areas.)

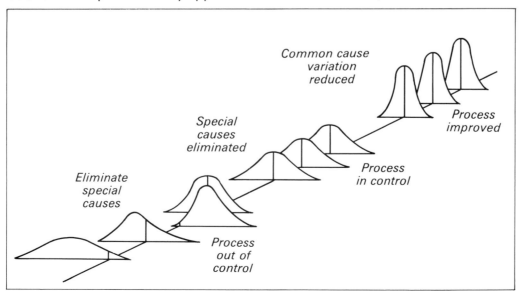

Fig. 6.6 Instability, control and reduced variability

Fig. 6.7 Process improvement loop

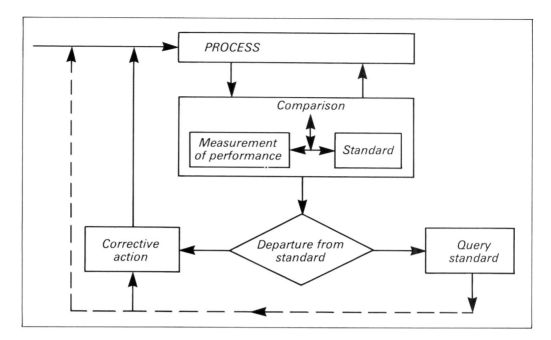

Having obtained the measure of performance it is then compared with a standard. If there is a difference, then it is advisable to check on the standard. (Was it changed? Has the calibration been correctly carried out?) Similarly, is the measuring equipment used of an adequate standard to monitor the performance? In one organisation using electronic gauging equipment an operator input was required in the form of a thump at a particular place on the cabinet to provide the correct digital display of the reading. A case like this does not reflect well on the management concerned. The operators can see that, as usual, the purchasing department has bought on the basis of cheapest is best. The shop floor staff concerned would consequently have little faith that a new way of life is around the corner.

If there is a real difference between the measurement of performance and the standard then action must be taken. If the special cause results in a deterioration in the process performance then the cause must be sought and eliminated. On the other hand, if the special cause represents improvement, then the reason for that cause should be found and integrated into the system. It will be seen how these two approaches are represented in the way that control charts themselves are interpreted.

6.4 Measuring the process

If the process is to be improved it must first be measured. How can this be done? What cannot be done is to monitor the whole output. It is retrospective and in any case is just not practical for many production processes. Production rates are such that one would be surrounded with data even within minutes – literally buried under cans of dog food, connectors or cigarettes. An alternative could be to take a large sample daily, hourly, or in certain cases every 15 minutes for example. From this data a histogram could be constructed. But again, any information is available too late to take appropriate action. In any case the histogram does not provide the whole story, as will be seen in Chapter 7. The remaining option is to take samples from the process at regular intervals and use the characteristics of the sample as an indication of how the process is performing. Sample size is considered in Section 6.13. As the sample size changes then it will require appropriate calculations which will be reflected in corresponding changes in the control limits.

Two measures of the process are critical: its position and spread. To monitor the position of the process the sample mean (or sample median) is used as an indicator of the process mean (Fig. 6.8). As the process mean changes, then so does the sample mean. Similarly, the sample range is taken as an indication of how the process spread is changing (Fig. 6.9). The sample standard deviation is really a better measure to use, but there are strong reasons for using the range. These will be covered later in the chapter. Process shape is not usually a major cause for concern. There are ways of monitoring this but they are adequately covered by some of the more technical statistical texts.

Thus for most applications across industry in general, sample means and sample ranges are plotted over time, and this is done on the control chart.

6.5 The control chart

All control charts follow the same pattern in that there are three main sections. One allows for the recording of data and the calculations that go with it. A second block provides a grid on which the appropriate results can be plotted. A third block allows for the basic administrative information to be recorded. Typical control charts appear in Appendix H. More detailed consideration of some of these issues appears in Chapter 8.

It is necessary now to plot some values to gain appreciation of the patterns that are generated on control charts and of how statistically based rules can be applied to detect the presence of special causes.

6.6 Calculating and plotting

Fig. 6.10 shows a section from a typical control chart corresponding to the initial stage of taking samples, and then calculating and recording.

Twenty samples of size 5 are recorded. The readings are measured in units from some standard value and could apply to a host of situations. The case study which

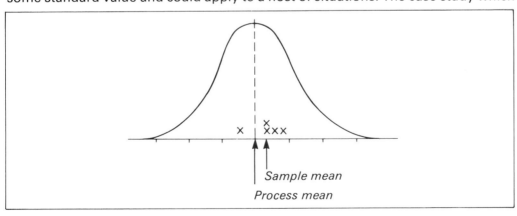

Fig. 6.8 Sample mean and process mean

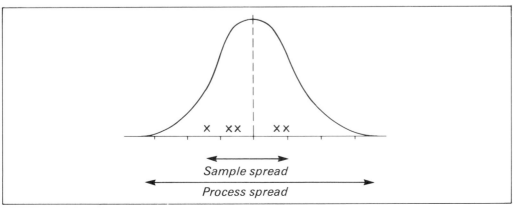

Fig. 6.9 Sample variability and process variability

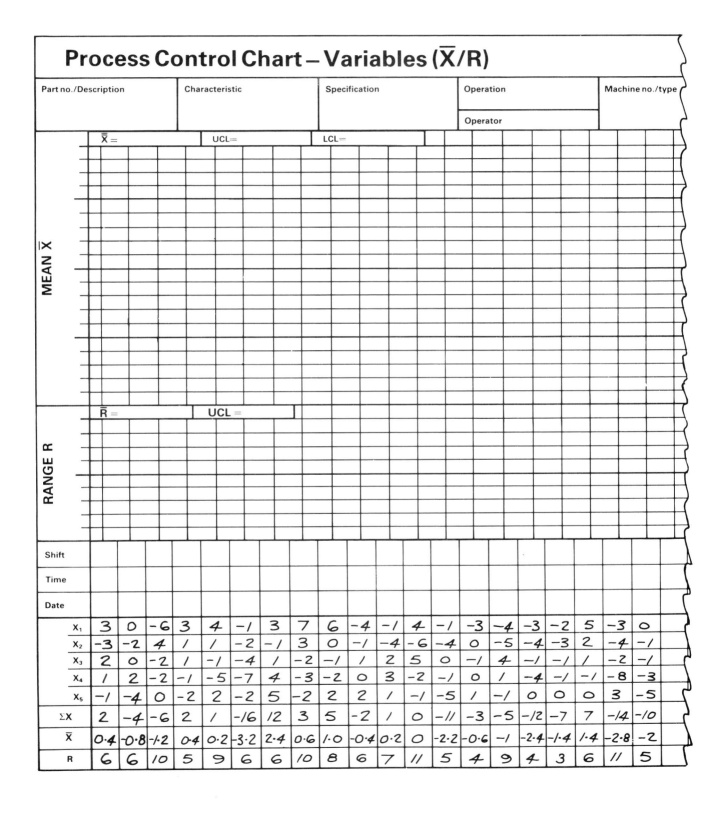

Process Control Chart – Variables (X̄/R)

Part no./Description	Characteristic	Specification	Operation	Machine no./type
			Operator	

X̄ =   UCL=   LCL=

MEAN X̄

R̄ =   UCL =

RANGE R

Shift																				
Time																				
Date																				
x_1	3	0	−6	3	4	−1	3	7	6	−4	−1	4	−1	−3	−4	−3	−2	5	−3	0
x_2	−3	−2	4	1	1	−2	−1	3	0	−1	−4	−6	−4	0	−5	−4	−3	2	−4	−1
x_3	2	0	−2	1	−1	−4	1	−2	−1	1	2	5	0	−1	4	−1	−1	1	−2	−1
x_4	1	2	−2	−1	−5	−7	4	−3	−2	0	3	−2	−1	0	1	−4	−1	−1	−8	−3
x_5	−1	−4	0	−2	2	−2	5	−2	2	2	1	−1	−5	1	−1	0	0	0	3	−5
ΣX	2	−4	−6	2	1	−16	12	3	5	−2	1	0	−11	−3	−5	−12	−7	7	−14	−10
X̄	0·4	−0·8	−1·2	0·4	0·2	−3·2	2·4	0·6	1·0	−0·4	0·2	0	−2·2	−0·6	−1	−2·4	−1·4	1·4	−2·8	−2
R	6	6	10	5	9	6	6	10	8	6	7	11	5	4	9	4	3	6	11	5

Fig. 6.10 Section from control chart

follows shows a specific application. The 20 samples reflect the guideline that in carrying out the initial process study at least 20 samples are required, although certain local conditions may require 25, 30 or even more samples. These guidelines hold for any type of chart, not just the mean and range chart. These 20 samples would be taken over a sufficient period of time to allow the natural variability to express itself. It may be measured in hours, days or even months, depending on the production rate, but common sense will indicate that the greater the production rate the greater the benefits to be obtained from taking a sample of 5 as often as possible. A typical sampling frequency would be five per hour. For each sample of 5, the mean and range can be calculated and recorded. These values can then be plotted. It can be seen that the sample mean is recorded on one section of the control chart and the sample range on a second section.

In setting up the control charts an appropriate scale should be used. It is here that some difficulty arises. Should the sample results be plotted before the control limits are calculated, or vice versa? In practice a useful rule is to calculate control limits first, because they provide a good guide when determining the scales used on the charts. However, because the concern at this stage is understanding how the chart works, the first step here will be to plot sample means and sample ranges.

Some rules for determining the appropriate scales are available. One suggestion is to choose a scale for the $\overline{X}$ chart so that the difference between the extremes on the scale is twice the difference between the greatest and least values for $\overline{X}$. For the R chart, a scale should be chosen which extends from 0 to twice the maximum value of R.

For the set of values given in Fig. 6.10 the least value of $\overline{X}$ is -3.2 and the greatest value is 2.4. This gives a difference of 5.6 and the corresponding scale on the $\overline{X}$ chart has therefore been chosen to span twice this value, i.e. 11.2. At the same time the nominal for the process should be set centrally on the vertical scale. Similarly, the maximum value of R is 11 and the scale on the R chart has been chosen to span twice this value, i.e. 22 units, with a minimum value of 0.

The appropriate scales are shown in Fig. 6.11, where the $\overline{X}$ and R values for the 20 sample readings have been plotted. In plotting the mean and range, the vertical line corresponding to the centre of the corresponding data column should be used. A cross is drawn where this line intersects the horizontal line corresponding to the $\overline{X}$ value, and similarly for the R value. The crosses are joined by a series of straight lines which helps to bring out the pattern represented.

What do these plots say about the process? Is the mean or range under statistical control? Are any changes in the pattern evident, such as trends, high or low spots, or any other unusual features? Are there some underlying causes which are influencing the pattern?

It is not possible at this stage to determine whether any features which may seem apparent on the two charts are due to chance or to the presence of some underlying factor. The patterns show a reasonable enough situation on the basis of experience of just looking at a time series of points. There may be a suspicion in places of untoward happenings, but without some guidelines based on statistical reasoning no real conclusions would be justified. Some information is therefore needed on how to set up control limits and central lines for the charts so that sounder judgement can be made.

6.7 The basis of control limits

The guidelines for control charts for variables are based on the properties of the normal curve. These were discussed in some detail in Chapter 5, the important point here being that well over 99% of the values involved fall within a band width of 6 standard deviations – plus and minus 3 standard deviations measured outwards from the central line (Fig. 6.12).

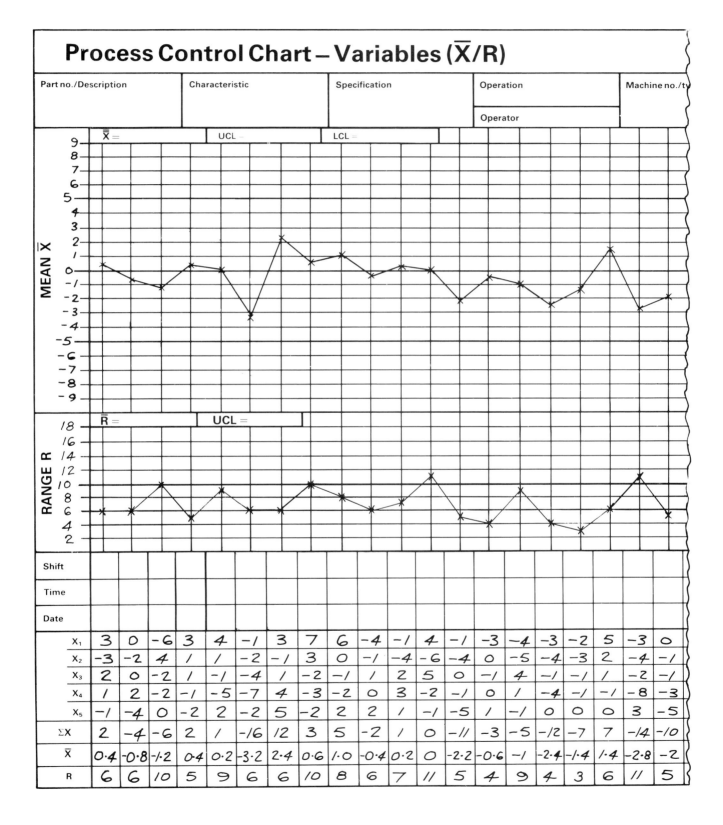

Process Control Chart – Variables (X̄/R)

Part no./Description	Characteristic	Specification	Operation	Machine no./ty
			Operator	

X̄ =   UCL =   LCL =

R̄ =   UCL =

Shift																				
Time																				
Date																				
X₁	3	0	-6	3	4	-1	3	7	6	-4	-1	4	-1	-3	-4	-3	-2	5	-3	0
X₂	-3	-2	4	1	1	-2	-1	3	0	-1	-4	-6	-4	0	-5	-4	-3	2	-4	-1
X₃	2	0	-2	1	-1	-4	1	-2	-1	1	2	5	0	-1	4	-1	-1	1	-2	-1
X₄	1	2	-2	-1	-5	-7	4	-3	-2	0	3	-2	-1	0	1	-4	-1	-1	-8	-3
X₅	-1	-4	0	-2	2	-2	5	-2	2	2	1	-1	-5	1	-1	0	0	0	3	-5
ΣX	2	-4	-6	2	1	-16	12	3	5	-2	1	0	-11	-3	-5	-12	-7	7	-14	-10
X̄	0.4	-0.8	-1.2	0.4	0.2	-3.2	2.4	0.6	1.0	-0.4	0.2	0	-2.2	-0.6	-1	-2.4	-1.4	1.4	-2.8	-2
R	6	6	10	5	9	6	6	10	8	6	7	11	5	4	9	4	3	6	11	5

Fig. 6.11   Chart with scales and plotted values

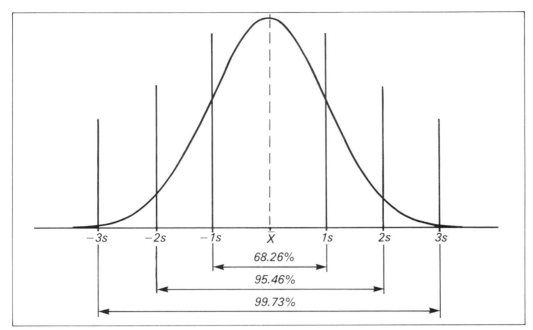

Fig. 6.12 Areas and multiples of s

It has been established that for a variety of reasons the individual values which generate the population cannot be used for the control chart. Instead samples are used, so the knowledge of sampling distributions accumulated from Chapter 5 needs to be reinforced. Fig. 6.13 shows the build-up of information as more and more samples of the process are taken. For each sample there is a value which is representative of that sample and at the same time an indicator of how the process (i.e. the parent population at that time) is performing. To illustrate the reasoning only the pattern of sample means will be considered. A similar argument, however, holds for other quantities.

It is known that in the same way as individual readings build up to form a normal curve, then similarly the sample means form a normal distribution but with a narrower spread than that for the individual readings, as shown in Fig. 6.14. For any normal distribution, 99.73% of the readings fall within plus and minus 3 standard deviations measured outwards from the central line. Hence, since it is now sample

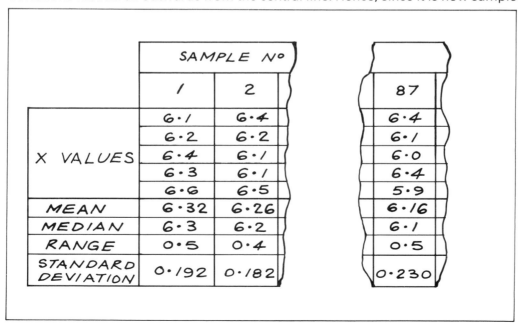

Fig. 6.13 Data from samples

means that are being considered, then equally 99.73% of the sample means satisfy this requirement. In other words, the chance of a sample mean falling outside the lines shown in Fig. 6.15 is extremely small. This particular normal curve therefore becomes the basis of the control chart for sample means.

Conventionally, the line of symmetry of the normal curve is made to be horizontal. The normal curve in Fig. 6.15 is therefore rotated, which rotates as a result the lines indicating 3 standard deviations and the central line. The control limits of the control chart for sample means then appear as in the last diagram in Fig. 6.16. The

Fig. 6.14   Individuals and sample means

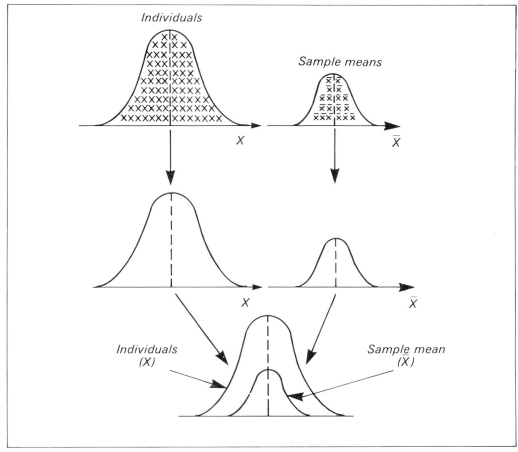

Fig. 6.15 Normal curve for X̄

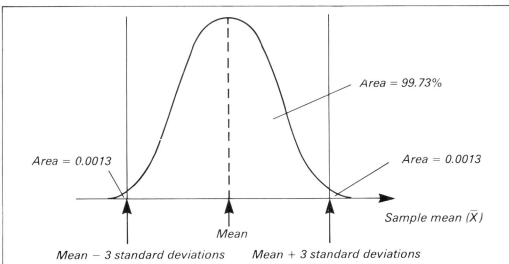

probability of a point falling outside one of the control limits is then equal to 0.0013, and is so slight that if it does occur a special cause can be assumed to be present.

In order to monitor the process fully, it is logical for a central line also to be drawn on the control chart. It will be seen that this central line corresponds to the overall process mean (and, by implication, if the process is to be correctly set, the nominal). The control chart for sample means will therefore have three lines, as shown in Fig. 6.17. Typically, continuous lines are used for the control limits and a broken line for the central line.

Control charts for the range follow the same pattern, as indeed do other charts such as those for standard deviation, moving mean and attributes. It may be that the lower control limit is not always present, for reasons to be discussed in Section 6.8.3. That will not invalidate the principle of using the performance-based chart to determine the presence of special causes which can be identified using rules to be considered shortly.

6.8 Setting up the charts

It has been seen how the control limits originate. What is now required is to go through the steps in obtaining numerical values for control limits and central lines from a given set of data. The necessary information is already available in Fig. 6.10.

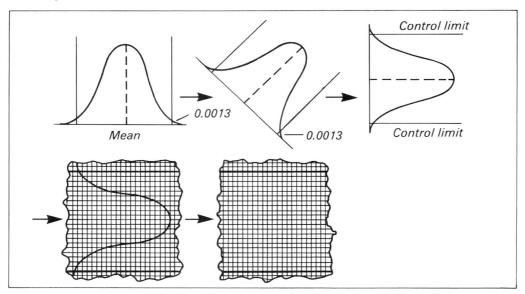

Fig. 6.16 Generating the control limits

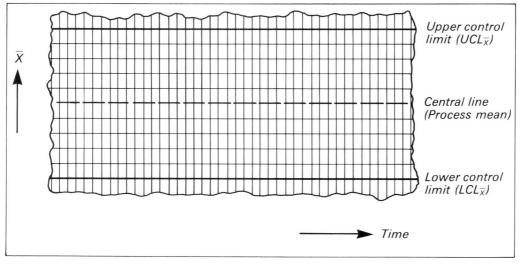

Fig. 6.17 Lines on $\bar{X}$ chart (US)

6.8.1 Central lines

In order to determine the control limits, the values of the central lines for both $\bar{X}$ and R are needed.

(a) *Central line for $\bar{X}$ chart*

The line corresponds to the mean of all sample means. It is denoted by $\bar{\bar{X}}$ and, as introduced in Chapter 4, is given by:

$$\bar{\bar{X}} = \frac{\Sigma \bar{X}}{k}$$

where k is the number of samples.
Ideally $\bar{\bar{X}}$ and the nominal, ϕ, should be the same, but this is unlikely to occur on the basis of the first 20 samples. Using the given data:

$$\bar{\bar{X}} = \frac{0.4 - 0.8 + \dots -2.8 - 2}{20}$$

$$= -0.57$$

(b) *Central line for R chart*

The line corresponds to the mean of all the sample ranges. It is denoted by $\bar{R}$ and is given by:

$$\bar{R} = \frac{\Sigma R}{k}$$

where k is the number of samples.
Therefore

$$\bar{R} = \frac{6 + 6 + \dots + 11 + 5}{20}$$

$$= {}^\cdot 6.85$$

6.8.2 Control limits for $\bar{X}$ chart

The control limits (upper control limit [UCL] and lower control limit [LCL]) are to be set at 3 standard deviations $(3\sigma_{\bar{x}})$ away from the grand mean $\bar{\bar{X}}$, i.e.

$$UCL_{\bar{x}} = \bar{\bar{X}} + 3\sigma_{\bar{x}}$$

$$LCL_{\bar{x}} = \bar{\bar{X}} - 3\sigma_{\bar{x}}$$

How is the standard deviation $\sigma_{\bar{x}}$ determined in such a case? Use is made of the fact that there is a relationship between the standard deviation of the individual readings in the population and the mean range $\bar{R}$ of samples of a given size taken at regular intervals from that population. This relationship is such that $3\sigma_{\bar{x}}$ is equivalent to $A_2\bar{R}$, where A_2 is a constant which depends only on the sample size n. Hence

$$UCL_{\bar{x}} = \bar{\bar{X}} + A_2\bar{R}$$

$$LCL_{\bar{x}} = \bar{\bar{X}} - A_2\bar{R}$$

Tables of constants are available which enable A_2 values to be obtained for given values of n (see Appendix C). For those who are interested, an explanation of how the A_2 values have been determined is provided in Appendix E.

For a sample size of 5 it is seen that the corresponding value of A_2 is 0.577. Hence, for the readings being considered,

$$UCL_{\bar{X}} = -0.57 + 0.577 \times 6.85$$

$$= 3.382$$

Similarly

$$LCL_{\bar{X}} = -0.57 - 0.577 \times 6.85$$

$$= -4.522$$

Limits can now be drawn on the appropriate section of the control chart together with a line corresponding to $\bar{\bar{X}}$. These are shown in Fig. 6.18. For reference the data boxes have also been completed.

6.8.3 **Control limits for R chart**

The distribution of sample range is not normal but skew, and particularly so for a sample size of 5. This means that it is not valid to think in terms of the mean ± 3 standard deviations as a basis for generating the control limits. Nevertheless it is still statistically possible to determine other constants corresponding to A_2 which will again enable the limits to be fixed at probability levels of approximately 1 in 1000. The statistical reasoning behind the determination of these constants is beyond the scope of this book (and is really not necessary in any case). All that is needed are the values of these new constants, D_3 and D_4, which when multiplied by $\bar{R}$ give the appropriate limits.

Hence

$$UCL_R = D_4\bar{R}$$

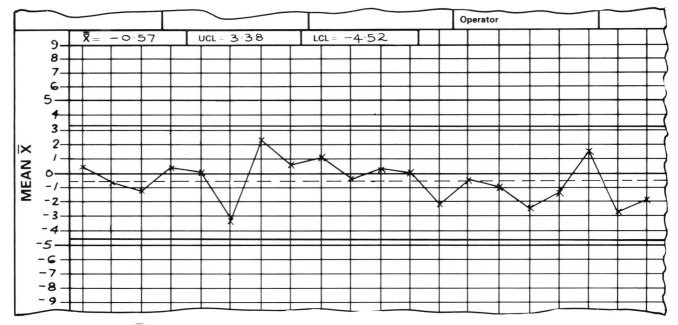

Fig. 6.18   Section of $\bar{X}$ chart

For a sample of size 5, the value of D_4 from Appendix C is 2.114, and for the set of readings being considered, $\bar{R}$ is 6.85. Therefore

$$UCL_R = 2.114 \times 6.85 = 14.48$$

Again, lines can be drawn on the R section of the control chart and the data boxes completed, as shown in Fig. 6.19.

D_3 is the constant corresponding to the lower limit LCL_R, and it can be seen from Appendix C that for samples of size 6 or less D_3 is zero (or at least as sufficiently close to zero to be treated as such). The explanation for this is associated with the fact that the distribution of R is skew. As the sample size increases, the distribution becomes less skew (and more symmetrical) and the value of the LCL changes. At a critical sample size of 7 the LCL can be registered on the control chart as a line distinguishable from the horizontal axis. Fig. 6.20 indicates the reasoning by looking at the distributions (and corresponding control limits) for sample sizes of 2, 5, 7, 10 and 20, using an $\bar{R}$ value of 6.85 in each case.

The LCL should always be drawn on the range chart if the sample size allows for it to be present, as it can be used to determine the presence of a special cause indicating improvement which can then be incorporated as part of the system. The use of the control chart for R in this way to trigger off process improvement is considered in detail in Chapter 7. It is a very important aspect of using control charts as a means to continuous improvement.

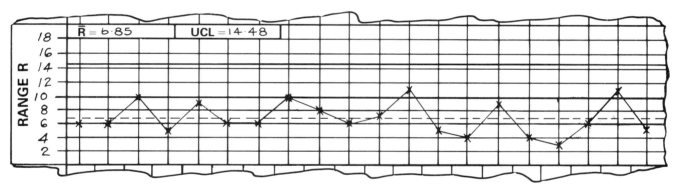

Fig. 6.19   Section of R chart

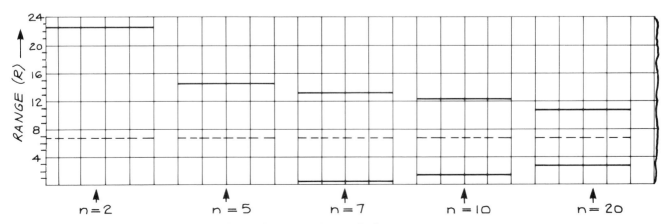

Fig. 6.20   Effect of sample size on R chart

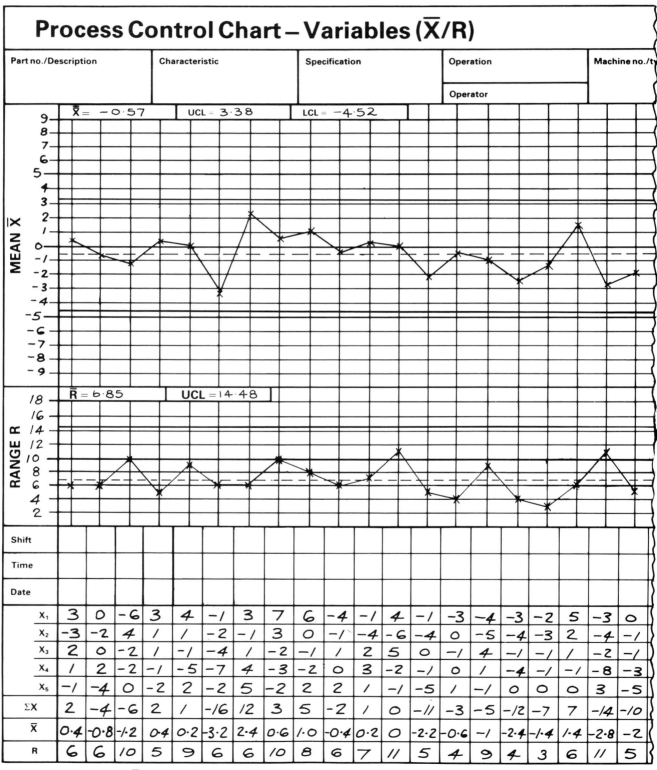

Process Control Chart – Variables (X̄/R)

Part no./Description	Characteristic	Specification	Operation	Machine no./ty
			Operator	

X̿ = −0·57   UCL = 3·38   LCL = −4·52

R̄ = 6·85   UCL = 14·48

Shift																			
Time																			
Date																			
X₁	3	0	−6	3	4	−1	3	7	6	−4	−1	4	−1	−3	−4	−3	−2	5	−3	0
X₂	−3	−2	4	1	1	−2	−1	3	0	−1	−4	−6	−4	0	−5	−4	−3	2	−4	−1
X₃	2	0	−2	1	−1	−4	1	−2	−1	1	2	5	0	−1	4	−1	−1	1	−2	−1
X₄	1	2	−2	−1	−5	−7	4	−3	−2	0	3	−2	−1	0	1	−4	−1	−1	−8	−3
X₅	−1	−4	0	−2	2	−2	5	−2	2	2	1	−1	−5	1	−1	0	0	0	3	−5
ΣX	2	−4	−6	2	1	−16	12	3	5	−2	1	0	−11	−3	−5	−12	−7	7	−14	−10
X̄	0·4	−0·8	−1·2	0·4	0·2	−3·2	2·4	0·6	1·0	−0·4	0·2	0	−2·2	−0·6	−1	−2·4	−1·4	1·4	−2·8	−2
R	6	6	10	5	9	6	6	10	8	6	7	11	5	4	9	4	3	6	11	5

Fig. 6.21   Completed (X̄, R) chart

The completed control chart for the first 20 samples is now as shown in Fig. 6.21. The control limits reflect the natural variability of the process and provide a basis for determining whether the process is under statistical control. They also provide the necessary framework for the rules used in determining the presence of special causes.

6.9 Rules for special causes

Four rules are typically used for determining the presence of special causes. At this stage they are applied to the first 20 samples in carrying out the initial process capability study.

6.9.1 Rule 1: Any point beyond the control limits

The positions of the control limits have been determined on the basis of a probability level of approximately 1 in 1000. In other words, for every 1000 values of, say, $\bar{X}$ that are measured, with no change in the process, then only one of those values would be expected by chance to fall outside either the UCL or the LCL. Hence if there is a situation as represented in Fig. 6.22(a) or (b), then it means that a special cause is present and must be sought.

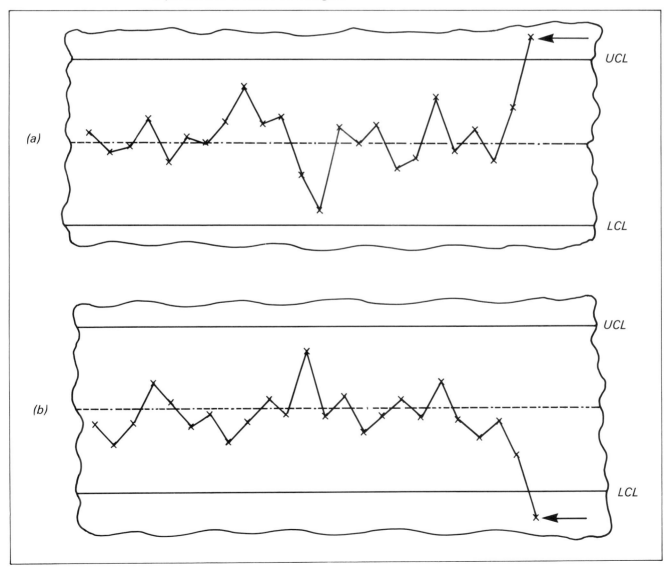

Fig. 6.22   Rule 1: Any point beyond the control limits

If the reason for this special cause is not found, then it must be accepted that the sample value is part of the system which generated the limits. If the reason for the special cause is found, then that reading is removed from the group of 20 and the control limits recalculated using the remaining 19 points.

The use of the control chart in this way reflects a change from past practice. Traditionally, the 20 sample means were used, as now, to calculate the control limits. It was then argued that there was no reason to plot the results on the chart because, since the points had been used to generate the limits, then the process must be under control. This does not necessarily follow. Plotting the points as well as drawing the lines increases the likelihood of eliminating readings which have an excessive influence on the calculation of control limits.

If there is a point outside LCL_R, then this corresponds to improvement not deterioration. The presence of such a point in the first 20 samples heralds a potential change for the better by reflecting a reduction in the variability. If possible the special cause should be found and incorporated as a permanent feature of the system.

6.9.2 **Rule 2: Rules of seven**

The rules of seven apply when there are seven consecutive points either all on one side of the mean, or all increasing, or all decreasing. These various alternatives are represented in Fig. 6.23(a), (b), (c) and (d).

The number 7 is chosen because the probability of a sequence of seven occurring by chance is small. (It is equivalent to obtaining seven heads in succession in seven tosses of a coin: a probability of 1 in 128.) If such a sequence does occur, therefore, a special cause can be suspected. However, there is less certainty associated with such a cause than there is with rule 1 causes because of the difference in probability levels.

As with the first rule, the reason for the special cause should be found and that factor either eliminated or introduced permanently, depending on its effect on the process.

Rules 1 and 2 are more statistically based than rule 3.

6.9.3 **Rule 3: Unusual patterns or trends within the control limits**

It is possible to have control chart patterns which may not be affected by rules 1 and 2. For example in Fig. 6.24 a cyclic pattern is suggested. It should not be too difficult to find the reason for this. Similarly in Fig. 6.25 there is a cluster of points near one of the limits. This is known as bunching and could well be due to mixing materials from different suppliers. A variety of other non-random patterns can be obtained. Whatever the pattern, it is imperative that the reasons for the particular format of points are established, so that a random pattern is allowed to develop.

6.9.4 **Rule 4: Middle third rule**

If the number of points in the middle third of the overall distance between the control limits is either much greater or much less than two-thirds the total number of points present, then the middle third rule applies. Fig. 6.26 shows such a case.

This rule is based on the properties of the normal distribution. As a first step, the distance between the control limits is divided into six bands of equal width, and lines drawn on the chart corresponding to 1 band width on either side of the mean. Since the control limits are set at 3 standard errors on either side of the mean, then these lines correspond to 1 standard error on either side of the mean. It is known that in a normal distribution 68.3% (i.e. approximately two-thirds) of the readings lie within this part of the curve, and this is the basis of the rule.

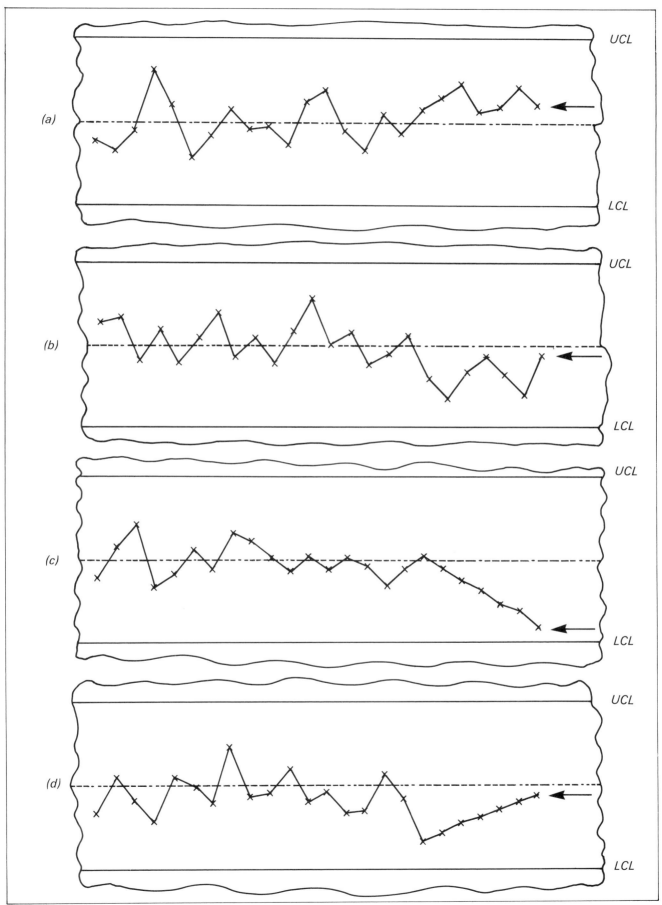

Fig. 6.23 Rule 2: Rules of seven

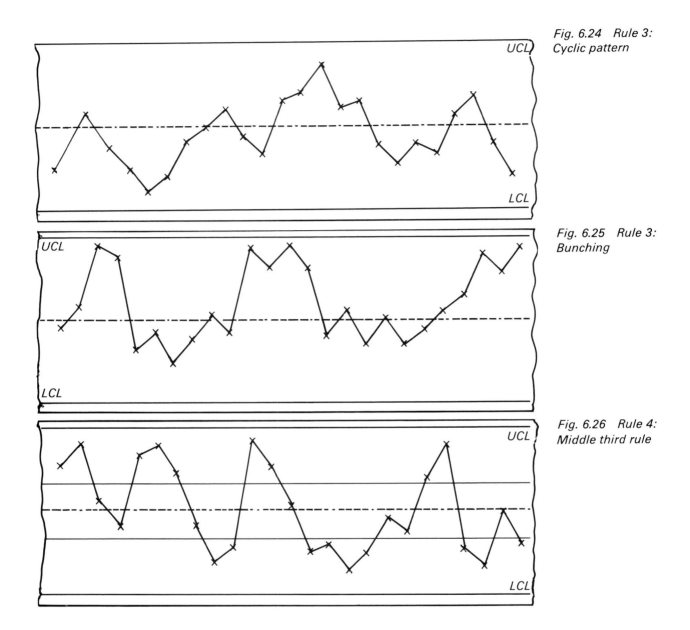

Fig. 6.24 Rule 3: Cyclic pattern

Fig. 6.25 Rule 3: Bunching

Fig. 6.26 Rule 4: Middle third rule

These four rules should be applied at the first stage of process control using the initial 20 samples. It is likely that some of the rules, e.g. rule 4, may have much more potential when used for ongoing process control with considerably more readings available. The middle third rule has to be applied with some care when distributions are skew, as in the range chart. This is because of the lack of symmetry and hence the uncertainty in applying a rule based primarily on the properties of the normal curve.

Other rules are available and can be applied as appropriate. They are often based on an extension of the middle third rule and require counts of points in zones other than the middle third zone. Reference to specialist sources is necessary for those interested in more sophisticated rules. It would be unwise to introduce too many rules at an early stage and hence the common usage of the four rules covered here.

On the basis of these four rules, therefore, the state of the process as represented in Fig. 6.21 can be checked. The pattern of points shows the process is under statistical control in that none of the rules applies.

6.10 **Ongoing control**

Once the initial 20 samples have verified a state of statistical control, the control limits and central line are used as the basis of ongoing control and are projected ahead. One approach is to use the nominal, φ, for the process as the central line for the X̄ chart from this point onwards. This results in control lines set at values of φ ± A₂ R̄ and emphasises the need to monitor against the nominal as part of continuous improvement. It is unlikely that the lines will be set about φ rather than X̄̄ until the process has generated sufficient readings. It is obviously beneficial, though, if the setting of the process can be on nominal as quickly as possible. This does suggest seeking out new techniques in this respect and a helpful reference is included in Appendix I

Typical guidelines suggest that ongoing control limits should be drawn in advance to cover some 25–30 future sample points. A further suggestion is that the limits are reassessed whenever a continuation control sheet is used. These issues will be covered in more detail in Chapter 7.

6.11 **Use of rules for ongoing control**

The rules for determining special causes apply in exactly the same way for ongoing control as they do for the initial stage of setting up the chart. There are some differences in acting on the signals however.

For example, if a sample falls outside the control limits (rule 1) typically a second sample would be taken straight away without waiting the specified time between samples. If the point was below LCL$_R$ then it may be wiser to continue to run the process for a short time before taking action in implementing the improvement.

The middle third rule has more scope for application as the process continues to run. As the number of points available increases then so the confidence in applying the rule with more valid conclusions increases.

6.12 **Alternative control rules**

The current wave of interest in control chart techniques brings with it understandable yet unnecessary confusion which stems from differences between national standards. Some historical background is necessary here.

The constant A₂ forms the basis of the control chart standards used in the USA. The basis of its formulation uses the fact that an area in the tail of the normal curve corresponds to a probability value of 0.0013 (as summarised in Fig. 6.27), and the relationship between standard deviation and range.

Why has there, then, been excessive usage in British organisations of the A₂ constant and the control limits that go with it? Why is more use not made of existing British Standards? The answer lies with the automotive industry and its associated American influence. The Ford Motor Company, for example, has adopted as part of its SPC documentation the control chart standards developed by the American National Standards Institute (ANSI). As a result, Ford of Europe in devising its own version of the Ford North American Operations SPC manual, has made extensive use of the existing Ford material.

In the UK the British Standards Institute (BSI) has been developing a series of control chart standards. These differ in interpretation from the ANSI standards in two respects at least:

(a) *Probability levels*

Instead of using control limits based on 3 standard deviations with a corresponding probability level of 0.0013, the British standard is based on a limit set at 3.09 standard deviations with a corresponding probability level of 0.001, i.e. exactly 1 in 1000. As a result instead of using A₂, as in the American system, a similar constant

$A'_{0.001}$ is used. Tables of $A'_{0.001}$ values are available for different sample sizes as for A_2, and are given in Appendix D. Thus for a sample size 5 the $A'_{0.001}$ value is 0.59 in comparison with the A_2 value of 0.577. Fig. 6.28 shows the basis of the British system of charting.

The difference basically reflects the influence of statistical philosophies. In practice the effect is negligible: 3 standard deviations or 3.09 standard deviations as the basis of the control limits is immaterial to the operators. The statistical niceties are irrelevant in the practical world of getting quality improvement operating on the shop floor. All the operator is interested in is whether the point being plotted is above or below a line set at a probability level of about 1 in a 1000.

(b) *Warning lines*

The British standard charts also use an additional pair of control limits based on areas in the tail of the normal curve of 0.025 (i.e. a probability level of 1 in 40) and corresponding to 1.96 standard deviations away from the mean (Fig. 6.29). Appendix D provides values of $A'_{0.025}$ which are used in conjunction with $\bar{R}$ (or $\bar{w}$), to obtain the warning lines. When samples are considered instead of individual readings, then the normal curve can be rotated as before, generating an additional pair of control limits for sample means set inwards from the original ones and at $1.96\sigma_x$ away from the mean (Fig. 6.30).

These lines are used as an alternative to the rules of seven. They are called warning limits as opposed to the action limits set at $3.09\sigma_{\bar{x}}$. The convention adopted is that a point falling outside the warning limits is an indication of the possible existence of a special cause. A further sample is taken immediately, and if that also falls outside the warning limits then a special cause is deemed to be present and the process is stopped. (The chance of two independent results falling outside the warning limits is $1/40 \times 1/40 = 1$ in 1600. This is an example of the multiplication law for probabilities.) Warning limits can also be generated for other types of chart, in particular the R chart and attribute charts in general.

Note also that a different convention for representing the lines is adopted – another area which can contribute to confusion in chart usage.

A possible objection to the BSI approach is that the introduction of an extra pair of lines on the chart as represented in Fig. 6.31 could be confusing to the operators. The American system , with its rules of seven, is more straightforward. Ultimately,

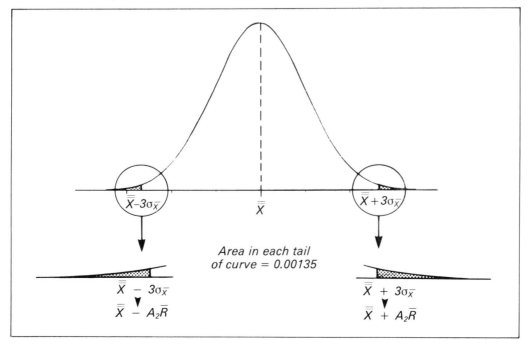

Fig. 6.27 Area in tail and A_2

Fig. 6.28 Area in tail and A'$_{0.001}$

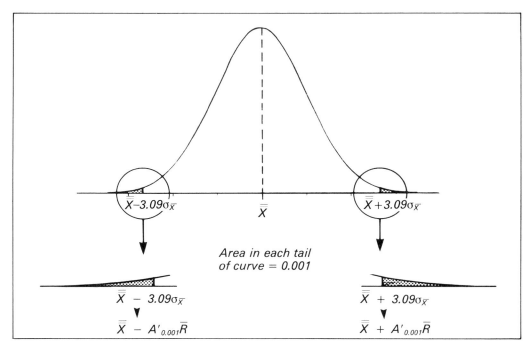

though, the really important factor is whether the control chart, of whatever variety, is being used as part of an SPC system with full management commitment. Provided suppliers have an effective system, then organisations looking for SPC programmes in their supply base are not going to insist that British systems as opposed to American systems are used.

It is the American system that will be used here. Those who wish to pursue the matter will find further details of the British approach in the BSI documentation.

6.13 Sample size

A sample size of 5 is typical but not sacrosanct, and samples of other sizes may be appropriate in some circumstances. In batch processes, for example, individual and moving range charts are used, utilising single readings corresponding to each batch. For example, one batch of paint gives a certain viscosity reading which will

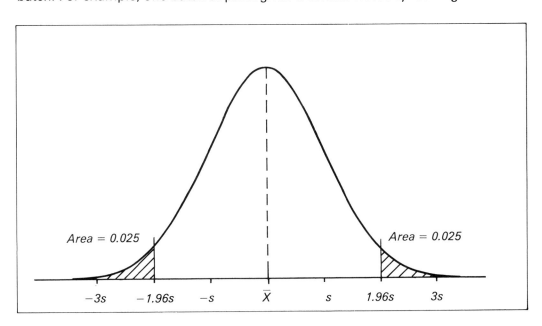

Fig. 6.29 0.025 probability lines

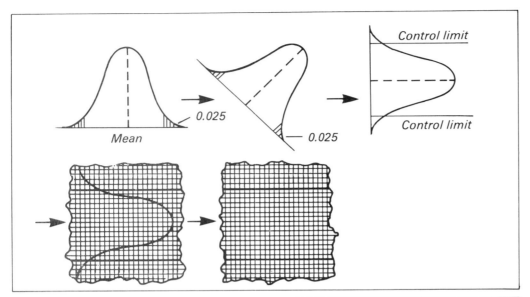

Fig. 6.30   Generating warning limits (UK)

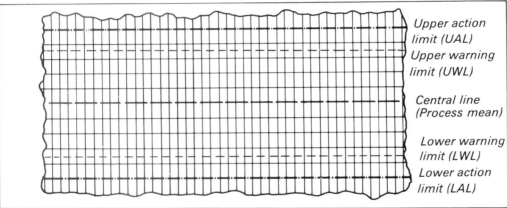

Fig. 6.31   Lines on $\overline{X}$ chart (UK)

vary from the reading provided by the next batch. These single readings can then be combined to give a composite sample size of 2, 3, etc.

A four-headed filling procedure would logically suggest that a sample of size 4 is used to monitor the machine as a whole. However, if it is desirable and in fact possible to check on the performance of each head, then a sample of 5, for example, could be used for each head separately.

A basic requirement, however, is that the sample size is between 2 and 12 when R is being used.

Similarly, and particularly with high production rates, a sample size of, for example, 10 may be utilised. However, in such cases the standard deviation should be used to monitor variability rather than the range. Improved automated measuring technology is assisting greatly in this area. If only one reading is obtained then clearly a range is impossible. If the sample size is greater than 12 then the validity of some of the statistically based constants is open to question. In a great majority of instances a sample size of 5 is used because calculating $\overline{X}$ is easier for this than any other sample size. In addition, and more importantly, the relevant statistical analysis indicates that a sample size of 4 or 5 is large enough to indicate important changes in the process yet at the same time not too large so that smaller, unnecessary changes are picked up.

One final point to be made is that once chosen, the sample size should be kept constant as far as possible whilst the process continues to be monitored. If there is some reason why the sample size must change, then the control limits must be recalculated to allow for this.

The following case study brings out some of the points introduced to date.

Case study 2

Fig. 6.32 shows the results for the first 20 samples of 5 obtained when measuring a particular bore diameter of a housing. The completed data boxes at the top of the chart provide relevant details, but some explanation may be useful.

The unit of measurement for the diameter is 0.0001in. and the setting ring which provides the zero on the vertical scale of the chart has been set at 2.1135in. Hence the individual readings shown in the appropriate data block are all multiples of 0.0001in. measured above 2.1135. The component has an upper specification limit of 2.1143 (8 coded units) and a lower specification limit of 2.1138 (3 coded units), so that the nominal for the process is 2.11405 (5.5 coded units).

Control limits and central lines have been calculated for both mean and range in the usual way and the results recorded in the boxes provided.

Note also that all other relevant information has been recorded. Chapter 8 will consider aspects relating to the administrative information which supports the sample data.

Both the mean and range chart clearly show an out-of-control situation. The excessive fluctuation on the $\overline{X}$ chart results in very few points in the middle third band. The situation is too severe to justify eliminating points outside the control limits and recalculating, even if the reason for the special cause in each case could be found. It is clear that a major factor is operating which is distorting the pattern from a random one. Whilst the range chart does not display such wild fluctuations, even so it cannot be satisfactory that three points out of the first 20 are outside the UCL.

An analysis of the process quickly resolved the problem: the air gauge used to measure the diameters was providing incorrect readings. As a result it was stripped, cleaned and reassembled. Following this modification a further 20 readings were taken and the results plotted. New control limits were obtained and an improved picture emerged. This is shown in Fig. 6.33 where the first 14 of the new set of readings are plotted, the remaining 6 being recorded on the next control chart in the time sequence.

The mean is now under control apart from the one sample reading of 7.8. As the measuring equipment is known to be operating satisfactorily, some other special cause must be present to account for this reading. It may well be that in this instance the machine setting had been affected or that the material was of a different standard for that particular housing. If the cause can be found then it should be eliminated and at the same time the control limits recalculated. The sample range is under statistical control.

It will be noted that imperial units, in this case inches, are used in this example. This reflects the uncertain position of British industry: considerable progress towards metrication was made in the 1970s but there are still organisations using the imperial system – as do American firms. Imperial units will therefore be used at times here to reflect the practical situation.

Summary

- SPC programmes are not just about filling in numbers on a chart.
- Resolve issues relating to charting very early on.
- Common causes relate to the system and reflect management input.
- Special causes are outside the system and indicate changes (for the worse or the better).
- A process is generally defined in terms of its position and spread.
- Mean (or median) measures position and range (or standard deviation) measures spread.

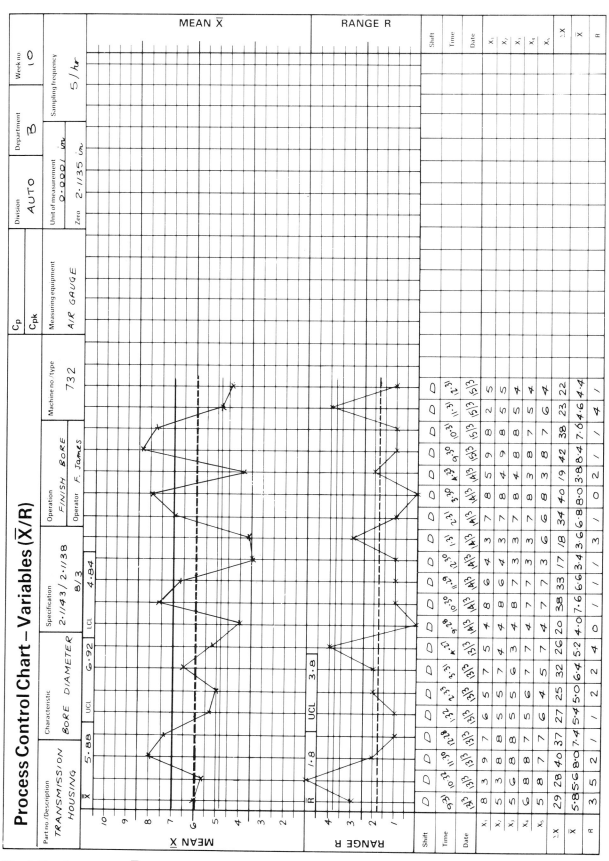

Fig. 6.32 Completed ($\overline{X}$, R) chart: initial study

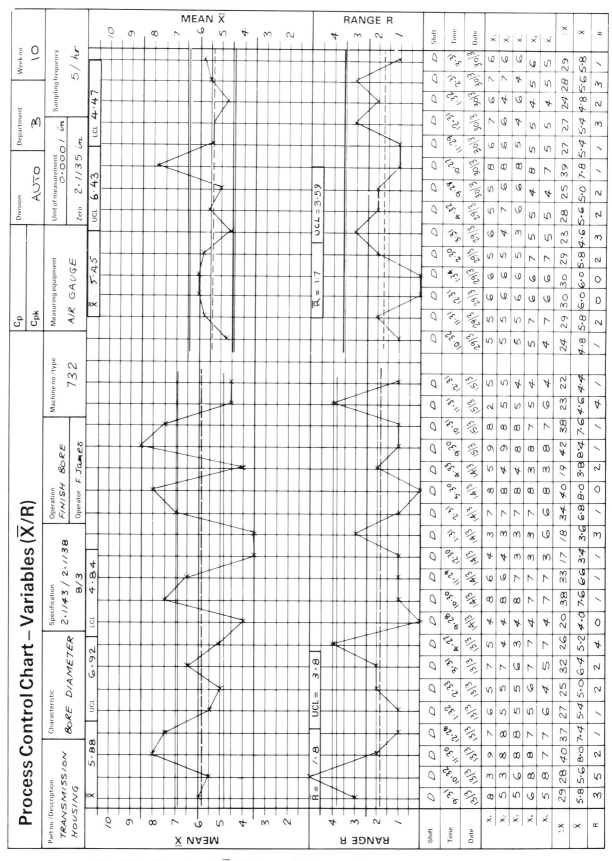

Fig. 6.33 Completed ($\bar{X}$, R) chart: further readings

- Control charts enable position and spread to be monitored.
- Typically, sample size is 5 and at least 20 samples are required to assess process performance.
- Guidelines are available for scaling charts.
- Control limits are based on the normal distribution and/or areas in the tail of the distribution corresponding to a probability level of approximately 1 in 1000.
- Typically, the $\overline{X}$ chart has two control limits and a central line and the R chart one control limit and a central line.
- There is no LCL_R for sample sizes of 6 or less.
- Constants, dependent on sample size, are available to enable control limits to be calculated.
- Charts can be used as a guide to improvement not just an indicator of problems.
- The presence of special causes is determined by using four standard rules.
- Control limits are used for ongoing control following the initial process study.
- American and British control chart standards vary slightly in interpretation.
- Sample size is typically 5, though not in moving mean and range charts.

Throughout this chapter there has been no reference to specification limits; the focus has been on determining whether or not a process is under statistical control. The process was assessed to see whether it was performing consistently and predictably irrespective of any specification limits which operate. But it is also important that while performing consistently and predictably, the process is also capable. There is no point in producing controlled scrap.

In fact, if the process is in control and incapable it is not difficult to make it capable, and then yet more capable. Capability therefore needs to be measured, not only so that the customer is satisfied, but in the long term to enable comparison of the process with a standard at a given time. This can be used as a reference point from which to measure improvement relative to both past achievements and competitors.

The following chapter on capability and control shows how tolerances are of limited relevance – and then only as a basis for measuring improvement. The control chart is yet again the key, and in particular the R chart takes on a new significance in the drive for improved performance.

Chapter 7 Capability and control

7.1 Introduction

In Chapter 6 the emphasis was on the steps to be covered in setting up the initial control charts in order to obtain a state of statistical control. Whether or not the items being processed were within specification was not discussed. It was control that was considered, not capability. The main message was that SPC requires the three stages of data collection, control and improvement. Specification limits, however, cannot be ignored in the short term in the practicalities of industrial life. It is a requirement to have capable processes as well as ones which are under control. At some stage in the sequence indicated in Fig. 7.1 the process will move from an incapable one to a capable one – and the earlier the better.

Control and capability, therefore, are quite distinct features. Various combinations of the two are possible, as shown in Fig. 7.2. Organisations obviously have a long-term objective of having all processes in the 'Under control/Capable' box.

This chapter deals with the ways of measuring capability. Even though capability and control are quite distinct, it will be seen that there is a close association between the two because of the role of the control chart for range in assessing capability.

7.2 The histogram

Chapter 6 described how 20 samples of 5 were taken to generate control limits for sample mean and sample range, as in Fig. 7.3. A first attempt at measuring capability is to assess the relative positions of each of the 100 individual readings with respect to any specification limits provided. To test this, all 100 readings are used to generate a tally diagram, with the representation as shown in Fig. 7.4 (This pattern was discussed earlier in Chapter 5, and illustrated in Fig. 5.28(a).) For convenience, this tally diagram can be rotated and related to the same scale as that used to monitor the sample mean $\overline{X}$; Fig. 7.5 shows the result.

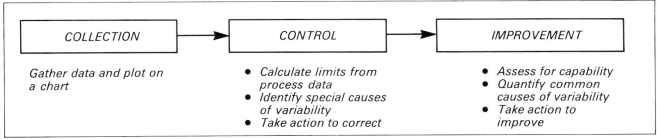

Fig. 7.1 Stages of process improvement

Out of control	Out of control
Incapable	Capable
Under control	Under control
Incapable	Capable

Fig. 7.2   Control and/ or capability

Some control charts are designed to allow for the recording of individual readings in this way and it is a useful addition to the chart. However, it does have the disadvantage that it cuts into plotting space and also requires the use of a scale on the right-hand side of the chart as well as the left. Organisations implementing SPC programmes need to take these issues into consideration at the control chart design stage.

The specification quoted for the items being plotted is ± 8.5, in coded units from the nominal. (It is recommended that imaginary lines are drawn on the chart at -8.5 and $+8.5$ to make the position clearer. The lines have not actually been drawn in Fig. 7.5 because with performance-based systems it is no longer an acceptable practice to confuse the issue by retaining specification limits as lines on the chart.) The process is seen to be under statistical control and as none of the individual readings falls outside these specification limits a natural assumption is that the process is capable.

Or is it? Things are not always what they seem. Even though there are 100 readings which give a good picture of the process, there is still a normal distribution in force which relates to the readings as shown (Fig. 7.6).

It is known that the limits of natural variability of a process correspond to 6 standard deviations, or $\pm 3s$ on either side of the mean. Using the 100 readings in question, s works out to be 2.945. Thus $6s = 6 \times 2.945 = 17.670$. Now the tolerance for the process is $2 \times 8.5 = 17$. Since this value is less than $6s$, it must be concluded that the process is incapable.

This analysis shows the danger of making decisions regarding capability on the basis of insufficient data. Capability must be determined on the natural spread corresponding to the process and this requires using a normal distribution as shown. Using only tally marks, or the related histogram, can be misleading when assessing capability accurately. The histogram is of even less value as a control feature (see Section 7.6).

Fig. 7.6 also illustrates the point that the distance between the control limits for sample mean is considerably less than the $6s$ value corresponding to the process performance. This is because the distribution of sample mean has a narrower spread than does the distribution of individual readings, as discussed in Chapter 5. This tightening of the operational limits often causes problems on the shop floor when charts are introduced against a background of 'working to tolerance'. There is misguided belief that more is being demanded of the operators than before – the

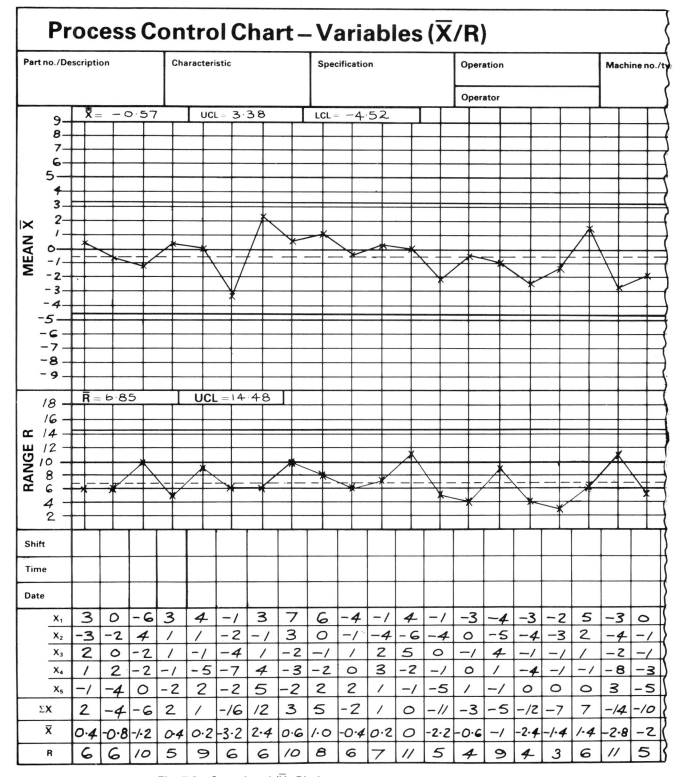

Fig. 7.3   Completed ($\overline{X}$, R) chart

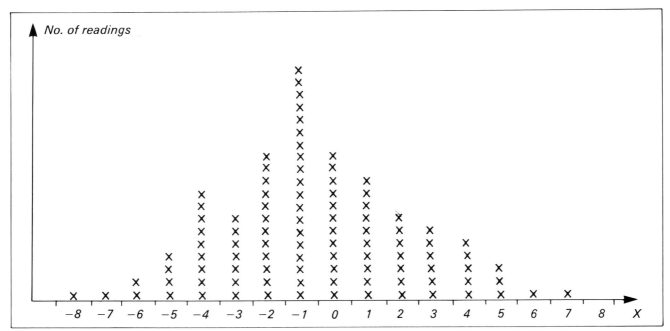

Fig. 7.4   Tally diagram for X

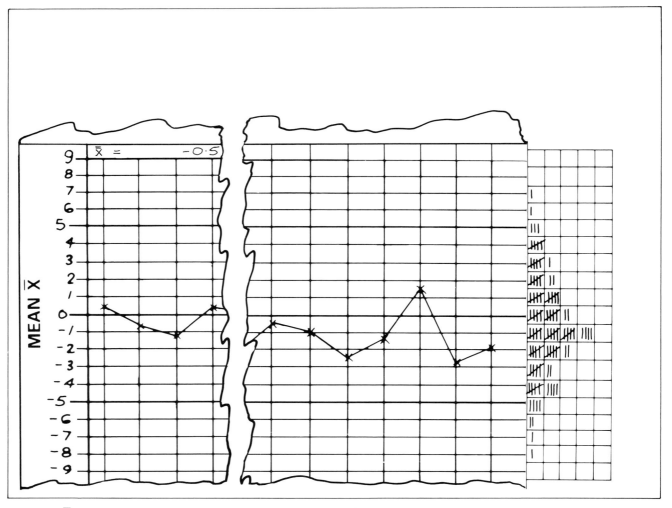

Fig. 7.5   X̄ chart plus tally diagram for X

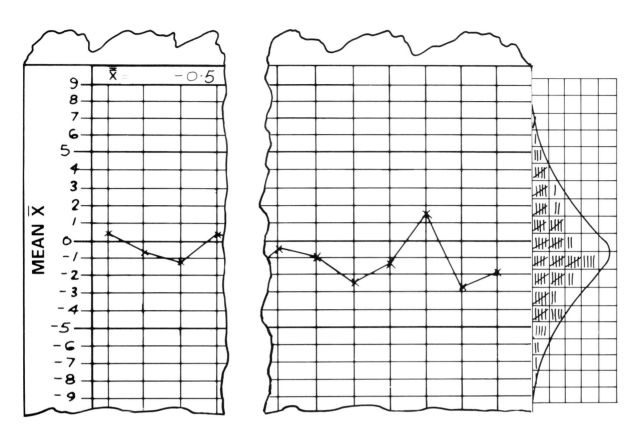

Fig. 7.6   X̄ chart plus normal curve for X

tolerance has been cut, in other words. In fact there should be little difference in decision-making. Instead of using specification limits, and plotting individual readings against those, performance-based limits are used which are naturally narrower because they reflect the variability in the mean, not in the individual readings. If the specification limits continue to be drawn on the chart as performance-based systems are introduced, then indecision and uncertainty will result.

Having worked out the capability using the calculated standard deviation based on individual readings, it is evident that the procedure is cumbersome and time-consuming. It is not really feasible in practice to work out s in this way. There has to be an easier method to determine capability and the statisticians have provided it as usual. The range chart is used to resolve the problem.

7.3 The R chart and capability

In Chapter 4 it was explained that the expression

$$s = \sqrt{\frac{\Sigma(X-\overline{X})^2}{n-1}}$$

is used to provide the best estimate of the variability of the parent population, i.e. $\hat{\sigma}$. In fact, this expression was used to obtain the result of 17.670 in the previous section. In the same way, it can be shown that $\hat{\sigma}$ can also be obtained from $\overline{R}/d_2$, where d_2 is another mathematical constant which varies with sample size. $\overline{R}$ is known from the range values.

A sample size 5 has been used in the analysis and reference to Appendix C gives a d_2 value of 2.326. The process performance is then given by

$$6 \times \frac{\overline{R}}{2.326} = 2.58\ \overline{R}$$

Fig. 7.7 provides a summary of this neat relationship between the range chart and process performance, again using a sample size 5. In the example above $\bar{R}$ is 6.85 and the process performance is therefore $2.58 \times 6.85 = 17.673$. This shows very good agreement with the value of 17.670 obtained in Section 7.2.

Two reasons account for the small difference that does exist. One is the fact that the 100 readings do not provide an ideal normal fit, as reference to Fig. 7.6 shows. Secondly, the relationship $\hat{\sigma} = \bar{R}/d_2$ is only an approximation, but nevertheless it is sufficiently accurate to justify being used. There are circumstances in which the accuracy lost in using a less robust result is of little concern in contrast with the considerable benefits in time and convenience gained by using a valid approximation. Hence the expression $\hat{\sigma} = \bar{R}/d_2$ is invaluable in control charting. Of particular value is the fact that for a sample of size 5, $6s = 2.58\bar{R}$. This relationship is used extensively, but it must be remembered that it only holds true for a sample size 5. Other sample sizes require the use of the correct d_2 constant in first calculating $6/d_2$ to obtain the number which is equivalent to 2.58.

7.4 **Measuring improvement**

7.4.1 **Variability: C_p**

Assessing the capability of a process by comparing $6s$ with the tolerance is a tedious method when using the individual readings to obtain σ. In Chapter 4 the measure C_p was introduced, C_p being defined as the ratio Tolerance/$6s$ as summarised in Fig. 7.8.

It has already been seen how, for a sample size 5, $6s$ can be replaced by $2.58\bar{R}$. Thus C_p can be defined as Tolerance/$2.58\bar{R}$ for a sample of size 5. Hence for the set of readings being considered,

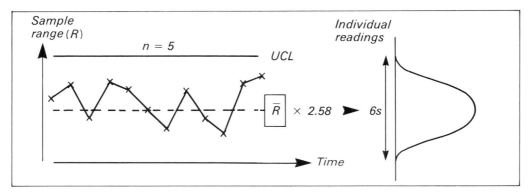

Fig. 7.7 *R chart and process performance*

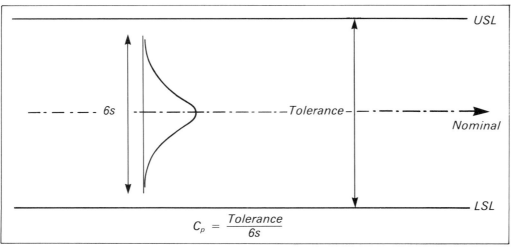

Fig. 7.8 C_p

$$C_p = \frac{2 \times 8.5}{2.58 \times 6.85} = 0.96$$

The fact that this is less than 1 is simply another way of confirming that the process is inherently incapable and is the numerical equivalent of stating, as in Section 7.2, that the tolerance is less than the process performance 6s.

Unfortunately a C_p value does not provide all the answers. It can be seen why by referring to Fig. 7.9, where processes with different settings all have the same C_p value. This is because C_p is purely a measure of the inherent ability of the process to match the specifications. It means that even though the process could be running at different levels of performance, as shown in Fig. 7.9, the variability at any setting is constant and 6s is always less than the tolerance.

C_p in itself is therefore not a satisfactory definition of capability. Another index is required to assess the position of the process with respect to the nominal. This quantity is denoted by C_{pk}. The expression for calculating C_{pk} takes two possible forms, as discussed below.

7.4.2 **Variability and setting: C_{pk}**

In addition to the two specification limits USL and LSL, the C_{pk} value involves the process mean $\overline{\overline{X}}$. The C_{pk} value is used to measure a change in the process setting from the nominal position.

Fig. 7.10 represents the optimum position with the setting on nominal. The peak of the curve is now analogous to the bow of a boat which is to be kept on the central

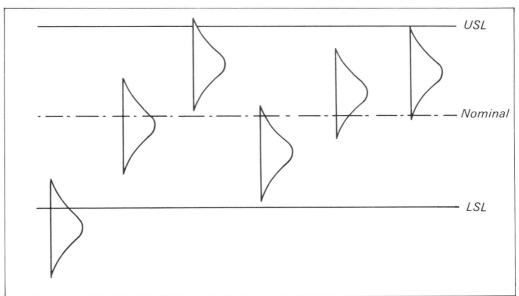

Fig. 7.9 Constant C_p with varying settings

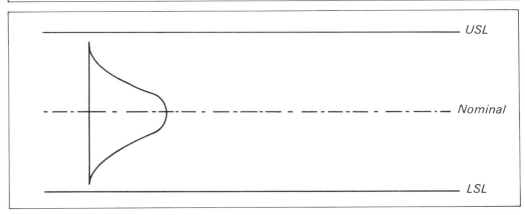

Fig. 7.10 Optimum process setting

line midway between the banks of a river. As soon as the bow moves away from this central line the movement needs to be detected. In terms of the process, this is equivalent to keeping the line of symmetry of the normal curve on nominal at all times. Any change from nominal is then detected by means of the appropriate C_{pk} expression as defined in Fig. 7.11.

It is seen that these expressions make use of the specification limit nearest to the $\overline{\overline{X}}$ value. It will not be clear at this point why the specification limits are used rather than the nominal, ϕ. This point will be discussed further after the relationship between C_p and C_{pk} has been considered.

For the set of readings being used, it is known from Fig. 7.5 that the value of $\overline{\overline{X}}$ is below nominal and therefore nearer to the lower limit. Hence the expression to be used is

$$C_{pk} = \frac{\overline{\overline{X}} - LSL}{3s}$$

Again, it is known that for a sample of size 5, 6s is 2.58 $\overline{R}$ and therefore 3s is 1.29 $\overline{R}$. Therefore

$$C_{pk} = \frac{\overline{\overline{X}} - LSL}{1.29\,\overline{R}}$$

$$= \frac{-0.57 - (-8.5)}{1.29 \times 6.85}$$

$$= 0.897$$

The significance of this figure may not be obvious. It can no doubt be appreciated, though, that for a process to be capable C_{pk} must be greater than 1 in the same way that C_p must be. However, a comparison of C_p and C_{pk} values also provides useful information. Some further discussion on the interpretation of C_{pk} values will be useful, therefore.

Interpreting C_{pk}

Rather than contend with both forms of the C_{pk} result the analysis will be based on the use of the form using USL, i.e.

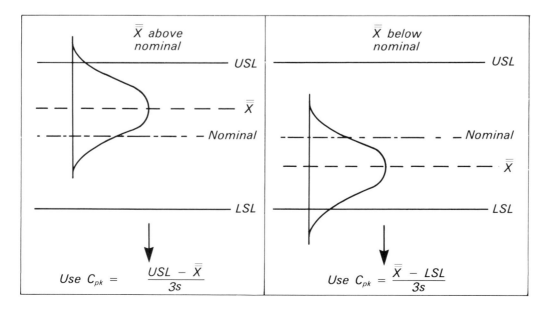

Fig. 7.11 *Alternative expressions for C_{pk}*

$$C_{pk} = \frac{USL - \overline{\overline{X}}}{3s} .$$

Using the LSL form would provide the same interpretation equally well.
As $\overline{\overline{X}}$ varies, then so does C_{pk}. Two initial values of $\overline{\overline{X}}$ are of interest:

(i) $\overline{\overline{X}}$ = USL (as in Fig. 7.12(a))

Here C_{pk} must equal zero because USL $- \overline{\overline{X}}$ equals zero.

(ii) $\overline{\overline{X}}$ = Nominal (as in Fig. 7.13)

In this case,

$$C_{pk} = \frac{USL - Nominal}{3s}$$

$$= \frac{Tolerance/2}{3s}$$

$$= \frac{Tolerance}{6s}$$

But Tolerance/6s, by definition, is the C_p value. Hence when the process is running on nominal, $C_p = C_{pk}$.

As the process mean moves the other side of the nominal towards LSL the C_{pk} value becomes smaller again until it is once more 0 when $\overline{\overline{X}}$ = LSL (as in Fig. 7.12(b)).

Note that using USL/LSL rather than the nominal in the definition of C_{pk} results in

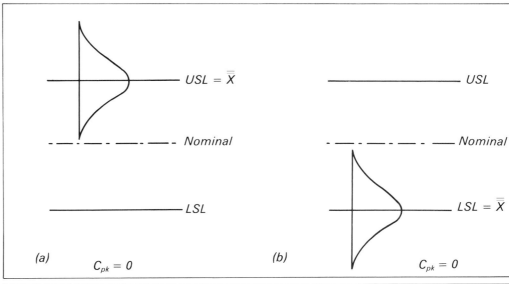

Fig. 7.12 Setting and USL or LSL

(a) $C_{pk} = 0$

(b) $C_{pk} = 0$

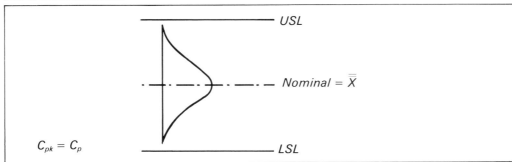

Fig. 7.13 Setting on nominal

$C_{pk} = C_p$

C_{pk} becoming bigger as $\overline{\overline{X}}$ moves towards nominal. This is consistent with the notion of the index value getting bigger as the setting improves. (If the nominal was used as a measure the reverse would be true.)

If $\overline{\overline{X}}$ were at a level outside either USL or LSL, then C_{pk} would be negative. Whilst this is technically feasible, it should not really happen in practice. If it were the case then it would indicate the need for immediate action in process setting on the long road of quality improvement.

Fig. 7.14 shows how C_{pk} values can vary, using the mean value of 6.85 for $\overline{R}$ and a sample size 5. The maximum possible value of C_{pk} is C_p and this occurs when the setting is right. No improvement in the C_{pk} value is therefore possible by tinkering with the setting. The difference between C_p and C_{pk} indicates that $\overline{\overline{X}}$ is offset relative to the nominal ϕ. Continuous improvement means ongoing reduction in the variability with the setting on nominal.

Fig. 7.15 shows a range of C_p / C_{pk} values. It illustrates the concise way of reporting on the capability status of a process. A quick scan of the readings results in a rearrangement (Fig. 7.16) which ranks the processes in terms of capability (from worst to best). For completeness, a statement on the setting is now also provided.

It can be seen that the C_{pk} value does not in itself indicate whether the $\overline{\overline{X}}$ value is above or below the nominal. One approach which has been adopted in such cases is to use a convention whereby C_{pk} is defined as positive if the formula $(USL - \overline{\overline{X}})/3s$ is used and defined as negative if the formula $(\overline{\overline{X}} - LSL)/3s$ is used. This may not be a common practice but it indicates how one organisation is attempting to remove some of the mystery regarding the index and at the same time adopting a

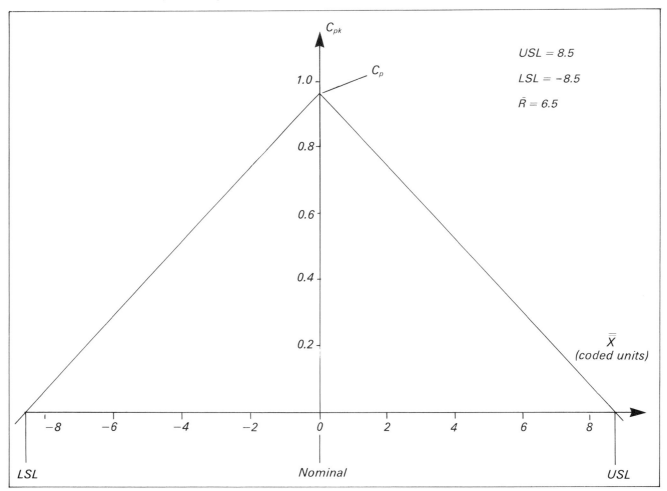

Fig. 7.14   Graph of C_{pk} values

Fig. 7.15 Table of C_p and C_{pk} values

C_p	C_{pk}
0.7	0.65
3.5	3.5
2.0	0.8
2.7	0
1.5	−0.2

Fig. 7.16 Process assessment based on C_p and C_{pk} values

C_p	C_{pk}	Setting
0.7	0.65	Close to nominal
1.5	−0.2	Outside specification limit
2.0	0.8	Between specification limit and nominal
2.7	0	On specification limit
3.5	3.5	On nominal

company-specific approach. However, care must be taken in this case to make sure that this convention is only used when $\overline{\overline{X}}$ is between USL and LSL, otherwise the 'negative' value of C_{pk} could be confused with a genuine negative value.

C_p and C_{pk} are just two of several similar indices which have been developed primarily in Japan. Other indices measure, among other things, skewness and kurtosis (i.e. the degree to which the distribution is 'peaked'). However, the basic indices of C_p and C_{pk} are quite sufficient in the main. They should form part of the common language of SPC and their significance should be generally understood throughout the organisation. They provide a means of breaking down communication barriers between departments.

7.5 Continuous improvement

7.5.1 Introduction

If the process is incapable, as in the example being considered, then action has to be taken. But action by whom? The operator, supervisor, management?

The responsibility for improving processes rests primarily with the management. This is because reduction in variability is the result of improved maintenance, better suppliers, more consistent materials, etc. – and decisions on these factors are outside the authority of operational staff. Thus a C_{pk} value less than 1 requires management action. However, a C_{pk} value showing 'offset' of $\overline{\overline{X}}$ demands operator action so that the process is reset or adjusted. This will make $C_{pk} = C_p$ but will still leave the process incapable if the C_{pk} value is less than 1.

In another sense, the control chart can be seen as a reflection of Deming's 85% / 15% division of responsibilities. Decisions based on the $\overline{X}$ chart are very much in the hands of the operator, whereas decisions in connection with the R chart are in the hands of management. Fig. 7.17 illustrates this. This does not mean there is an absolute division of responsibilities, simply that management must recognise its major responsibility for process improvement and operators should work with management in achieving this. Hence the importance of problem-solving groups – not quality circles as generally understood, although the techniques used might well be the same, but rather groups working on solving problems which have been raised as a result of charting.

As the variability is reduced then this will be indicated by a reduction in the range values. A sequence of seven points below $\overline{R}$ heralds a special cause (concerned with improvement, not deterioration) and would indicate that reduced variability is likely. In practice, a total of 20 samples would be taken in order to substantiate the new level of performance and then new control limits would be calculated for R and, consequently, for $\overline{X}$. Fig. 7.18 shows the link between the two charts.

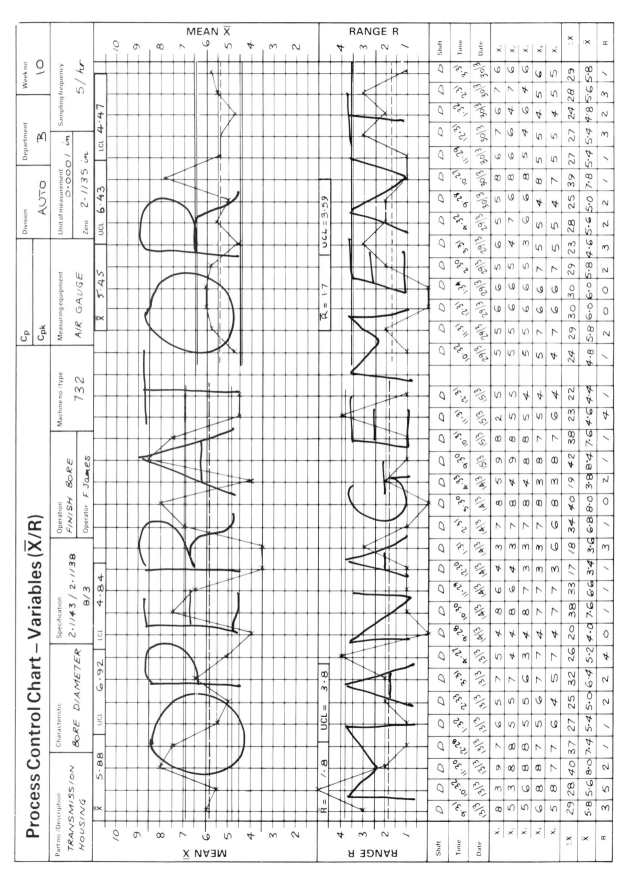

Fig. 7.17   Control chart responsibilities

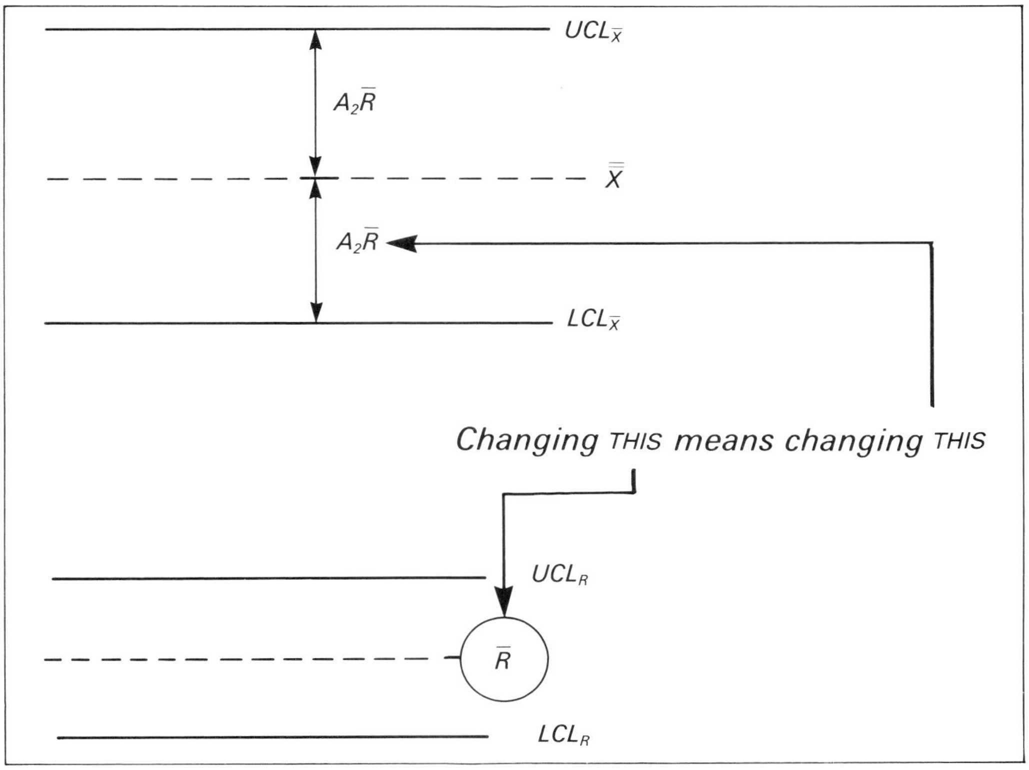

Fig. 7.18 $\bar{R}$ and the $\bar{X}$ limits

The sequence does not end there however. Continuous improvement means that the procedure is repeated, as shown in Fig. 7.19. The R chart is the one that can influence a major improvement in performance. Not only does it provide a quick means of determining capability (unlike the $\bar{X}$ chart, which has nothing to do with the capability of the process), it also provides the evidence of sustained improvement which can lead to further improvement, and so forth.

How far down the road of improvement for any one process should an organisation go in an SPC programme? It has to be a matter of judgement, based on economics. A stage will be reached, no doubt, when the time and effort involved in further reducing the variability are not justified. Continuous improvement is a company-wide activity just as much as it is an activity concentrating on one process alone. Therefore, more immediate benefits will be gained if resources are switched to other processes which have not yet been analysed rather than obsessionally trying to squeeze all the variability out of one process, ideal as that goal may be.

In any event, continuous improvement does require a positive way of thinking. Old habits die hard, and the elimination of tolerance based ideas does not come easily.

7.5.2 **Matching the competition**

A typical problem arises when the initial variability of the process is small in

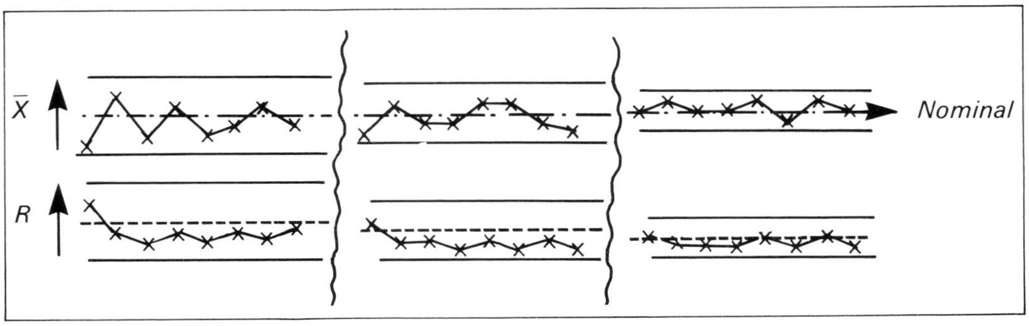

Fig. 7.19 Control limits and continuous improvement

comparison with the tolerance specified. This situation is represented in Fig. 7.20. Traditionally one would be tempted to argue that the process is being controlled unnecessarily tightly in the light of the demands of the customer. After all, the process is well 'within spec'. Why waste time and effort holding the process tightly around nominal? Why not relax things? One traditional approach in such cases is to use 'modified control limits' whereby performance-based limits are set inwards from the specification limits. In so doing, the mean is allowed to vary a lot more before action is taken.

Such an approach is inconsistent with the philosophy of continuous improvement because effectively a tolerance-based method is still being used, with all its associated limitations. Far from letting things slide by allowing greater variability of the mean, the correct course of action would be, as a minimum, to keep the control limits at the current level with a firm intention of improving on the level of performance.

It would be inappropriate to suggest that no other options are possible, and much would depend on customer/supplier relationships regarding the viability of the existing tolerance, the nature of the product being produced, and so on. There can be no denying the fact, however, that competition in the marketplace is fierce and will become more intense. A competitor may already be operating at a reduced level of variability, as shown in Fig. 7.21. All things being equal, business with the common customer will be lost unless the performance of the competitor can be matched. Resorting to modified limits may alleviate a short-term crisis, but such an approach will not help in the long term.

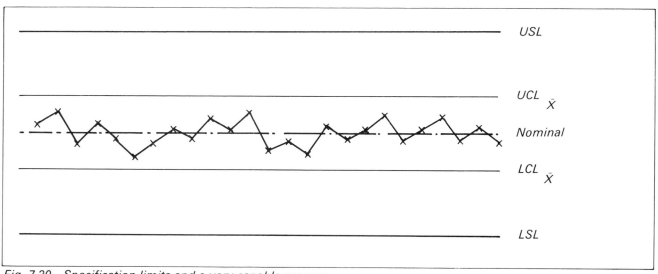

Fig. 7.20   *Specification limits and a very capable process*

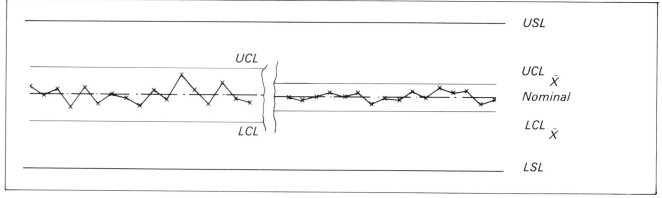

Fig. 7.21   *Effect of reduction in variability*

Where does capability fit into this sequence? It may be that the process is initially incapable in that the specification limits are too tight to be matched by 6s. Gradually closing up the control limits around nominal by using the range chart to trigger the need for recalculation will result in the process becoming capable at some stage. Ideally, the process should be capable initially, so that continuous improvement, as shown in Fig. 7.22, means that satisfying the specification limits becomes less and less of a problem. If it is felt necessary to apply a new specification limit at a later stage, this can be done with confidence because there is now sound information available which, perhaps for the first time, enables process performance to be matched to any specification limits which may be required.

Care is required in utilising C_p and C_{pk} values. They depend on using the tolerance applied when process performance was being measured initially. The gradual improvement in C_p and C_{pk} values over time is masked if, at the same time, the tolerance is changed.

Measuring improvement by using C_p and C_{pk} values can be applied over a wide range of processes. For example Fig. 7.23 shows how SPC control plans can indicate overall performance over time.

One final point regarding C_p/C_{pk} is worth mentioning. If control charts are being used to control a process dimension rather than a product dimension a high C_{pk} value can sometimes lead to a change in the sequence of operations. This can result in cost savings with no detrimental effect on the quality of the final product received by the customer.

Fig. 7.24 refers to the ongoing example which formed the basis of Fig. 6.21. Previously calculated limits have been projected ahead for the first control sheet. A further series of readings has been taken. These are recorded in sequence as a continuation of those already plotted on the control chart. These readings follow the action taken on the process by management, which recognised that without such action the process may well have remained in statistical control but no improvement was possible.

The effect of the process change is illustrated by the fact that the first seven points on the R chart all fall below $\bar{R}$. This supports the assertion that introduction of a special cause (i.e. a known process change) is then reflected by the 'rules of seven' coming into operation. As further readings are taken they substantiate the new, improved situation. According to the guidelines proposed, a total of 20 sample readings is taken to provide the information which enables new control limits to be worked out. Note that the 'middle third' rule on the $\bar{X}$ chart also applies. It is to be expected that if R is running at a consistently lower level then the points on the $\bar{X}$ chart should cluster more closely around the $\bar{X}$ value over time.

On the basis of the latest set of 20 points, the following results are obtained:

$$\bar{\bar{X}} = -0.49$$

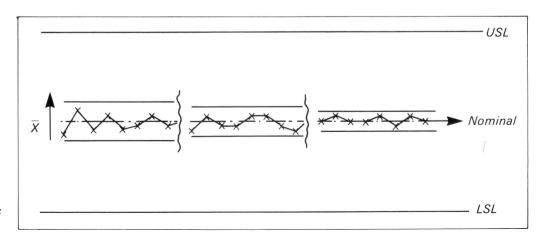

Fig. 7.22 Continuous improvement

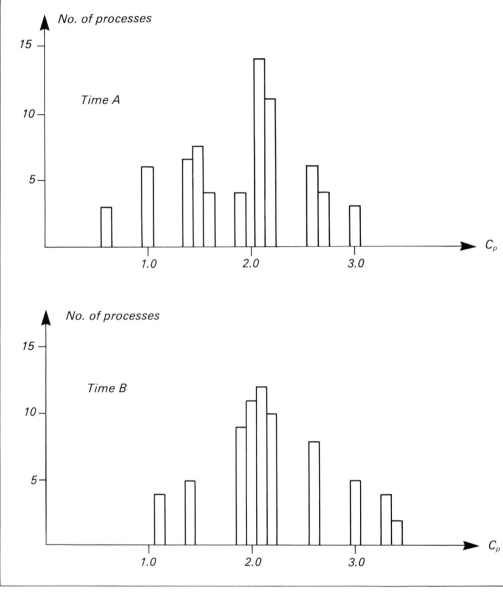

Fig. 7.23 Process improvement over time

$$\bar{R} = 4.15$$

$$\begin{aligned} UCL_{\bar{x}} &= \bar{\bar{X}} + 0.577\,\bar{R} \\ &= -0.49 + 0.577 \times 4.15 \\ &= 1.905 \end{aligned}$$

$$\begin{aligned} LCL_{\bar{x}} &= \bar{\bar{X}} - 0.577\,\bar{R} \\ &= -0.49 - 0.577 \times 4.15 \\ &= -2.885 \end{aligned}$$

$$\begin{aligned} UCL_R &= 2.114 \times \bar{R} \\ &= 2.114 \times 4.15 \\ &= 8.77 \end{aligned}$$

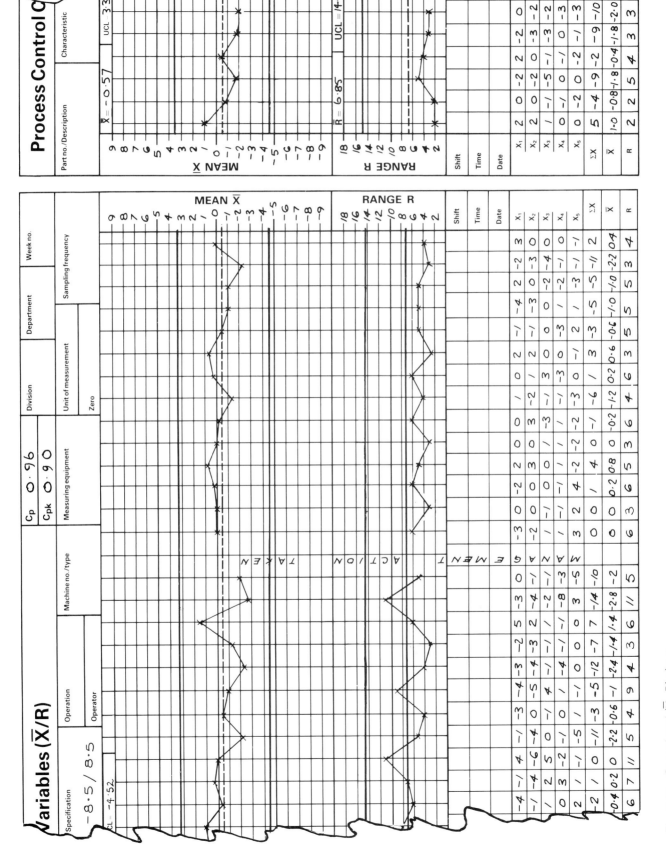

Fig. 7.24 Completed ($\bar{X}$, R) chart

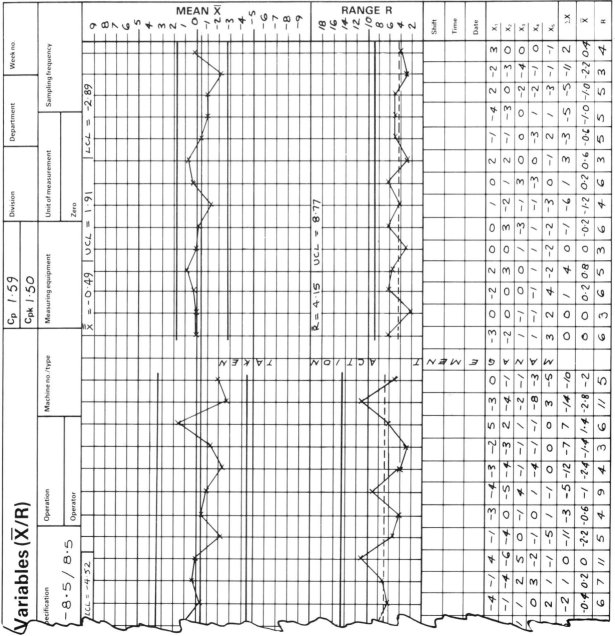

Fig. 7.25 $(\bar{X}, R)$ chart with recalculated limits

New control limits are drawn for the new set of 20 readings covering both control charts, as a check that no other special causes are present. The situation is then as shown in Fig. 7.25.

The new situation is under statistical control and following the usual guidelines the limits are projected ahead to monitor the process, and improve it again if possible, over the next series of readings as they become available (Fig. 7.26).

The improvement can be emphasised by recalculating C_p and C_{pk}. C_p is 1.59 and C_{pk} is 1.50, so the process is now capable. There is still a difference between C_p and C_{pk}, however, reflecting the fact that the setting is not on nominal. An adjustment is necessary to improve the setting whilst at the same time progress is maintained on continually reducing the variability.

In this example a specific management action was taken to improve the process, the result of which was reduced variability. In practice it may be that improvement in variability is being detected by the R chart because of some unknown special cause. Detective work may be required to identify this. Ultimately the cause will be integrated as a permanent feature of the system. This means that a record of events relating to chart usage must be kept. This will be considered in Chapter 8.

7.6 **Limitations of the histogram**

The histogram has been used at various places in this text. Whilst it is a useful visual indicator of the performance of a process, it does have the serious handicap of providing information too late to take action.

Look again at the histogram generated by the first set of sample means used in the current example. The tally chart represented in Fig. 7.27(a) can form the basis of a histogram of sample performance as shown Fig. 7.27(b). How much confidence can be placed in this histogram as a true indicator of process performance?

In terms of capability analysis the histogram does give a reasonable representation of how process performance relates to the specifications. Even here, however, caution is needed because it has been seen that when the distribution is fitted to the histogram a different interpretation normally results. A process which is assumed capable on the basis of a histogram is seen to be incapable when a normal curve is fitted (as was found in Section 7.2).

The histogram does not really provide the answer when used as an indication of process performance. If more sample readings are taken and a histogram drawn for the $\overline{X}$ values, then a typical picture would be as shown in Fig. 7.28(a). What pattern of points, on a time scale, contributed to this format, or equally to the equivalent but more visually striking diagram in Fig. 7.28(b)? It is not possible to say. Many variations are feasible, of which Fig. 7.29 illustrates four typical ones.

The histogram on its own is of little use, therefore, as a means of controlling the process. It represents a picture of performance, but any variations over time are not revealed. The only way to monitor a process effectively is to use the control chart, and hence the constant reference to plotting over time against performance-based limits. These limits are clearly related to a histogram but unlike the histogram they provide the necessary means of controlling the process and then improving it.

Summary

- Capability and control are two distinct features.
- Measuring items against specification limits does not provide a true indication of capability.
- Using performance-based limits as a means of control does not mean that operators have to work to tighter limits.
- For a sample size 5, process performance = $2.58\,\overline{R}$.
- C_p measures capability; C_{pk} measures capability and setting.

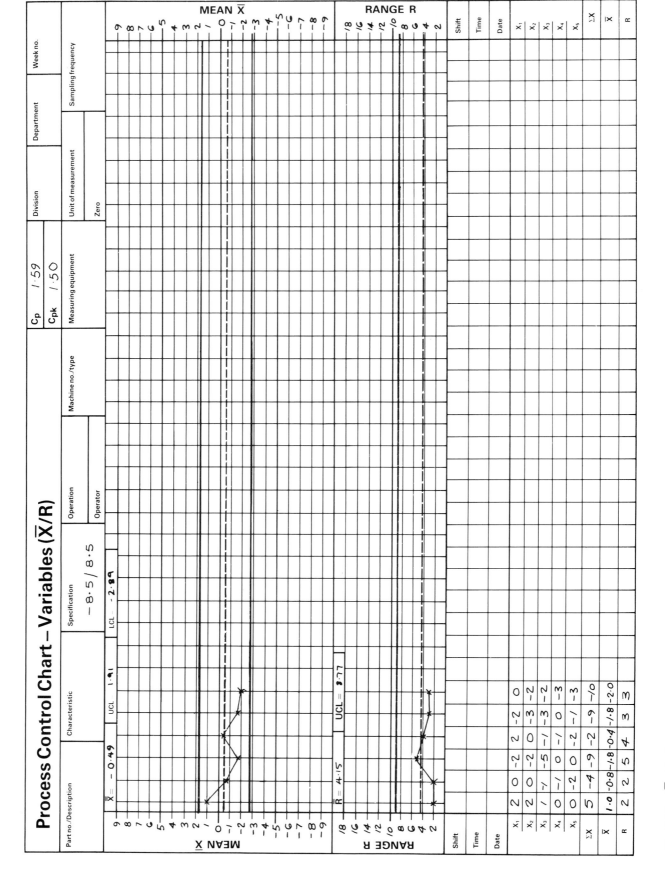

Fig. 7.26 (X̄, R) chart prepared for further readings

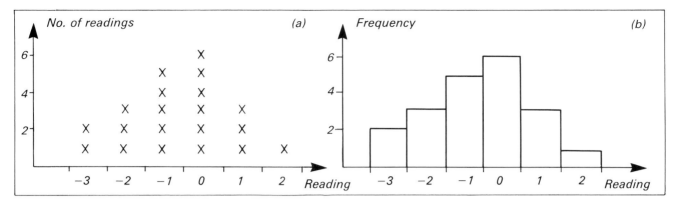

Fig. 7.27   Pattern of $\bar{X}$ results

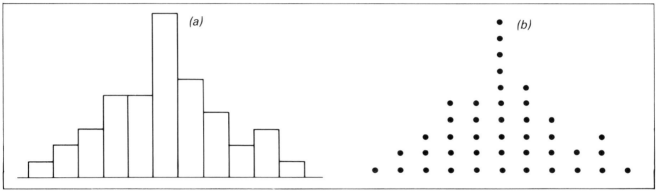

Fig. 7.28   Histogram and practical equivalent

- The $\bar{X}$ chart corresponds to operator action; the R chart corresponds to management action.

- Problem-solving groups are a means of providing joint action to identify special causes on the R chart.

- The R chart provides the key to process improvement.

- Modified limits have no place in quality improvement programmes.

- C_p and C_{pk} values provide means of communication as well as effective process measures in the programme of continuous improvement.

- The histogram is retrospective and should not be used for control purposes.

These last two chapters on charts for variables have provided information on the basic ideas of control and capability. The mean and range chart has been used as the working example. It has been seen how special causes can be recognised by using the four rules. The various patterns that the charts display have not yet been considered. The use of the graphical and database section of the control chart has been covered but not the need to document adequately other factors besides those directly connected with measuring, recording and plotting.

The following chapter, which is the final one on charting variables, provides practical detail on an area sadly neglected at times. Quality improvement programmes require thought and commitment enough without being handicapped by lack of care in the design and use of the control chart: hence the need to consider issues which may seem trivial to some. For those at the work-face of implementing control charts, however, the material to follow could be as relevant as any in the book.

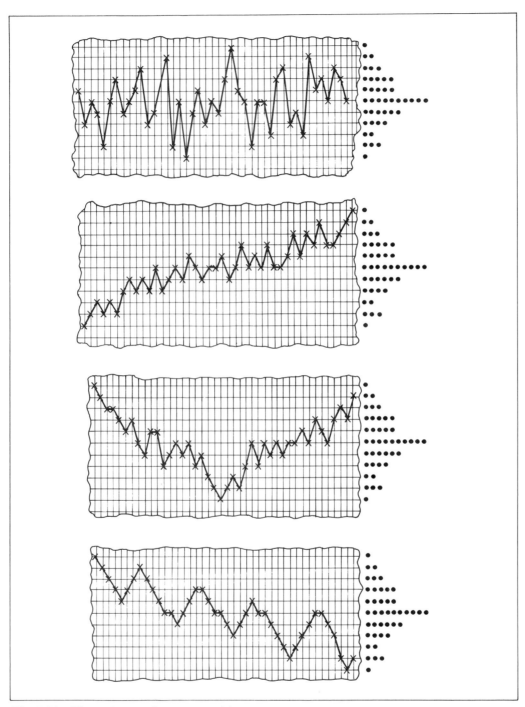

Fig. 7.29   Time plots and a constant histogram

Chapter 8 Interpreting patterns and setting up charts

8.1 **Introduction**

The previous two chapters have referred constantly to the need to interpret charts. There has been an emphasis on using the four rules to determine the presence of special causes which should then be eliminated, or permanently incorporated, depending on their nature. Fig. 8.1 illustrates the types of pattern which indicate improvement in the process. Patterns other than these are connected with special causes which adversely affect the current situation and would therefore result in poorer process performance.

There is a need to be able to recognise patterns so that full benefit can be gained from the information provided by the use of the rules. Understanding the four rules is only a start. It is also necessary to understand in a practical sense the relationship between the pattern produced by the chart and the behaviour of the process on which the chart is inherently based.

These considerations will be dealt with in this chapter, together with issues relating to the practical use of the control chart itself, such as the design of the chart and the information which needs to be recorded. These latter points tend to be overlooked, but it is vital that the chart is designed as a practical working document. Failure to do so could mean failure of the SPC programme itself.

8.2 **Recognising patterns**

Familiarity with recognising patterns in conjunction with the four rules is vital. These rules enable the presence of special causes to be determined from the control charts. Here are the rules again:

Rule 1 Any point beyond the control limits.
Rule 2 Rules of seven, i.e. a run of seven consecutive points that are either all on one side of the mean, or all increasing/all decreasing.
Rule 3 Unusual patterns or trends within the control limits.
Rule 4 Middle third rule, i.e. the number of points in the middle third of the overall distance between the control limits differs greatly from two-thirds the total number of points present.

It goes without saying that it is the machine operators, the clerical staff preparing the invoices, the cleaning staff who empty the waste-bins, and others closest to the operation concerned, who are in the best position to advise on process improvement. SPC now provides them with statistical tools so that decisions can be based on data which everyone can see and understand. Understanding control chart

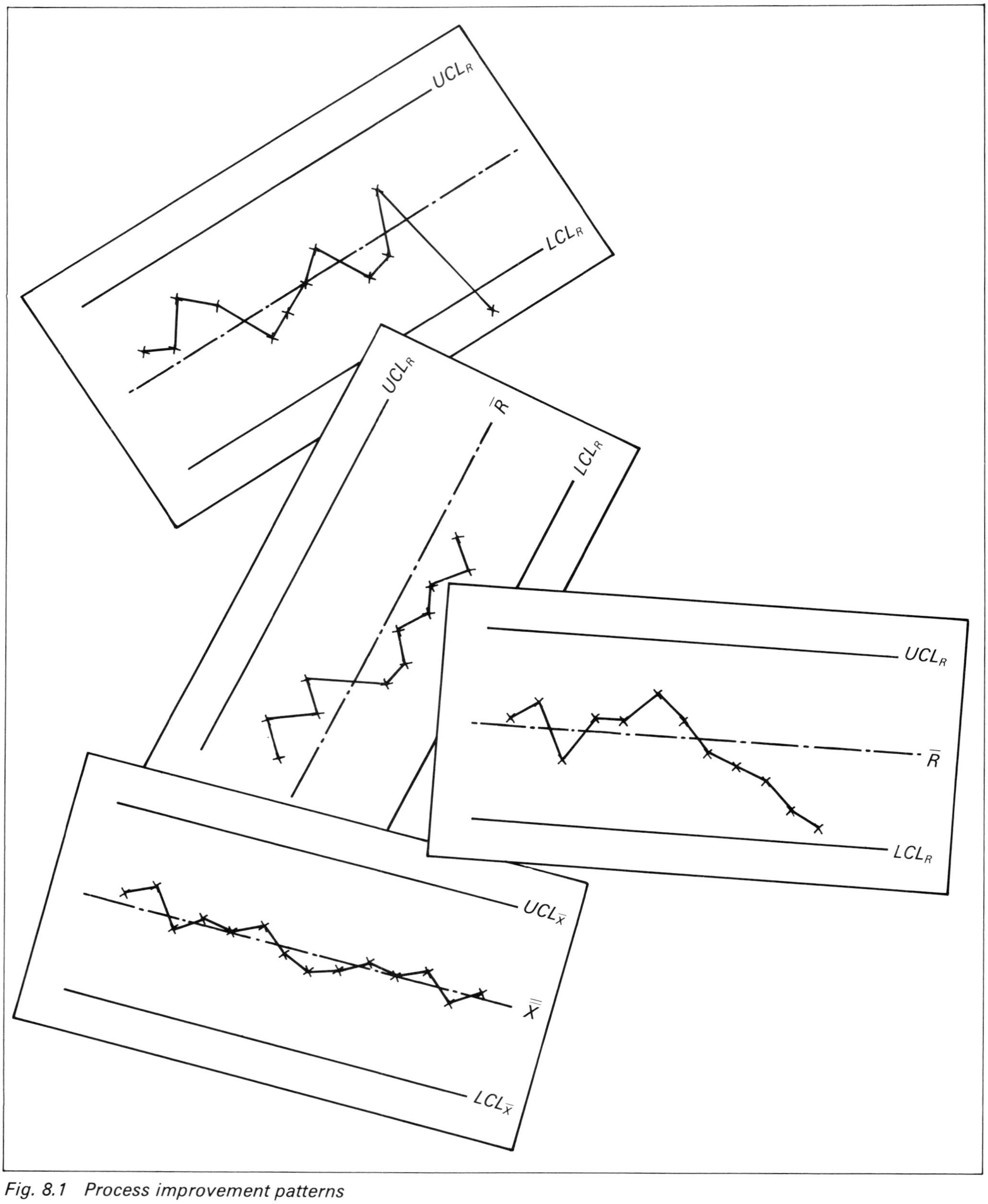

Fig. 8.1   Process improvement patterns

patterns at the operational level is therefore critical. Amongst other things it enables the right action to be taken with confidence, whether by manager or operator.

For any chart, what is looked for initially is a natural random pattern as represented in Fig. 8.2. It is hoped that improvement will be seen, such as that indicated in Fig. 8.3. Here the setting is being monitored and it is clear that there is reduced variability about the nominal, as verified by the application of rule 4. As a result the control limits should be recalculated to reflect the improved performance. There should be little difficulty seeing how rule 1 applies in Fig. 8.4, or seeing the various rules of seven, relating to rule 2, as they apply to Figs. 8.5 and 8.6. It is rule 3, the presence of unusual patterns, which often causes difficulty.

Sometimes the position is relatively straightforward, such as the bunching shown in Fig. 8.7, the cyclic pattern in Fig. 8.8 and the drift indicated in Fig. 8.9. Regular jumps appear in Fig. 8.10; on a range chart these may reflect the use of measuring equipment with a limited level of accuracy. At other times the position is

Fig. 8.2 Random pattern

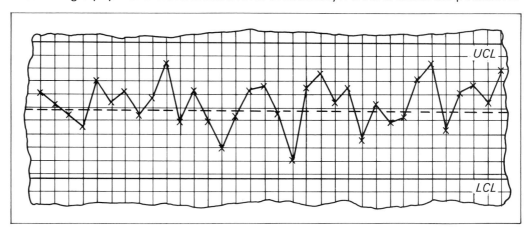

Fig. 8.3 Process improvement

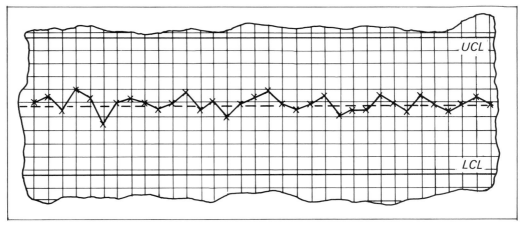

Fig. 8.4 Special cause: Rule 1, any point beyond the control limits

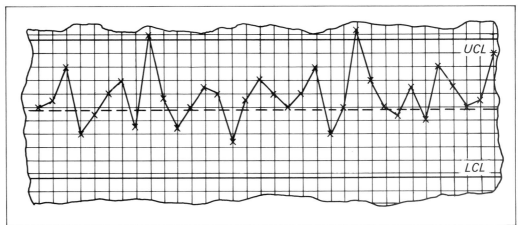

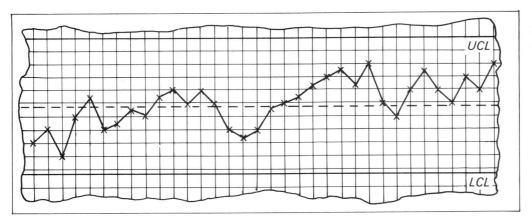

Fig. 8.5   Special cause: Rule 2, rules of seven

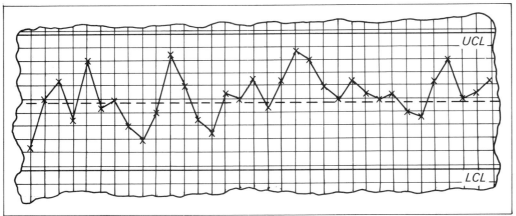

Fig. 8.6   Special cause: Rule 2, rules of seven

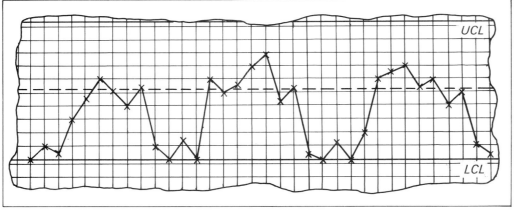

Fig. 8.7   Special cause: Rule 3, bunching

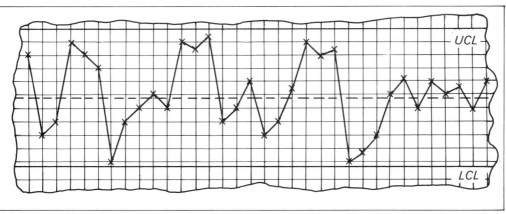

Fig. 8.8   Special cause: Rule 3, cyclic pattern

Fig. 8.9   Special cause:
Rule 3, drifting

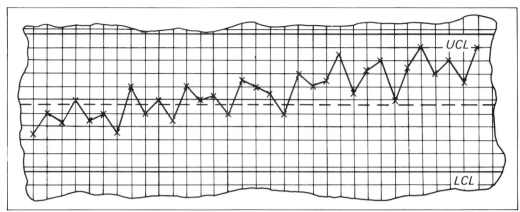

Fig. 8.10   Special
cause: Rule 3, jumps

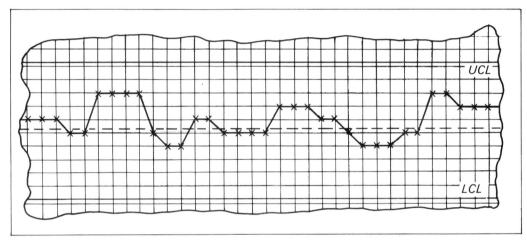

Fig. 8.11   Special
cause?

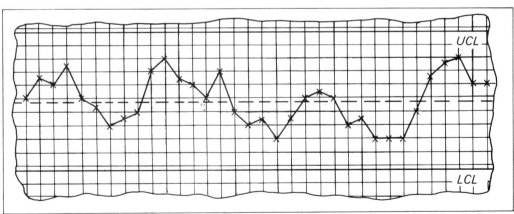

not as clear cut. For example Fig. 8.11 suggests various changes in the pattern, but it would need some experience to be sure of what was really happening.

There is not only a need to recognise the way in which these patterns communicate information about the process; there is also the further, and much more important requirement of finding out what caused the pattern and doing something about it.

It is here that shop floor experience is vital. It is evident that over the years the operators in industry have been neglected when it comes to using their knowledge and familiarity with processes, whether in manufacturing or administrative functions. Decisions have been taken by managers usually often working in offices detached from the machine or process. A recent training course for operators involved obtaining responses to a series of illustrations based on patterns typical of those discussed above. In each case the group was able to suggest the probable reason for the special cause. Management would be foolish not to respond positively to the enthusiasm of such operators. They should be encouraged to channel their efforts into constructive problem-solving rather than being employed in activities which they know involve the generation of waste and the recycling of unacceptable material. The same argument holds for clerks, secretaries and typists. Pinpointing the special causes, and doing something about them, is not an easy process. It requires patience, and a deep sense of commitment to change. Problem-solving groups are a necessity, and hence the need for familiarity with the techniques covered in Chapter 3. But in order to find the reason for the special causes, a knowledge of the relationship between the chart pattern and the process on which is was based is also required.

8.3 **From chart to process**

As the process changes so does the chart pattern. Because the sample mean is an indicator of the process mean, then an erratic process, as was seen in Fig. 6.5, results in the control chart for the sample mean also showing instability, as represented in Fig. 8.12. Since the variability is also changing over time, then it similarly follows that the range chart also indicates instability. The aim is predictable processes, as shown earlier in Fig. 6.3. This results in the corresponding sample mean chart showing stability (Fig. 8.13). Continuous improvement can then follow.

Having recognised the connection between the movement of the process and a corresponding pattern on the chart, it must also be appreciated that in practice the reverse is happening: the control chart once set up duly reflects process performance. It is important, therefore, that changes in the chart can be identified with factors which influence the performance of the process. There are three possible alternatives:

● Changes in $\overline{X}$ chart, i.e. a change in process setting with no change in process variability.

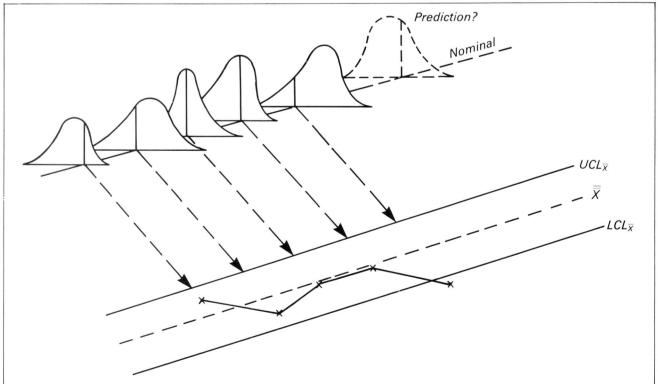

Fig. 8.12 Unstable process

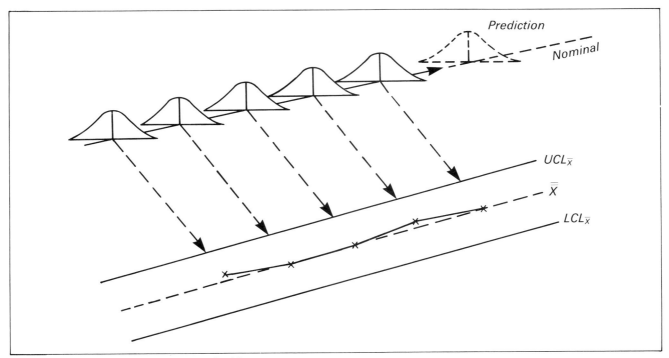

Fig. 8.13   Stable process

- Changes in R chart, i.e. a change in process variability with no change in process setting.
- Changes in $\overline{X}$ and R chart, i.e. a change in process variability together with a change in process setting.

In considering the three options use will be made of some of the patterns illustrated in Figs. 8.4–8.11.

8.3.1   Changes in $\overline{X}$ chart

(a) *Instantaneous change*

If there are seven points above the central line on the $\overline{X}$ chart, as in Fig. 8.6, then this indicates that the process setting has moved to a different level, as shown in Fig. 8.14. The sample range has remained unchanged over the same period.

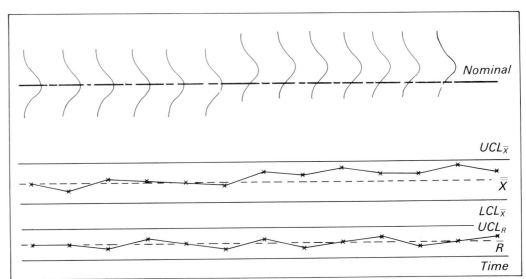

Fig. 8.14   ($\overline{X}$, R) chart: instantaneous change in $\overline{X}$

(b) *Trend*

If there are seven points in succession moving upwards on the $\bar{X}$ chart, as in Fig. 8.5, then this shows that the process setting is gradually increasing, as shown in Fig. 8.15. Again the sample range has remained unchanged.

(c) *Irregular changes*

If there are irregular changes indicated on the $\bar{X}$ chart, as Fig. 8.10 , then this reflects a similar feature in the process performance, as shown in Fig. 8.16. $\bar{X}$ is now moving in response to the process setting whilst the R chart once more reflects stability. Other features on the $\bar{X}$ chart, such as cycles or out-of-control points, reflect an associated pattern in the performance of the process.

The reasons for these features depend very much on the process concerned, and hence the need for utilising the experience of the operators. Some typical reasons, in a manufacturing environment, for change in the process setting without changes in the process variability are:

- Material
- Machine setting

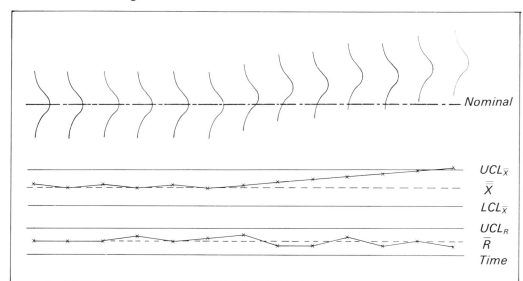

Fig. 8.15   ($\bar{X}$, R) chart: drifting $\bar{X}$

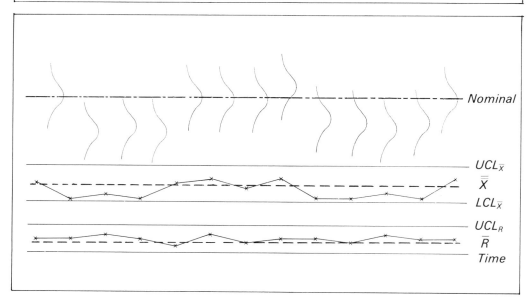

Fig. 8.16   ($\bar{X}$, R) chart: irregular changes in $\bar{X}$

- Hardness of stock
- Supplier
- Mould or cavity dimension
- Tool wear
- Moisture content
- Operators

The operator is expected to adjust the process accordingly in the light of the out-of-control signals. However, given the typical list of causes above, it is evident that without the necessary management action, getting the setting as near to nominal as possible is still a short-term solution. In the typical Western organisation factors such as material and supplier are not normally ones over which the operator has much control. As the use of control charts increases, then so does the need for increased operator involvement in many of the decision-making processes of the organisation. This has nothing to do with shop floor revolution in a political sense, but a great deal to do with utilising operators' knowledge for the benefit of the whole organisation.

8.3.2 Changes in R chart

Changes in the R chart similarly reflect changes in the process variability. Instantaneous changes, trends, irregular changes, cyclic patterns, bunching and other features can all be present on the R chart and indicate corresponding features in the process itself. The example in Fig. 8.17 shows the effect of a gradual increase in variability while the process mean remains stable.

Some typical reasons, in a manufacturing environment, for an increase in the process variability without changes in the process setting are:

- New operator
- Lack of maintenance
- Unstable test equipment
- Loose fixtures
- Careless handling
- Broken bolt
- Non-uniform materials
- Lack of housekeeping
- Poorly trained operator
- Worn threads on screws

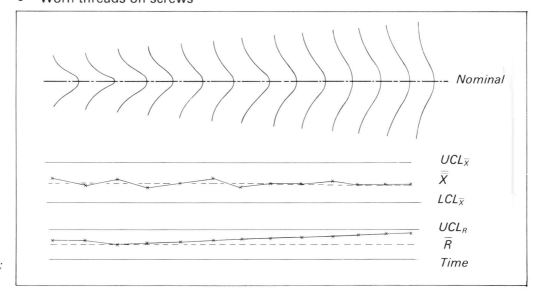

Fig. 8.17 (X̄, R) chart: drifting R

It is not difficult to relate many of these factors to an administrative environment:

- New clerk
- Filing cabinet worn
- Papers lost in transit
- Photocopier not maintained
- Pens and pencils not of uniform standard
- Typewriter support broken
- Desks and offices untidy
- Invoice clerks untrained
- Pencil sharpeners not working efficiently

A statistically based control chart may be too refined a tool to remedy many of the faults, and a performance indicator graph will probably suffice. The principle is nevertheless valid throughout an organisation: operators, manual or clerical, generally work in a system which is not conducive to them doing as good a job as they would wish. At the same time they have no influence to change the system within which they operate.

8.3.3   Changes in $\bar{X}$ and R charts

There are occasions when $\bar{X}$ and R patterns are related. For example, there may be causes which are in reality associated with the R chart but also influence the $\bar{X}$ chart. These form indirect causes and appear on the $\bar{X}$ chart only as a reflection of the change on the R chart. For example, a mixture of two materials, freaks in the data and other abnormalities show up clearly as special causes on the R chart and are also likely to influence the level or pattern on the $\bar{X}$ chart.

It is not the intention here to cover in detail the various aspects of special cause analysis, and the reader is referred to the relevant references in Appendix I for further information. The crucial point is that whatever the chart, and whatever the pattern being displayed, the reason for the special cause must be determined and then acted upon. There is no substitute for hard-won experience in this respect, and each organisation will build up its own case histories. These will usually require hands-on knowledge of the process, and hence the necessity for involving the operators. The role of problem-solving groups has been dealt with in Chapter 3, and their importance cannot be overstated.

The control chart can be used to solve long-term problems. With a management committed to support the SPC programme by providing, for example, ready finance to buy a piece of equipment for which many requests have been made over the months, it is not surprising that barriers start coming down and working relationships improve for the benefit of all. The operators find less aggravation in their jobs, they have greater involvement, they can see management being supportive, and they see the organisation becoming more effective.   All this stems from a control chart which is providing the correct information on which decisions can be based.

These decisions are worthless, however, if the information on the chart is open to question. The system around which the chart operates will be looked at in Chapter 15 when the implementation of SPC is discussed. There is a need, however, to consider the control chart itself as a document, and for this the ($\bar{X}$, R) chart will be taken as an example. All other charts follow basically the same pattern, but any variations which relate to other types of chart will be included as appropriate.

8.4   The control chart

Section 6.5 referred to the fact that the control chart has three main sections: one for recording data and calculations, one for plotting the results, and one for recording administrative information relevant to the process, product, sampling frequency, operator etc. In practice the sections are not quite as clearly defined as this, but they are nevertheless a good starting point. In the schematic diagram in Fig. 8.18, which is based on a typical control chart for mean and range, the three sections have been labelled 'Recording and calculation', 'Plotting' and 'Process data'. Each of these needs to be considered in some detail.

8.4.1 Recording and calculation

This section of the control chart provides for the recording of the sample readings and the calculation of the sample mean and sample range. An extra box is provided for entering the total of the individual readings in the sample, which is needed if the mean is to be calculated mentally.

Not all organisations will adopt the same policy when it comes to the plotting of the sample mean. There is a natural inclination to succumb to electronic gauging. After all, quick and accurate answers avoid tedious arithmetic and they can be linked to impressive display units. The advantages are many and are repeatedly pressed home – far too often by companies whose only interest in SPC is the quick financial return they can make from gullible customers. There is a danger that the reputation of SPC as a proven means of quality improvement is being tarnished in some instances by unjustified claims regarding the use of electronic gauging. There is also the problem of unnecessary accuracy. By displaying a numerical result to excessive decimal places, the operator is confused rather than helped. Such illusions of accuracy work against effective operation of SPC at shop floor level.

It should not be thought that electronic gauging is always inappropriate, however. For components with, say, eight or ten dimensions to be monitored, then there is no alternative but to measure by electronic means. Otherwise, the operator is spending more time in recording calculations and plotting readings than in being involved with the actual manufacturing.

Sophisticated electronic gauging equipment is worthless, however, unless operators have been adequately trained in manual charting methods. Some of the best examples of SPC in action on the shop floor occur when the operator is directly involved with taking sample readings, recording them, mentally working out the mean, recording it, plotting it, and then interpreting the chart. The chart should therefore be designed so that there is adequate space for recording the various results and the calculations that go with them.

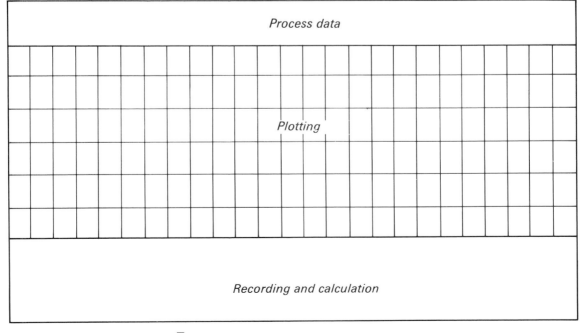

Fig. 8.18 ($\bar{X}$, R) chart: schematic diagram

A further feature associated with the recording of results is the need to indicate the shift, date, and time of sampling. The first two are straightforward, but time of sampling may be a problem. Fig. 8.19 shows a typical section from a control chart. It reflects the fact that samples of 5 are being taken hourly, or at least near-enough hourly. Bearing in mind that SPC brings with it a need to be specific, there would be some justification in querying the phrase 'near-enough'. The intention in sampling, though, is to avoid the samples being taken exactly on the hour, every hour, because there may be some cyclic feature operating which would distort the results. Hence the variation around the exact hourly interval.

It is suggested that the five sample readings are taken, as far as possible, as a sequence. They then represent an instantaneous picture of the process at that time. There may be cases where the sample taken is a random one based on the previous hour's output. This poses a problem as to how representative the sample is. It could be that there is still a dependence on laboratory staff to assess the sample results, and that a random sample of 5 based on the hour's output has always been used. This raises questions regarding the role of the Quality Control Department in SPC programmes, and some comments on this are given in Section 8.5. Whatever the situation in the past, continuous improvement dictates that a sequence of 5 at a predetermined time is preferable, as otherwise the 5 readings are representative of a much wider spread (Fig. 8.20). More than that, because there is not likely to be any indication of when these 5 items were produced within the hour, they are not traceable in terms of production sequence. One of the related issues of an SPC programme is the emphasis on identifying times, materials, operations etc., and hence the need for SPC and Just-in-Time to be developed in tandem.

8.4.2 **Plotting**

The plotting of the sample mean and sample range seems straightforward enough: follow the vertical line corresponding to the centre of the sample data column and then look for the point of intersection with the two horizontal lines corresponding to the sample mean and sample range respectively. In practice, every effort should be made to eliminate possible difficulties in carrying this out.

Perhaps the major potential problem relates to the alignment of figures at the correct position on the vertical scale. Fig. 8.21 shows how confusion could arise if the sample mean values, or sample range values, are not aligned correctly with the horizontal grid. In some instances, potential difficulties have been eliminated by enclosing the scale readings in boxes. Jaguar Cars, amongst others, use a boxed notation as shown in Fig. 8.22. It may also be preferable to continue the grid right across the page and add a second scale on the right-hand side. This certainly makes it easier to plot results, but there is possibly a disadvantage in that no room is now available for drawing a histogram. Organisations will need to assess priorities in such cases.

All this may seem trite and basic, but it tends to be the small things which irritate and confuse when introducing SPC. If some thought is given to these and similar

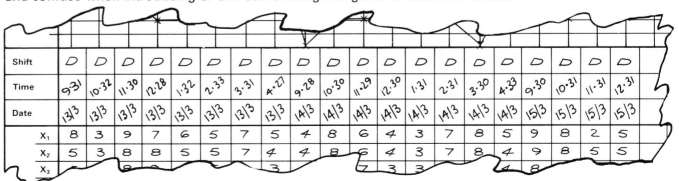

Shift	D	D	D	D	D	D	D	D	D	D	D	D	D	D	D	D	D	D	D	D
Time	9·³¹	10·³²	11·³⁰	12·²⁸	1·³²	2·³³	3·³¹	4·²⁷	9·²⁸	10·³⁰	11·²⁹	12·³⁰	1·³¹	2·³¹	3·³⁰	4·²³	9·³⁰	10·³¹	11·³¹	12·³¹
Date	13/3	13/3	13/3	13/3	13/3	13/3	13/3	13/3	14/3	14/3	14/3	14/3	14/3	14/3	14/3	14/3	15/3	15/3	15/3	15/3
X_1	8	3	9	7	6	5	7	5	4	8	6	4	3	7	8	5	9	8	2	5
X_2	5	3	8	8	5	5	7	4	4	8	6	4	3	7	8	4	9	8	5	5
X_3								3				7	3	3			8			

Fig. 8.19 Section from ($\overline{X}$, R) chart

Fig. 8.20 Sampling
implications

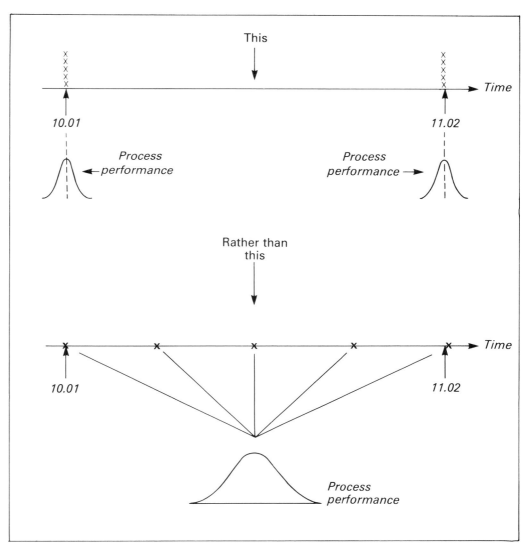

Fig. 8.21 Scaling
difficulties

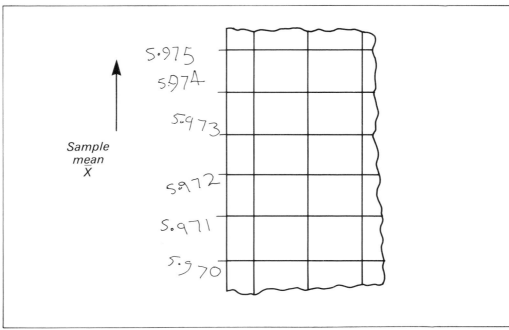

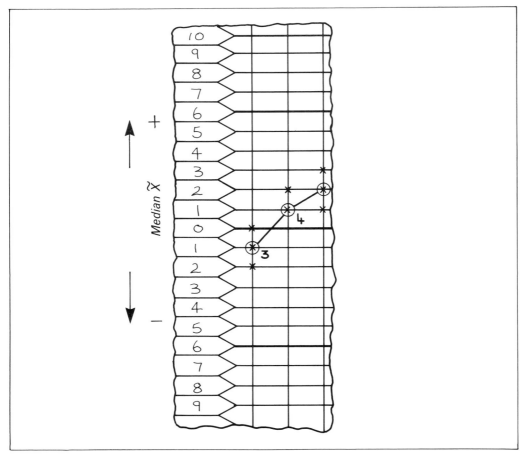

Fig. 8.22 Minimising scaling problems

issues, operators can do a better job, with consequent improvement in productivity and quality.

The position and thickness of the lines on the grid also needs to be considered. Sauer Sundstrand have recognised that this is an important area in chart design – to such an extent that they have introduced a series of ($\bar{X}$, R) charts which incorporate different grid markings. Fig. 8.23 shows a selection of some of the different charts which are available.

It is also worth while considering the density of the lines. Some need to be bolder than others. Those used in the grid should be bold enough to allow for ease of plotting but not so bold as to mask the pattern of the points themselves.

8.4.3 Process data

The main administrative section of the chart, which is usually, but not necessarily, placed at the top of the sheet, allows for details of the process or product under consideration to be recorded. Fig. 8.24 shows a typical format for this section of the control chart. The exact format will depend on specific requirements. There is no point in re-inventing the wheel, however, and if a suitable standard chart is available then there is no reason not to use it. For example, the Ford Motor Company charts are used extensively within the company's supply base.

Most of the reference boxes in Fig. 8.24 require no explanation. There are some points worth pursuing, however.

(a) C_p and C_{pk} boxes

A relatively recent addition to the standard control chart format is the box which allows the updated C_p and C_{pk} values to be noted. This enables the process perform-

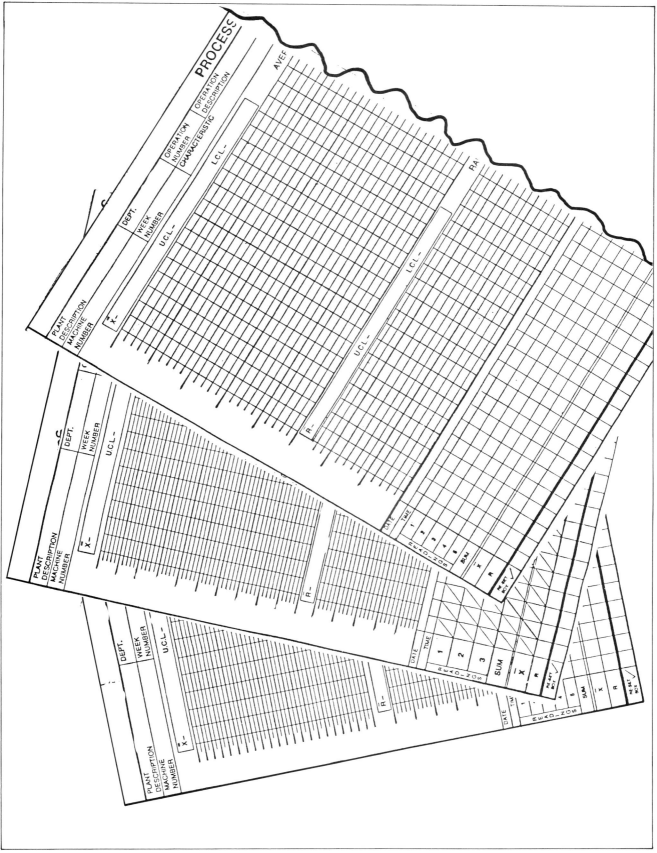

Fig. 8.23 Grid options

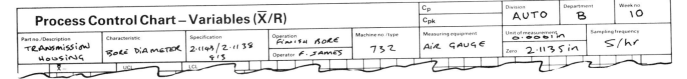

Fig. 8.24 (X̄, R) chart: administrative section

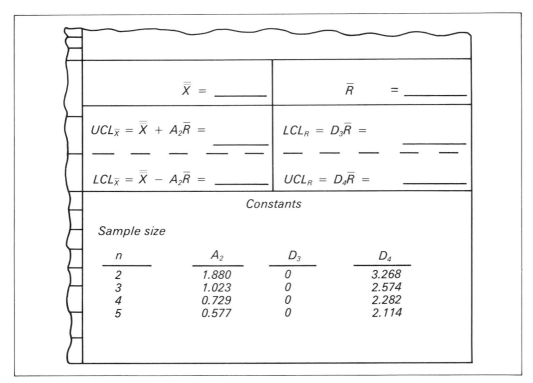

Fig. 8.25 (X̄, R) chart: control limits reference box

ance to be recorded in a way which makes it easier to monitor process improvement from chart to chart.

(b) *Control limits and central lines*

Spaces are allocated on the chart grid for noting the values of the various lines to be drawn. In some cases, as in the Ford chart, an extra section is included which provides a condensed version of the table of control chart constants. There is also a box for recording calculations involved with working out the control limits and central lines. Fig. 8.25 shows the appropriate section on the Ford chart.

In designing its own control chart, an organisation may prefer to allocate this information to the rear of the chart. The presence of this information on the front of the Ford chart is a reflection of its view that the control limits will be re-assessed for each new page. That may not be a policy followed by others. In addition, the introduction of standard reference data does limit, to a certain extent, the amount of current data which can be recorded and plotted.

Ease of use of the chart by the operators who will be plotting the data is clearly of paramount importance, and their views on chart design should be taken into account.

(c) *Rules for interpreting charts*

Some companies have included on their charts the rules for determining special causes. Fig. 8.26 shows the appropriate section of the Jaguar chart.

Fig. 8.26 Rules for
special causes

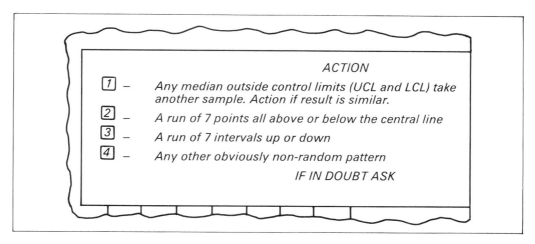

ACTION

1 – Any median outside control limits (UCL and LCL) take
another sample. Action if result is similar.

2 – A run of 7 points all above or below the central line

3 – A run of 7 intervals up or down

4 – Any other obviously non-random pattern

IF IN DOUBT ASK

8.4.4 Process log chart

A typical procedure is to use the reverse side of the control chart as a process log sheet. An example of a log sheet is shown in Appendix H and Fig. 8.27 illustrates a section from such a log.

This sheet provides the means for recording any changes which have taken place in the process. It cannot be stated too often that the control chart itself is worthless unless the necessary supportive information is recorded. The information is useful in two ways:

- It can be used to reference known changes in the process, such as tool change or material change, so that the corresponding changes in the control chart patterns can be identified. To save unnecessary repetition, a code can be used to indicate standard routine changes. Fig. 8.28 illustrates the Jaguar approach.

- Equally, and perhaps more usefully, there is the opportunity of tracing a change in process pattern to a particular change which took place in the operation. There may well be a time delay between noting the feature and seeing an effect

		PROCESS LOG SHEET		
DATE	TIME	COMMENTS	DATE	TIME
FEB 12	11-05	CHANGE OF CASTING CODE	MAR 17	8-15
	14-00	COOLANT LEVEL TOPPED UP		3-30
FEB 13	12-05	CUTTERS ALTERED	MAR 19	9-21
	13-30	CUTTERS RAISED	23	10-50
FEB 14	8-30	COOLANT LEVEL TOPPED UP	24	15-20
	15-00	CHANGE CASTING FROM 284 TO 513	31	10-30
FEB 20	11-15	CUTTERS ALTERED	APR 4	9-20
	14-45	CUTTERS RAISED	5	15-03
	16-05	CHANGE OF CASTING CODE	7	9-31

Fig. 8.27 Completed log sheet

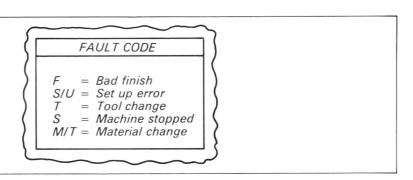

Fig. 8.28 Typical fault code

on the chart, and for this and other reasons the task of tracing the reason for the special cause is not an easy one. Without a formal record of events, the task could be reduced to a lottery.

The Japanese traditionally write process changes directly on the control chart, the operators being encouraged to write comments near the suspicious points. It makes the chart more of a live document and almost a personal diary of the process and the operator. A typical Japanese control chart appears in schematic form in Fig. 8.29. The design of the chart allows for the section 'Notes and Comments' and encourages the recording of comments as near to the source as possible. Western organisations are similarly encouraging their operators to use the chart in this way and experience has shown that the shop floor staff react positively to the approach by writing on the chart near to the point in question.

The fact that the operators may have oily hands and that the chart is smudged or greasy is of no consequence. If cleanliness is a prerequisite for some reason, then ways can be found of satisfying that. Management that suggests the Inspection Department maintain the records so as to keep the charts neat and tidy exposes its lack of understanding of the wider implications of SPC.

This section has referred to some mundane, basic and yet very necessary aspects of control charts. Designing the chart, printing it and updating it raises many issues. Chapter 15 will include further information on these as part of the discussion on implementing SPC.

Without a properly designed chart, complete with information recorded in full by the operators, no SPC programme will succeed. The charts will provide true working documents on the basis of which appropriate decisions can be made. They form a natural part of the everyday activity of the operators and they are not the

Process data
Recording and calculating
Plotting
Notes and comments

Fig. 8.29 Typical Japanese control chart: schematic diagram

prerogative of the Quality Control Department. An explanation of this statement follows.

8.5 **The role of the Quality Control Department**

The introduction of an SPC programme causes changes throughout the organisation. The daily pattern of work of each employee will gradually alter for the better. There will be greater involvement, less aggravation, better use of time. Staff involved with the Quality Control (QC) function face change in the same way as do maintenance, production or invoicing staff. However, because of their traditional involvement with quality matters it is understandable that the QC staff in some instances react negatively.

Historically the QC Department has been the sole arbiter on quality matters. The QC Manager would report to the General Manager and if the quality of the product was unacceptable in his view then he would demand that no shipment be made. This would cause difficulties on the production side because of the need to satisfy quotas, keep the line going and get products out to the customer. The resulting all too familiar conflict between Production and Quality would require a higher-level decision to resolve the impasse.

Because the QC Department has been in possession of the many quality-related tools and techniques which form the basis of SPC, and has been responsible for monitoring quality, it tends to assume that these techniques belong almost exclusively to the QC staff. It is not really the fault of the QC Department that such a situation has arisen. In the past there may have been an attempt to spread the responsibility for quality throughout the organisation, but by and large these attempts clearly failed, for reasons well known by now.

It is not surprising, therefore, that a picture is emerging of some QC staff viewing SPC with apprehension. The reasons are logical. QC inspectors have traditionally taken readings. QC laboratories have taken measurements and provided the results. QC technicians have performed the calculations relating to any charts that were operating. The advent of SPC programmes could now be seen as very threatening. Changing roles, with operators being trained in the new techniques which at one time were the prerogative of QC staff only, is always unsettling. Even worse, it could mean redundancy.

In practice, properly handled, the position is not so bleak. Overmanning is avoided by natural wastage and/or early retirement schemes. QC staff are retrained in new skills which enable them to play a role in supporting production staff, and others, in the new programme.

There are bound to be difficulties at times. In one case a member of the QC staff insisted on visiting the line and removing the relevant control chart in order to display it in the QC laboratory. The SPC facilitator, in turn, then replaced the chart on the shop floor location. The point was rightly made that the chart belonged to the operator and the machine. It was being used by the operator to control a process and as such had absolutely nothing to do with the QC Department. Such misunderstandings do occur and they do, unfortunately, reflect the rather traditional protectionist role of a minority of QC staff. It is for these reasons, among others, that the QC Department may not be the best base from which to choose a facilitator for the SPC programme. This point will be considered further in Chapter 15.

On the positive side, QC Departments have much to offer in terms of SPC. They are familiar with the techniques, and the staff involved have experience of the problems of other departments. Given the commitment of top management, enlightened QC Departments should see SPC as an opportunity rather than a threat. It is a powerful vehicle for changing the role of QC in a positive sense as well as helping the Department to influence other departments for the better.

Summary

- Operator experience should be utilised to the full in recognising chart patterns.
- Recognising patterns is not sufficient in itself.
- Chart patterns follow the movement of the process.
- Reasons for changes in the process must be determined.
- Organisations will tend to build their own case histories of special cause analysis.
- Control charts have three main sections that need to be completed.
- Electronic gauging and computerised displays should only be introduced following adequate experience of manual charting.
- All relevant administrative detail such as time and shift must be provided on the control chart.
- Samples should be taken sequentially, not randomly over a longer period.
- Design the chart to make it easy for the operator to use.
- Encourage operators to write directly on the chart.
- The Quality Control Department has a positive role to play in SPC programmes.

The mean and range chart has been referred to a great deal in this chapter and earlier ones. The following chapter will therefore give some detail on other charts that can be used for variables, as in some companies they provide the main focus of SPC programmes.

Chapter 9 Alternative techniques for charting variables

9.1 Introduction

Throughout the previous three chapters there has been a heavy emphasis on the use of the mean and range, or $(\bar{X}, R)$ chart. The reason is twofold. Traditionally, any attempt to introduce control charts within an organisation usually resulted in the use of this well-tried chart, and therefore it is natural to introduce it first in any text on chart techniques. Furthermore, many of the subsidiary issues relating to control charts can be considered using the $(\bar{X}, R)$ chart as the example. That does not mean that the issues are specific only to that chart, however.

In using the $(\bar{X}, R)$ chart as the working reference, there is an understandable danger of suggesting that this chart, and this chart alone, will cover all possibilities. At a recent SPC conference the keynote speaker made the following statement: 'If the only tool you have is a hammer, then it is not surprising that everything looks like a nail.'

Useful as it is, the $(\bar{X}, R)$ chart still has limited applications. For example, organisations involved in batch processing, particularly processing chemicals, paints, oils, etc., will find the chart of little value. Organisations such as British Alcan make far more use of charts other than the $(\bar{X}, R)$ type in applying SPC to their own environment. In a similar fashion, median charts may be of more immediate value than mean charts, individual readings could provide more appropriate information than sample values. If computing facilities are available, then the standard deviation is obviously a better measure of spread to use than the range. Hence charts making use of medians, individual readings and standard deviations form part of the SPC tool kit. In the same way, the Cusum technique, to be covered in Chapter 13, provides an alternative when it is small changes from the nominal that need to be detected.

There are, therefore, a range of various techniques available which can be readily adapted to suit the particular process (Fig. 9.1), and this chapter will provide information on all of these except Cusum. There is no intention to comment in detail on every possible method available for charting values, though, as some of the alternatives to the $(\bar{X}, R)$ chart are considered sufficiently in other sources.

It is probably logical to start with the median chart.

9.2 Control chart for medians

The median, $\tilde{X}$, has already been defined as the central value of a set of readings arranged in order of magnitude. Fig. 4.12 (reproduced again as Fig. 9.2) shows a typical section of a median control chart.

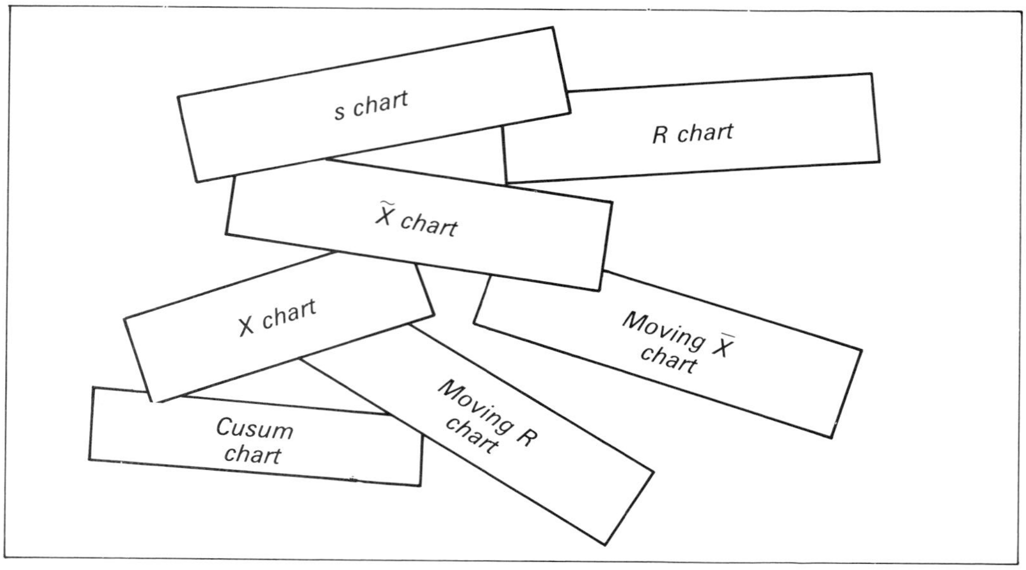

Fig. 9.1 Various charting techniques

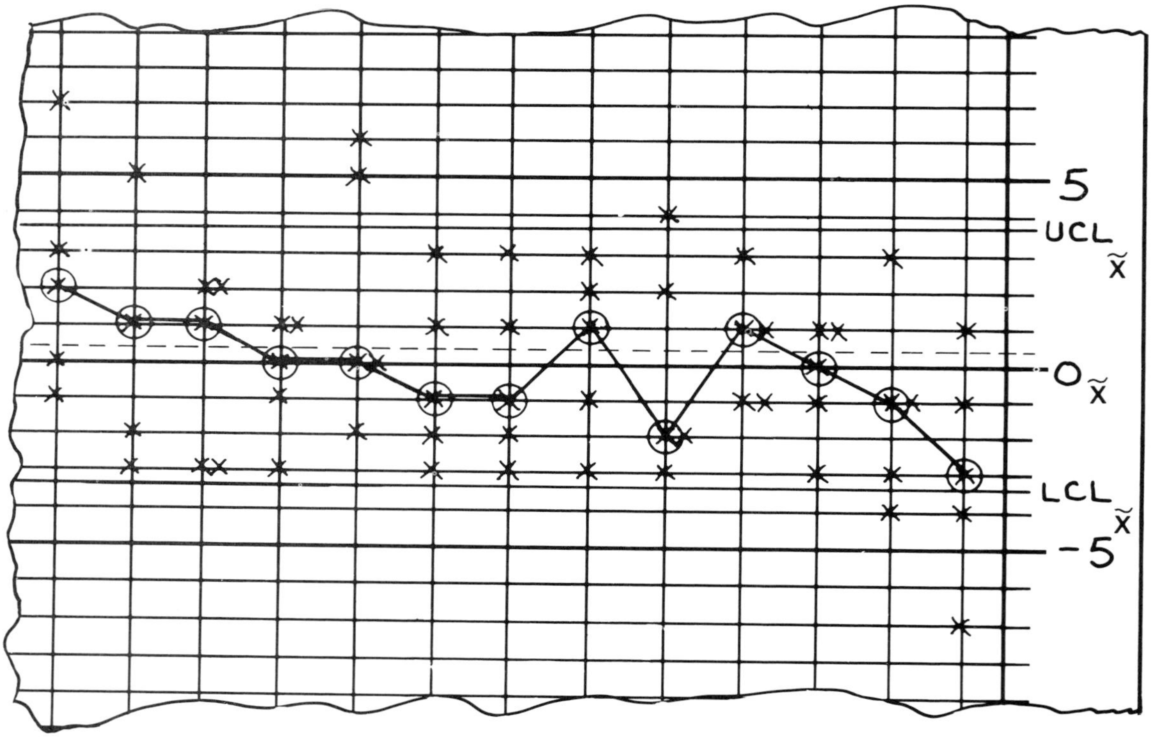

7	-3	1	1	0	1	-1	1	4	1	-1	3	-4	x_1
3	-2	-3	1	5	-3	-2	3	2	-1	-3	-1	-1	x_2
2	1	2	-1	0	-2	-3	2	-3	1	1	-3	-7	x_3
-1	1	2	-3	6	3	3	-1	-2	-1	1	-1	-3	x_4
0	5	-3	0	-2	-1	1	-3	-2	3	0	-4	1	x_5
2	1	1	0	0	-1	-1	1	-2	1	0	-1	-3	$\tilde{x}$

Fig. 9.2 Section from median chart

Some further explanation is required regarding collecting the data, setting up the chart, and monitoring the results.

9.2.1 Collecting, calculating and plotting

The guidelines relating to the collecting of initial samples apply equally to median charts as to mean charts. Twenty samples are normally taken, and once more the sample size is usually 5. As explained earlier, the sample size must be an odd number otherwise unnecessary difficulties are introduced. The twenty sample readings are recorded and plotted on the appropriate portion of the control chart in the usual way. The median is then ringed.

As a working example on the median chart, the same data which generated the $\overline{X}$ chart in Chapter 6 will be used. The information was detailed in the data boxes in Fig. 6.10. Fig. 9.3 shows the same basic information but this time for analysis in the form of a median chart.

The results of the first 20 samples corresponding to the initial process study have been recorded and plotted. The median has then been ringed and the medians connected by straight lines as is customary.

It is possible for a sample to contain more than one value equal to the median. For example, the second sample of 0, −2, 0, 2, −4 clearly has the median at 0, so there are thus two readings corresponding to the median. The way this is represented is to put two adjacent crosses on the horizontal line corresponding to the value of 0. There are slight variations on this convention but they do not fundamentally alter the standard procedure.

Since there are two readings corresponding to the median, it has to be decided

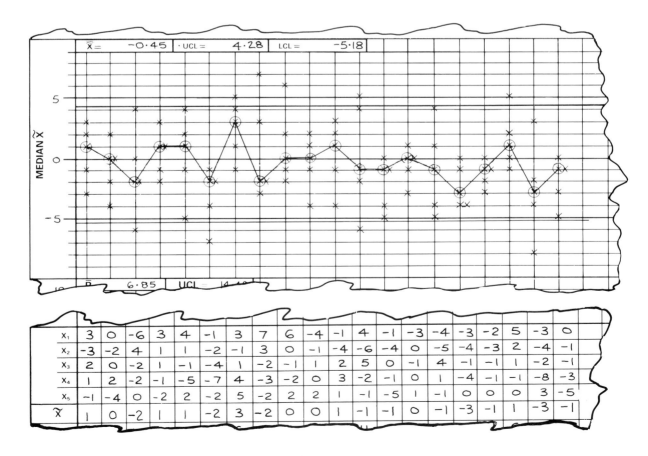

Fig. 9.3 Section from median chart

which of the two should be used as the median for charting purposes. Conventionally, the one lying on the vertical line corresponding to that sample is chosen. The reason for this is that points to be used as the basis of the final control chart for medians are then equidistant along the horizontal time axis, which improves the visual impact of the chart.

9.2.2 Control limits for $\tilde{X}$ chart

Control limits for the median chart are obtained in very much the same way as for the mean chart. First the central line is calculated. Having used the median for each sample, the grand mean, corresponding to the central line, is obtained by reverting to the definition of the mean rather than the median. The central line is therefore obtained as the mean of the sample medians and denoted by $\overline{\tilde{X}}$.

Hence

$$\overline{\tilde{X}} = \frac{\Sigma \tilde{X}}{k}$$

where k is the number of samples.
Therefore for the particular readings being considered

$$\overline{\tilde{X}} = \frac{1 + 0 + \ldots -3 -1}{20}$$

$$= -0.45$$

Understandably, this value is different from the grand mean $\overline{X}$ (which was previously determined as -0.57).

As with the mean chart, the control limits for $\tilde{X}$ are positioned at distances of 3 standard errors measured outwards from the grand mean. Once more, statisticians have provided an appropriate constant, $\tilde{A}_2$, so that the upper and lower control limits (UCL and LCL) for the sample median are given by

$$\text{UCL}_{\tilde{x}} = \overline{\tilde{X}} + \tilde{A}_2 \overline{R}$$

$$\text{LCL}_{\tilde{x}} = \overline{\tilde{X}} - \tilde{A}_2 \overline{R}$$

Tables of $\tilde{A}_2$ appear in Appendix C. For a sample size 5 the value of $\tilde{A}_2$ required is 0.691.

As might be expected, the mean range $\overline{R}$ is again required. For the 20 samples forming the basis of this analysis, the range values are as shown in Fig. 9.4. These range values are exactly the same as those used in setting up the mean chart and, as before, $\overline{R} = 6.85$.

X_1	3	0	-6	3	4	-1	3	7	6	-4	-1	4	-1	-3	-4	-3	-2	5	-3	0
X_2	-3	-2	4	1	1	-2	-1	3	0	-1	-4	-6	-4	0	-5	-4	-3	2	-4	-1
X_3	2	0	-2	1	-1	-4	1	-2	-1	1	2	5	0	-1	4	-1	-1	1	-2	-1
X_4	1	2	-2	-1	-5	-7	4	-3	-2	0	3	-2	-1	0	1	-4	-1	-1	-8	-3
X_5	-1	-4	0	-2	2	-2	5	-2	2	2	1	-1	-5	1	-1	0	0	0	3	-5
$\tilde{X}$	1	0	-2	1	1	-2	3	-2	0	0	1	-1	-1	0	-1	-3	-1	1	-3	-1
R	6	6	10	5	9	6	6	10	8	6	7	11	5	4	9	4	3	6	11	5

Fig. 9.4 Data box for median chart

Hence

$$UCL_{\tilde{x}} = -0.45 + 0.691 \times 6.85$$

$$= 4.283$$

$$LCL_{\tilde{x}} = -0.45 - 0.691 \times 6.85$$

$$= -5.183$$

The $\tilde{\tilde{X}}$, $UCL_{\tilde{x}}$ and $LCL_{\tilde{x}}$ values can now be recorded in the data boxes and lines drawn on the chart, with the results shown in Fig. 9.5.

9.2.3 Interpreting the $\tilde{X}$ chart

The control limits have been calculated in order to monitor the sample median. There may well be occasions, such as for the third sample in Fig. 9.5, when an individual reading falls outside the control limits. This does not matter as long as the median itself is under statistical control.

The usual four rules apply for determining the presence of special causes. It is clear that the rules show the process to be under statistical control.

One further aspect of plotting medians will be included here. It can be seen that the central line on the median chart is different to that on the mean chart. Also the control limits for medians are further apart than the corresponding control limits for the mean chart. Fig. 9.6 provides a direct comparison, based on differences between A_2 and $\tilde{A}_2$ for the same values of n and R. With exactly the same data, $\tilde{\tilde{X}}$ and $\bar{X}$ differ, as is to be expected. In addition the distance between $UCL_{\tilde{x}}$ and $LCL_{\tilde{x}}$ is greater than the distance between $UCL_{\bar{x}}$ and $LCL_{\bar{x}}$ for the same sample values. This is an indication of the fact that as a representative value of the sample, the median is not as good as the mean. The reasons for this were considered in Section 4.3.4 and are reflected in the statistics behind the determination of the $\tilde{A}_2$ values.

9.2.4 The R chart

The median chart should always be accompanied by a chart for controlling the variability. Typically this is the familiar R chart. The R values for the data being

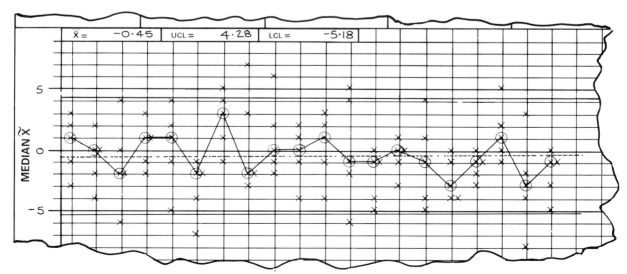

Fig. 9.5 Median chart with control lines

x_1	3	0	-6	3	4	-1	3	7	6	-4	-1	4	-1	-3	-4	-3	-2	5	-3	0
x_2	-3	-2	4	1	1	-2	-1	3	0	-1	-4	-6	-4	0	-5	-4	-3	2	-4	-1
x_3	2	0	-2	1	-1	-4	1	-2	-1	1	2	5	0	-1	4	-1	-1	1	-2	-1
x_4	1	2	-2	-1	-5	-7	4	-3	-2	0	3	-2	-1	0	1	-4	-1	-1	-8	-3
x_5	-1	-4	0	-2	2	-2	5	-2	2	2	1	-1	-5	1	-1	0	0	0	3	-5

$\bar{X}$ chart   (mean) $\tilde{X}$ chart   (median)

Fig. 9.6 Comparison of mean and median chart

considered are indicated in Fig. 9.4. From these $\bar{R}$ and UCL_R can be obtained in exactly the same way as in Section 6.8.3.

Since the same data is being used as in Chapter 6, then the control chart for ranges will be as shown in Fig. 6.19, reproduced again as in Fig. 9.7. The final control chart will then be as shown in Fig. 9.8.

9.2.5 Use of a template

It has already been stated that a major advantage of the median is its simplicity of use. This advantage may be further enhanced in that it provides the option of not plotting a range chart at all: instead the median chart can also be used to monitor the range.

This is done by making use of a transparent template, which is marked with two lines whose distance apart corresponds to the value of UCL_R. By superimposing this template on a set of sample readings, an out-of-control point will be readily picked up, as shown in Fig. 9.9. Clearly, the position of the reference lines on the template will vary depending on $\bar{R}$ (and the sample size).

A similar transparent graduated template can be used to measure the range directly from the individual plots on the median chart (Fig. 9.10). One advantage of this technique is that it provides an indirect way of familiarising operators with the problem of dealing with negative numbers when calculating a range. Also, use of a template removes the need for a data box: first the individual readings are recorded directly on the median chart, then the template enables the range values to be recorded directly. Direct plotting does have a possible disadvantage though. Cal-

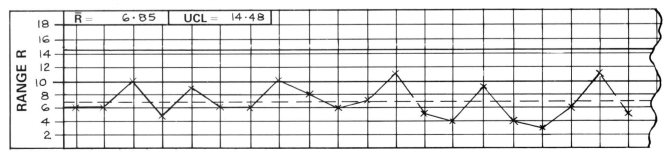

Fig. 9.7   Section of R chart

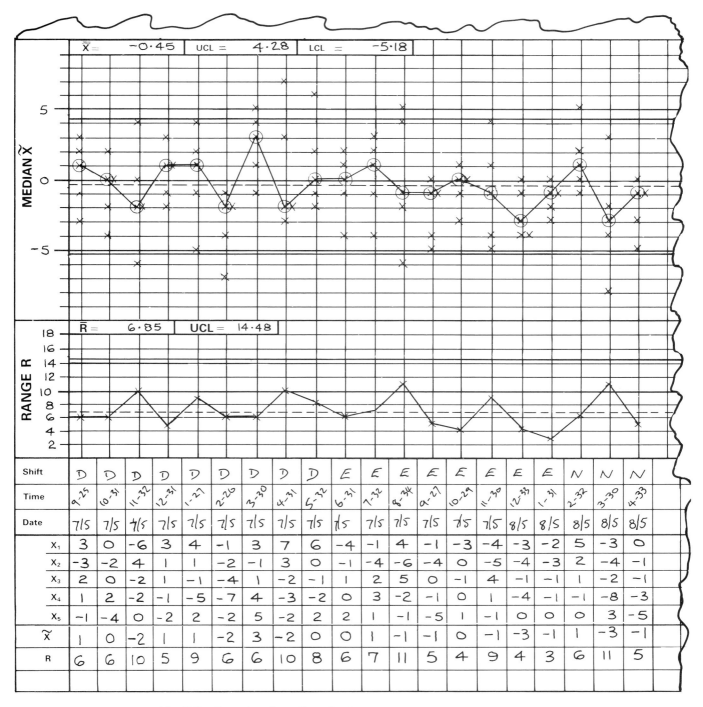

Fig. 9.8 Completed median chart

culating control limits is always easier if the numerical values are available in tabular form. Hence there may be a benefit in making use of a data box.

Case study 3: Jaguar cars

Jaguar cars has been making use of median charts in its Radford Engine Plant, Coventry, since 1983. For various reasons the company chose the median chart as the basis of operator control.

Extensive management and operator training in those areas directly or indirectly involved has resulted in a series of chart applications based entirely on machining

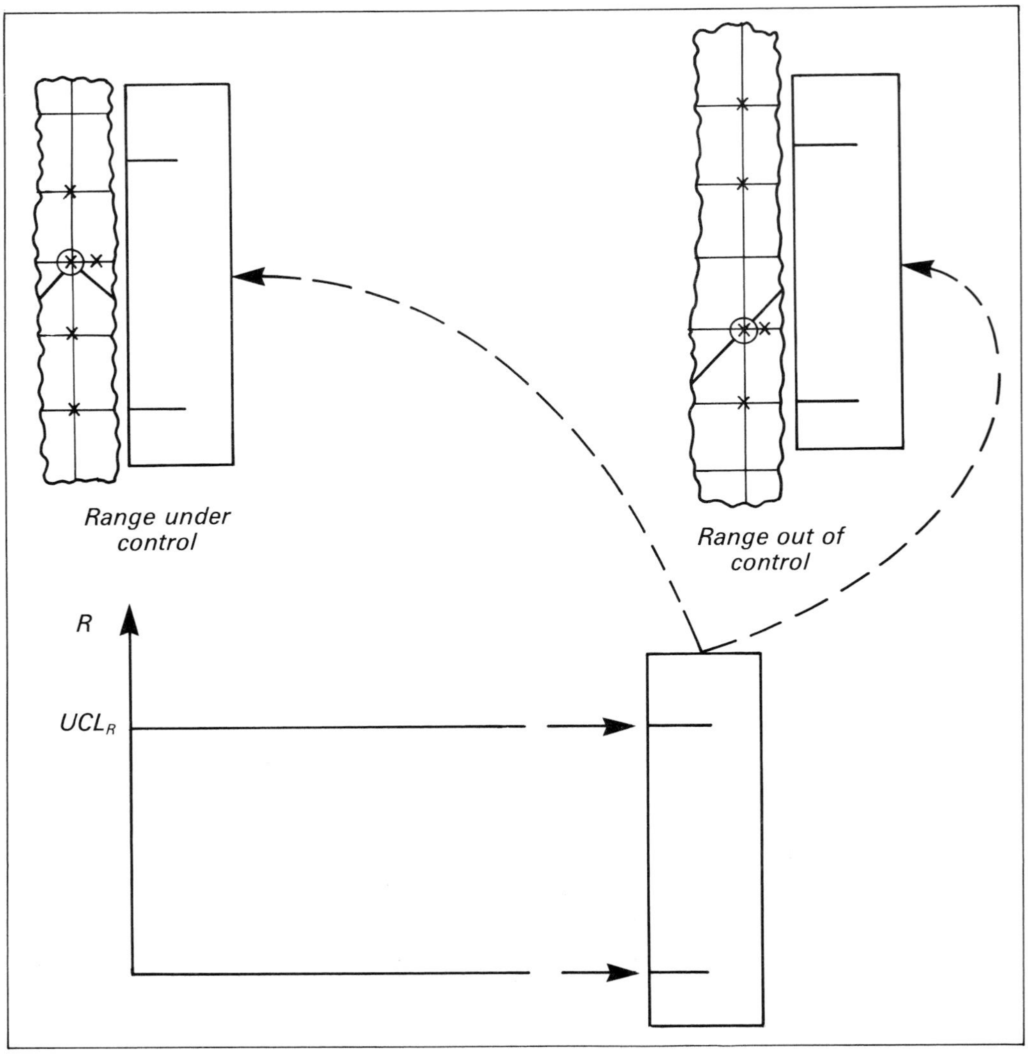

Fig. 9.9   Template for monitoring R on $\overline{X}$ chart

operations. Fig. 9.11 shows a typical chart. Note the convention used when recording the sample values on the chart. Two readings at the same value are indicated X2. In addition, the same control sheet is being used to monitor the results of two spindles.

With a bank of experience of manual charting behind it, Jaguar is introducing computerised SPC equipment. Each multi-dimensional operation will be linked by an electronic gauge and a computer. This will enable up to 35 dimensions from one operation to be monitored. The company has produced its own software package in conjunction with an outside supplier. Jaguar recognise that this sophisticated monitoring equipment can only be introduced after manual charting has been in operation for some time.

9.2.6   **Weights and measures legislation**

An interesting application of the median chart is in connection with the UK weights and measures legislation following the introduction of EEC directives.

Figure 9.12 shows a simplified median chart used in wine bottling. The references in Appendix I will provide further details on this chart and the more general statistical techniques which have formed the framework of various national standards on weights and measures legislation.

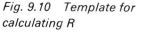

Fig. 9.10 Template for calculating R

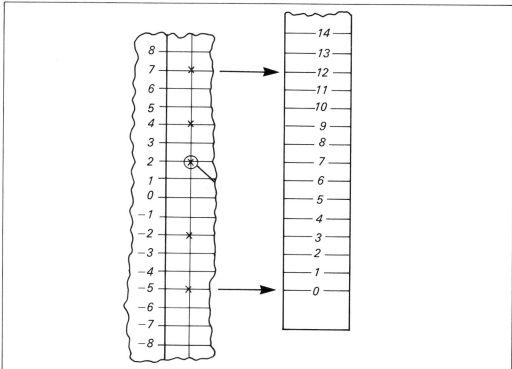

The median and mean chart are both based on taking a sample of a given number of discrete units at a particular time from the process. In many industrial practices this is not feasible. Hence the need to consider alternative methods of charting.

9.3 Moving mean/moving range chart

9.3.1 Introduction

Many organisations are involved in continuous processes, such as manufacturing steel, aluminium, paint, oil or chemicals. On the administrative side, invoices, shipments, shortages and losses often only provide one value for analysis. In such cases it is recognised that the $(\overline{X}, R)$ chart is inappropriate. Taking 5 samples from batches of oil going through the system does not make sense. Equally, coils of aluminium or steel do not lend themselves to being analysed in groups of 5, or groups of any other size for that matter.

Perhaps more important is the fact that in many of these applications the only effective way to monitor the activity is to control process variables such as pressure or temperature. This is the core philosophy of SPC in any case. Even with $(\overline{X}, R)$ charts, whilst it is the case that it is often the product being monitored, in the form of a dimension of a component, this is only because there is a direct relationship between the product and the process producing it. Controlling the former is then equivalent to controlling the latter.

In continuous processes extensive use is made of a moving mean/moving range chart.

9.3.2 Collecting, calculating and plotting

(a) *Moving mean*

The usual guidelines apply for collecting the data. At least 20 results are required. Fig. 9.13 shows the data box for a series of readings which are viscosity values for successive batches of paint being processed. Several options are now open when it

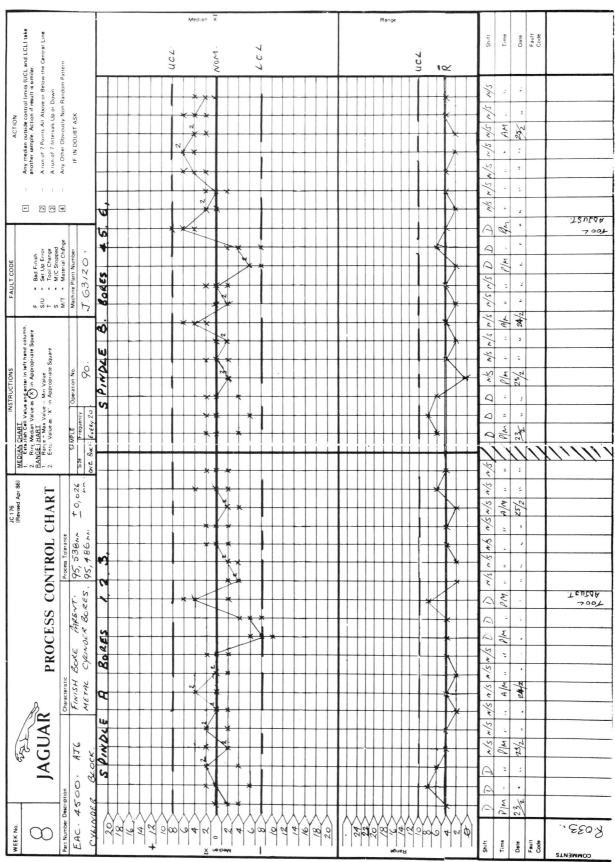

Fig. 9.11 Jaguar Cars: process control chart

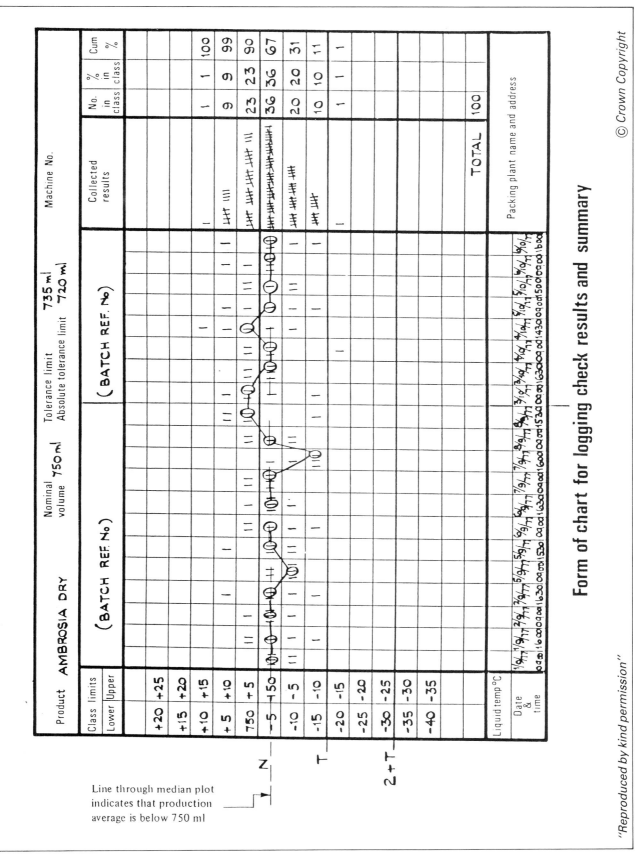

Fig. 9.12 Median chart used in wine bottling

comes to plotting the results. The readings could be plotted individually or as samples of 2, 3, 4, and so on.

Essentially, increasing the sample size, i.e. the number of individual readings considered as a group, smoothes out the form of the plot on the chart. A larger sample size therefore makes it easier to see the effect of any genuine changes in the process (in exactly the same way as increasing sample size makes the $\overline{X}$ chart more sensitive). Typically, a sample of size 3 is chosen, but there is nothing magical about the figure.

Section 4.3.1 described how to calculate the moving mean. For the first three readings of the set shown in Fig. 9.13 the moving mean is $(15.5 + 15.3 + 14.5) / 3 = 15.1$. This value is now recorded in the appropriate data box. The usual convention is to record the value in the box below the last of the grouped numbers, i.e. 14.5 in this case. The procedure is then repeated. At each step, the first of the previous group is dropped, and the next in sequence included. The moving mean of the next three readings is $(15.3 + 14.5 + 14.1) / 3 = 14.63$, and so on.

These moving means are then plotted. The completed data block, and the corresponding plot for moving mean, are then as shown in Fig. 9.14.

(b) *Moving range*

If a moving mean based on groups of three can be obtained, then it follows similarly that a moving range can also be calculated and plotted. The data block for ranges, and the corresponding moving R chart, appear in Fig. 9.15.

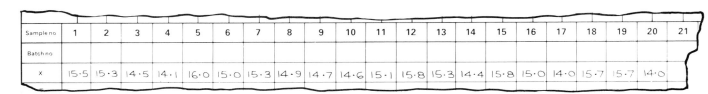

Sample no	1	2	3	4	5	6	7	8	9	10	11	12	13	14	15	16	17	18	19	20	21
Batch no																					
x	15·5	15·3	14·5	14·1	16·0	15·0	15·3	14·9	14·7	14·6	15·1	15·8	15·3	14·4	15·8	15·0	14·0	15·7	15·7	14·0	

Fig. 9.13   Section from data box of ($\overline{X}$, moving R) chart

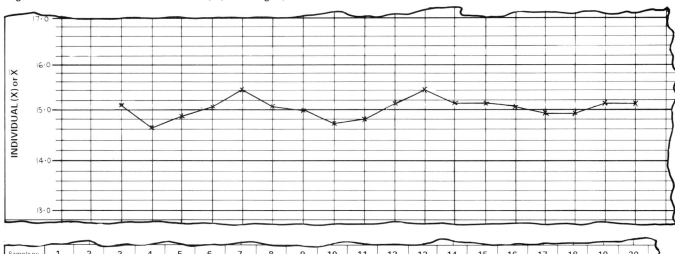

Sample no	1	2	3	4	5	6	7	8	9	10	11	12	13	14	15	16	17	18	19	20
Batch no																				
x	15·5	15·3	14·5	14·1	16·0	15·0	15·3	14·9	14·7	14·6	15·1	15·8	15·3	14·4	15·8	15·0	14·0	15·7	15·7	14·0
$\overline{x}$			15·1	14·63	14·87	15·03	15·43	15·07	14·97	14·73	14·80	15·17	15·40	15·17	15·17	15·07	14·93	14·90	15·13	15·13

Fig. 9.14   Plot of moving $\overline{X}$

9.3.3 Control limits for moving $\overline{X}$ chart

In defining control limits for a moving mean chart first the grand mean of the moving means needs to be calculated. This is obtained as $\Sigma\overline{X}/k$, where k, the number of samples, is now 18 not 20. Hence

$$\overline{\overline{X}} = \frac{270.7}{18}$$

$$= 15.04$$

Control limits are then obtained in the usual manner, using the now familiar relationship involving A_2 and $\overline{R}$. $\overline{R}$ is obtained as $\Sigma R/k$ where k is 18. Therefore

$$\overline{R} = \frac{21.8}{18}$$

$$= 1.21$$

The value of A_2, from Appendix C, is that which corresponds to a sample size 3. This is 1.023. Therefore

$$UCL_{\overline{X}} = \overline{\overline{X}} + A_2\overline{R}$$

$$= 15.04 + 1.023 \times 1.21$$

$$= 16.28$$

Similarly

$$LCL_{\overline{X}} = \overline{\overline{X}} - A_2\overline{R}$$

$$= 15.04 - 1.023 \times 1.21$$

$$= 13.80$$

9.3.4 Control limits for moving R chart

With a sample size 3, the corresponding value of D_4 from Appendix C is 2.574.

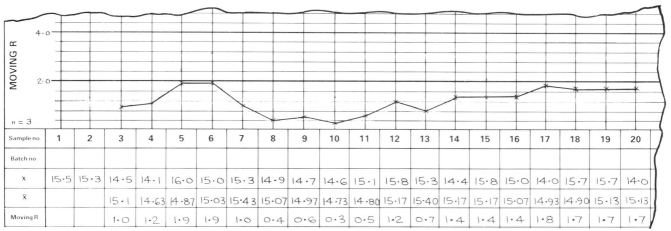

Sample no.	1	2	3	4	5	6	7	8	9	10	11	12	13	14	15	16	17	18	19	20
Batch no.																				
x	15·5	15·3	14·5	14·1	16·0	15·0	15·3	14·9	14·7	14·6	15·1	15·8	15·3	14·4	15·8	15·0	14·0	15·7	15·7	14·0
$\overline{X}$			15·1	14·63	14·87	15·03	15·43	15·07	14·97	14·73	14·80	15·17	15·40	15·17	15·17	15·07	14·93	14·90	15·13	15·13
Moving R			1·0	1·2	1·9	1·9	1·0	0·4	0·6	0·3	0·5	1·2	0·7	1·4	1·4	1·4	1·8	1·7	1·7	1·7

Fig. 9.15   Data block for moving R chart

Therefore

$$UCL_R = D_4\bar{R}$$

$$= 2.574 \times 1.21$$

$$= 3.11$$

The data boxes can now be completed and central lines and control limits for both $\bar{X}$ and R drawn on the respective charts. The completed moving mean/moving range chart is as shown in Fig. 9.16.

9.3.5 Interpreting the charts

$\bar{X}$ and R would seem to be under statistical control, apart from the fact that the R chart shows an out-of-control situation in that the last seven points are above the central line. However, it must be remembered that the points plotted are the means and ranges of groups of a given size: 3 in the example. Hence the sequential sample ranges (or sample means) are not independent, whereas every third sample range is. Fig. 9.17 provides a visual explanation.

The sample range corresponding to sample no. 14 is based on the readings for samples 12, 13, 14. The next sample range which is independent of those three is the one corresponding to sample no. 17 (based on the readings for samples 15, 16,

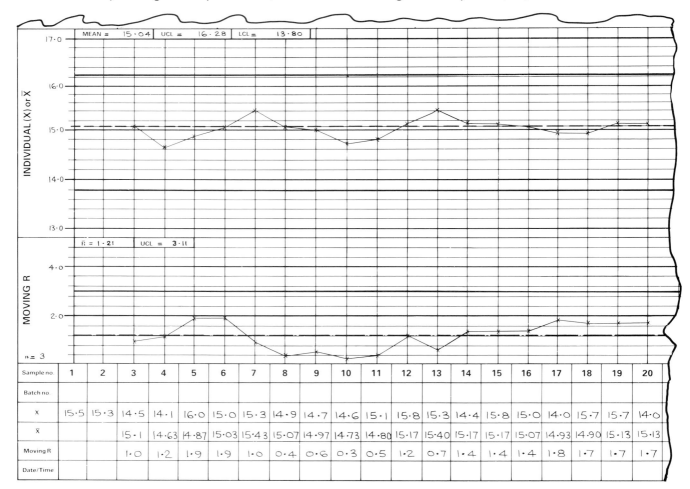

Fig. 9.16 Plot of moving R

17). Therefore if the process is genuinely out of control on the basis of the rules of seven it would be indicated by samples 14, 17, 20, 23, 26, 29 and 32 all having R values above the line. In such cases, therefore, the rules of seven imply that a minimum of 7n points is required, where n is the sample size. As a result, these rules are less applicable to this type of chart than others.

9.3.6 Varying the sample size

In Section 9.3.2 it was suggested that a sample size 3 was typical. Appendix H includes a copy of the control chart used by Century Oils. Note that individual readings are to be plotted here, not moving means. PPG Industries (UK) Ltd, along with others, also plots individual values but makes use of a sample size of 2 rather than 3. A moving range is still plotted, however, so as to make use of the constant A_2 in determining the control limits for individual readings. Appendix F provides an explanation of how the equations on the PPG chart have been obtained.

The main factor in determining the appropriate sample size is likely to be time. If there is a long delay between sample readings, then increasing the sample size will prolong the time before the chart indicates a need for action. As with all charts, the economies of commercial life must be taken into consideration. There is a balance to be found between using individual readings or a small sample, with less sensitivity in the chart, and utilising larger samples with a consequent delay in receiving signals from the chart.

It is of interest to note the effect of varying the sample size for the same set of data. Fig. 9.18(a–e) makes use of the 20 readings used previously. A chart of individual readings is shown, together with moving mean and range charts based on increasing sample sizes. The smoothing effect can clearly be seen.

9.3.7 Use of a mask

The calculation of moving mean and moving range is often confused by the visual presence of numbers outside the particular group chosen. To avoid this problem Sauer Sundstrand makes use of a simple mask. This reveals, horizontally, the values in the particular sample shown and, vertically, the blank data column in which ΣX, $\bar{X}$ and moving R values can be recorded. Fig. 9.19 shows a typical mask. Different masks are available corresponding to the sample size.

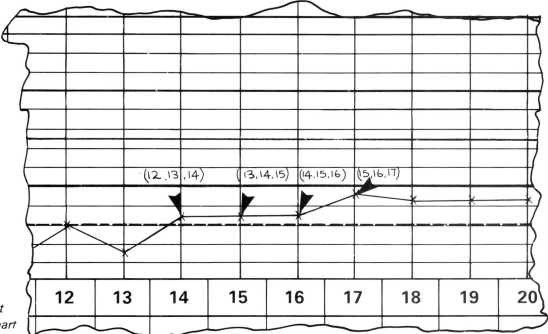

Fig. 9.17 Dependent samples for $(\bar{X}, R)$ chart

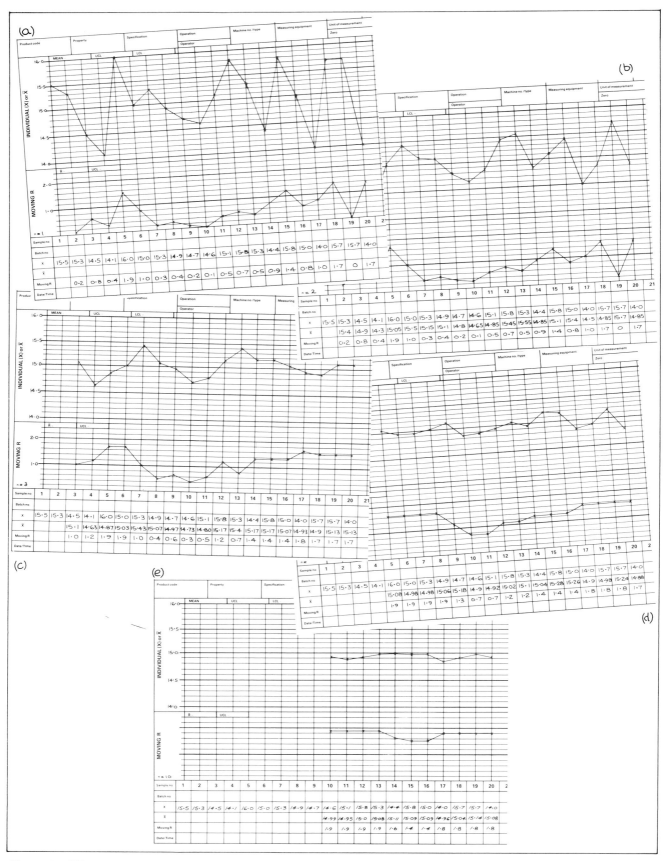

Fig. 9.18 Effect of increasing n

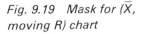

Fig. 9.19 Mask for ($\overline{X}$, moving R) chart

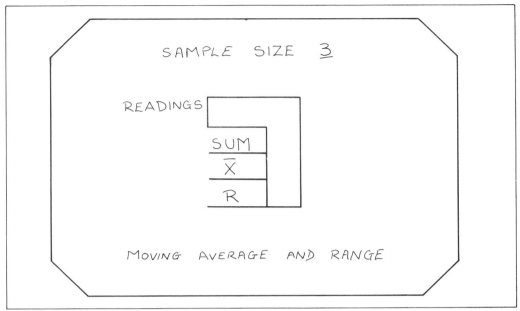

The use of such a technique reflects an essential aspect of SPC programmes. All possible steps should be taken to simplify things and remove the mystique. If the operators' involvement can be increased by using such simple practical resources, then there will be benefits to the organisation as a whole.

The following case study shows how an individual/moving range chart has been used effectively in the solution of a long-standing technical problem.

Case study 4: British Alcan (Rolled Products), Rogerstone

British Alcan (Rolled Products) has applied SPC charts in a variety of areas. Extensive use has been made of the individual/moving range chart as one which is far more applicable to its industry than the conventional ($\overline{X}$,R) chart.

The case study refers to the pretreatment line where a sulphuric acid process thoroughly cleans and prepares the surface of aluminium strip by creating an anodic film. This film should enable customers to lacquer and print the strip without carrying out a sizing operation prior to printing. However, inconsistent anodic film thickness from the pretreatment line resulted in the need for the customer to undertake sizing prior to printing. This meant an additional operation plus additional cost and could well have led to a potential loss of business to British Alcan.

On 7 January the company carried out an investigation into the operating parameters of the pretreatment line. The key factors to be considered included electrical characteristics, acid concentration, aluminium sulphate, and acid bath temperature, as well as the film thickness itself. A series of actions followed:

● The electrical characteristics were checked by Central Laboratories and were found to meet the requirements of the process. At the same time current density requirements were re-emphasised to line operators.
● Prior to the study, acid strength was checked once per shift by operators using a hydrometer and once per day by site laboratory staff. The operator checks would trigger the need to add acid if necessary. However, it was found that the sampling point used by the operators could be adversely affected by water additions. Water additions were made manually on a random basis and there was also an automatic system which added a minimum of 160 litres.

   Cross-reference between the hydrometer checks and the laboratory checks showed that the two methods generated different results. As a result, from 17 February, operator checks were discontinued, laboratory checks increased to twice per day and manual water additions stopped. The automatic water addition was also reduced from 160 litres to 60 litres and the acid usage consequently reduced from 74 litres to 30 litres per shift.

- The master bath temperature controller was reset. Work was put in hand to develop a reliable on-site check of film thickness to avoid the 2–3 week delay that resulted from sending samples to Central Laboratories.

In early March, Rogerstone Technical Dept carried out an investigation which concluded that the only variable which would have a significant effect on anodic film thickness was bath temperature. Following this report, the steam boiler was serviced on 10 March and from that date acid bath temperatures were recorded using individual and moving range charts.

Fig. 9.20 shows the improvement in the variability and also the consequent decreases in C_p value. The C_{pk} values reflected the continuing need to improve the setting. (The units are measured from a nominal of 0 for confidentiality purposes.)

The case study brings out different points regarding the use of SPC charts. In particular it shows the need for data to verify, or otherwise, decisions which would be made by instinct. In addition, the study provided an example of savings in material and time. This would result in a far greater chance of keeping a customer who might otherwise go elsewhere.

All charts considered so far in the text have concentrated on the range as the measure of variability. There will be situations where the standard deviation is preferable, and the next section describes how to set up the s chart and the accompanying $\bar{X}$ chart.

9.4 Control chart for $\bar{X}$ and s

9.4.1 Introduction

The main disadvantage in using s, particularly at operator level, is the fact that it is not so easy to understand and calculate as is the range. However, if there are calculators or computers available, particularly if these are linked to visual displays and electronic monitoring equipment, then the standard deviation is preferable. It

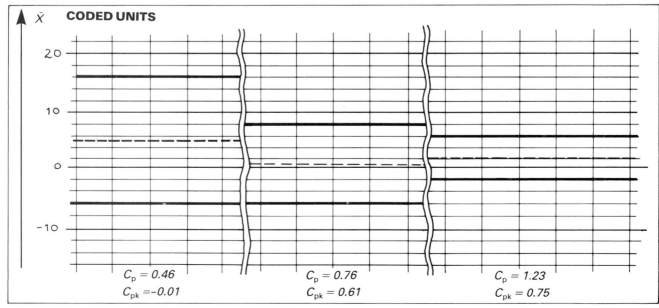

Fig. 9.20 Sequential charts showing process improvement: British Alcan

should only be utilised, however, after training followed by experience in manual charting.

9.4.2 Calculating and plotting

Twenty samples, as usual, are required. As an example the 20 sample readings from Chapter 6 will again be used. These are shown in the main data block in Fig. 9.21. The 20 sample means and sample standard deviations are then calculated, and Fig. 9.21 also shows these results.

A brief recap may be useful as a reminder of how the standard deviations have been obtained. In Chapter 4 the standard deviation was defined as

$$s = \sqrt{\frac{\Sigma(X - \bar{X})^2}{n - 1}}$$

It was stated there that when obtaining the standard deviation, the (n−1) form should always be used and the associated (n−1) key on the calculator. Section 4.4.2

	x_1	3	0	-6	3	4	-1	3	7	6	-4	-1	4	-1	-3	-4	-3	-2	5	-3	0
	x_2	-3	-2	4	1	1	-2	-1	3	0	-1	-4	-6	-4	0	-5	-4	-3	2	-4	-1
	x_3	2	0	-2	1	-1	-4	1	-2	-1	1	2	5	0	-1	4	-1	-1	1	-2	-1
	x_4	1	2	-2	-1	-5	-7	4	-3	-2	0	3	-2	-1	0	1	-4	-1	-1	-8	-3
	x_5	-1	-4	0	-2	2	-2	5	-2	2	2	1	-1	-5	1	-1	0	0	0	3	-5
ΣX																					
$\bar{X}$		0·4	-0·8	-1·2	0·4	0·2	-3·2	2·4	0·6	1·0	-0·4	0·2	0	-2·2	-0·6	-1·0	-2·4	-1·4	1·4	-2·8	-2
s		2·41	2·28	3·63	1·95	3·42	2·39	2·41	4·28	3·16	2·30	2·77	4·53	2·17	1·52	3·67	1·82	1·14	2·3	3·96	2

Fig. 9.21 Data block for $(\bar{X}, s)$ chart

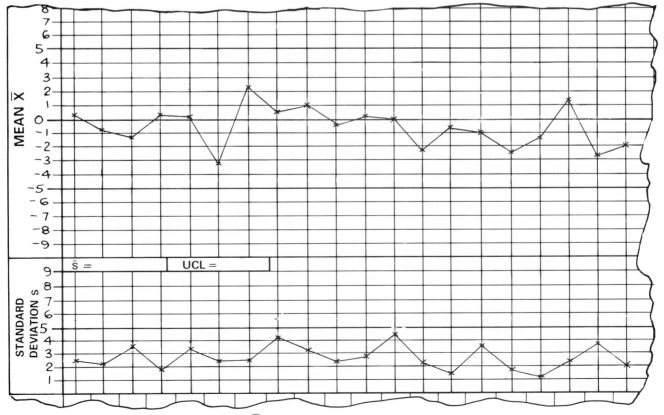

Fig. 9.22 Plot of $\bar{X}$ and s

described the calculation of the standard deviation for the first sample set of 3, -3, 2, 1, -1. It was found to be 2.408. The remaining 19 values of s in Fig. 9.21 were then obtained in a similar fashion.

The $\bar{X}$ and s readings can now be plotted in the usual way as represented in Fig. 9.22.

9.4.3 Control limits for s chart

As is to be expected, the first requirement when calculating control limits is the position of the central line. This is given by

$$\bar{s} = \frac{\Sigma s}{k}$$

where k is 20.

Hence for the readings being considered

$$\bar{s} = \frac{54.11}{20}$$

$$= 2.706$$

Using this value, the upper and lower control limits (UCL_s and LCL_s) can be determined. In the same way as constants D_3 and D_4 relate to the mean range $\bar{R}$, then similar constants are provided which relate to $\bar{s}$. These constants, B_3 and B_4, are multiplied by $\bar{s}$ to give LCL_s and UCL_s. B_3 and B_4 vary with sample size in accordance with the table in Appendix C. For a sample size 5, B_4 is 2.089. Therefore

$$UCL_s = B_4 \bar{s}$$

$$= 2.089 \times 2.706$$

$$= 5.653$$

Note that B_3 is zero for values of n less than 6, in comparison with D_3 which was zero for values of n less than 7. LCL_s is therefore zero in the case considered here.

9.4.4 Control limits for $\bar{X}$ chart

Since the same set of 20 samples as in Chapter 6 is being used, then the $\bar{X}$ chart will be as shown in Fig. 6.18. There is a problem, however, in that these control limits were based on a knowledge of $\bar{R}$. In the current analysis s has been used instead of R for each sample. Hence another constant is needed that relates to $\bar{s}$ in the same way as A_2 relates to $\bar{R}$. This constant is A_3, tabulated in Appendix C. As with the other constants, it varies with sample size. Using the appropriate A_3 value the control limits can be calculated from

$$UCL_{\bar{X}} = \bar{\bar{X}} + A_3 \bar{s}$$

$$LCL_{\bar{X}} = \bar{\bar{X}} - A_3 \bar{s}$$

$\bar{\bar{X}}$ has already been obtained in Section 6.8.1 as -0.57. For a sample size 5, A_3 is 1.427. Therefore

$$UCL_{\bar{X}} = -0.57 + 1.427 \times 2.706$$

$$= 3.291$$

$$LCL_{\overline{X}} = -0.57 - 1.427 \times 2.706$$

$$= -4.431$$

Central lines and control limits for both s and $\overline{X}$ can now be recorded in the appropriate data boxes and lines drawn on the respective charts. The completed ($\overline{X}$, s) chart then appears as in Fig. 9.23.

Now that the control limits for $\overline{X}$ based on s have been calculated, it is interesting to compare them with those obtained earlier using $\overline{R}$. Since $\overline{X}$ is common to both approaches, a more direct result is given by comparing $A_3\overline{s}$ and $A_2\overline{R}$ for n = 5.

$\overline{s}$ = 2.706 $\qquad\qquad$ $\overline{R}$ = 6.85

A_3 = 1.427 $\qquad\qquad$ A_2 = 0.577

$A_3\overline{s}$ = 3.861 $\qquad\qquad$ $A_2\overline{R}$ = 3.953

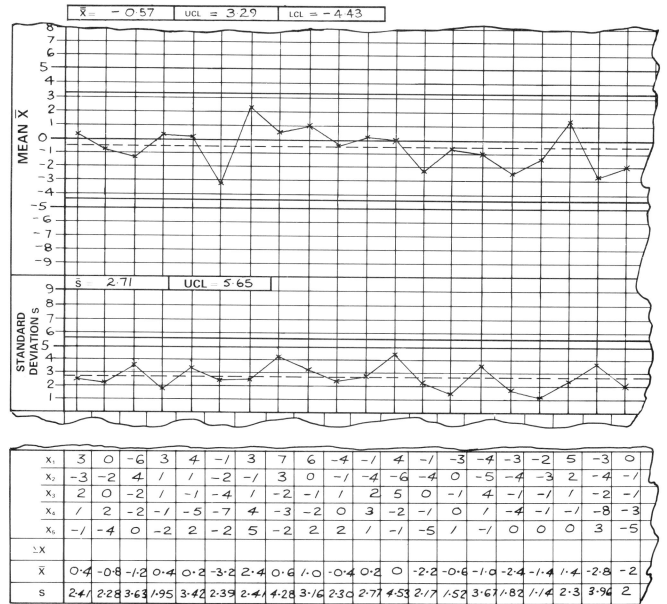

x_1	3	0	-6	3	4	-1	3	7	6	-4	-1	4	-1	-3	-4	-3	-2	5	-3	0
x_2	-3	-2	4	1	1	-2	-1	3	0	-1	-4	-6	-4	0	-5	-4	-3	2	-4	-1
x_3	2	0	-2	1	-1	-4	1	-2	-1	1	2	5	0	-1	4	-1	-1	1	-2	-1
x_4	1	2	-2	-1	-5	-7	4	-3	-2	0	3	-2	-1	0	1	-4	-1	-1	-8	-3
x_5	-1	-4	0	-2	2	-2	5	-2	2	2	1	-1	-5	1	-1	0	0	0	3	-5
ΣX																				
$\overline{X}$	0.4	-0.8	-1.2	0.4	0.2	-3.2	2.4	0.6	1.0	-0.4	0.2	0	-2.2	-0.6	-1.0	-2.4	-1.4	1.4	-2.8	-2
s	2.41	2.28	3.63	1.95	3.42	2.39	2.41	4.28	3.16	2.30	2.77	4.53	2.17	1.52	3.67	1.82	1.14	2.3	3.96	2

Fig. 9.23 Completed ($\overline{X}$, s) chart

This good agreement provides further evidence of the value of plotting the R chart. Not only is R much easier to use, it also loses little in accuracy in comparison with s when used as a basis of working out control limits for $\bar{X}$. Fig. 9.24 illustrates the alternatives. For all practical purposes there is little difference in the control chart for $\bar{X}$ irrespective of whether $\bar{R}$ or $\bar{s}$ is used.

9.4.5 Interpreting the charts

The ($\bar{X}$, s) chart is interpreted in exactly the same way as the ($\bar{X}$, R) chart.

9.4.6 The chart and capability

In Section 7.3 the relationship $\hat{\sigma} = \bar{R}/d_2$ was introduced. It might be expected that a similar relationship exists but this time making use of $\bar{s}$. The only index in Appendix C not yet introduced is c_4. This can now be used as a way of obtaining $\hat{\sigma}$ from $\bar{s} / c_4$. The value of c_4 again varies with sample size.

Hence, if the s chart is being plotted, then knowing $\bar{s}$ the process performance could be estimated as $6\bar{s} / c_4$. Similar modifications to the C_p and C_{pk} formulae would then enable these indices to be obtained from the $\bar{s}$ chart.

Case study 5: Ford of Europe.

Since the early 1980s the Ford Motor Company has been in the forefront in the use of SPC within both the organisation and the supply base. It has been responsible for significant improvement in performance in companies directly or indirectly involved with its activities by using SPC as the catalyst for change.

With some 60% of its product coming from outside its own locations, Ford recognises the need for long-term collaboration with suppliers who wish to work with the company in a partnership role. Progress in implementing SPC is a slow one, and both Ford and its suppliers understand that they are embarking on a long-

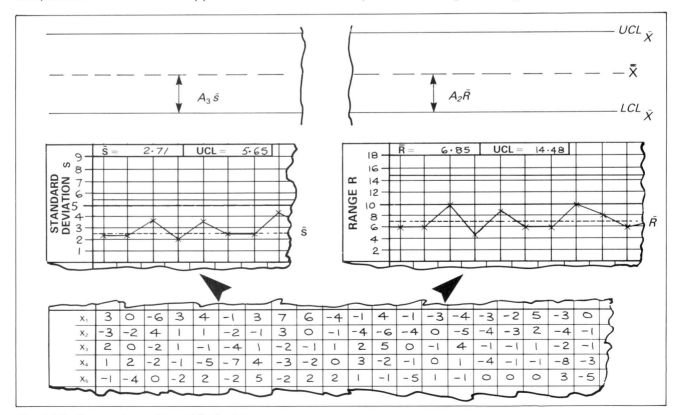

Fig. 9.24 Comparison of s and R charts

term programme of change. This does not mean that early successes are not possible, and internally Ford have built up a series of success stories. These reflect the company's Guiding Principles, which can be summarised as follows:

- Quality comes first.
- Customers are the focus of everything we do.
- Continuous improvement is essential to our success.
- Employee involvement is our way of life.
- Dealers and suppliers are our partners.

Quite apart from its application at operator level in a manufacturing environment, an important aspect of SPC implementation is the extension of its use to non-manufacturing areas and in management reporting. This case study illustrates how the control chart technique is being used in reporting Uniform Product Assurance System (UPAS) data, amongst others.

UPAS is an in-house system for monitoring the initial quality of vehicle and manufacturing assembly processes using customer-perceived criteria. The previous method of reporting UPAS results did not adequately reflect the normal process variation inherent in the manufacturing process, or those variations imposed by the UPAS system due to the small sizes often involved . An example is given in Fig. 9.25. The results are expressed in terms of concern per hundred vehicles (c/100) and the types of concern are defined as follows:

- High: The highest number of concerns on any one vehicle. (Because of a scale factor this value is multiplied by 100. Note that as a result of the UPAS rating system fractional concerns are possible.)

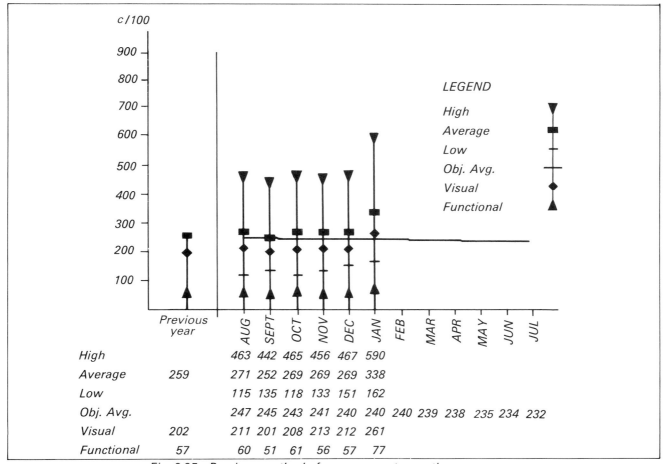

Fig. 9.25   Previous method of management reporting

- Low: The lowest number of concerns on any one vehicle.
- Average: The average number of concerns per vehicle.
- Obj. Av. (Objective average): The target value.
- Visual. A concern about the appearance of the vehicle.
- Functional: A concern about the function of the vehicle.

As can be seen from Fig. 9.25 the method can be misleading and may result in an over-reaction to events which could be within the normal process or sampling control limits.

In line with the corporate statistical approach to the monitoring of component and vehicle quality, a new method for reporting monthly UPAS results to management was therefore developed. This replaced the previous average and high/low vehicle graph with $(\bar{X}, s)$ charts based on sample sizes of 40 vehicles. Control limits are calculated and drawn on the charts, with adjustments being made as trends develop.

Each chart is an extract showing the last 24 plotted points from a continuous chart for a particular model from a specific plant. Breaks in continuity occur only at the introduction of major new model programmes. The new model will then continue to be reported on the same chart to allow comparison with the outgoing model. However, since the sample size of 40 vehicles will be divided by batching consecutive daily samples, the time period covered by the charts will vary from 6 months (for models produced at single-model plants where the daily sample is 8) to more than a year (for low-volume models produced at multi-model plants where the daily sample can be as low as 2).

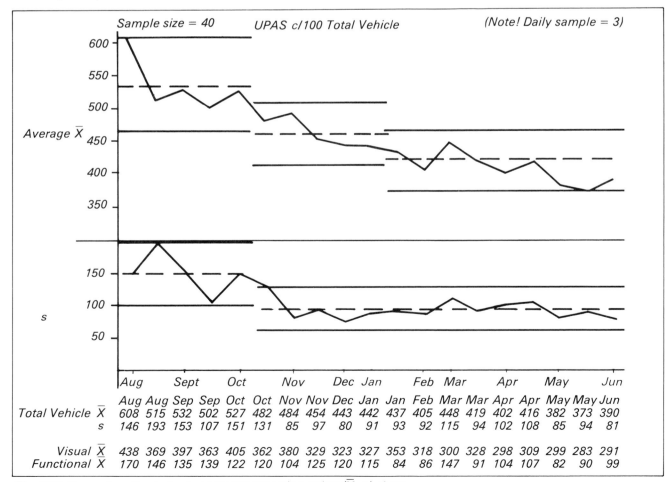

	Aug	Aug	Sep	Sep	Oct	Oct	Nov	Nov	Dec	Jan	Jan	Feb	Mar	Mar	Apr	Apr	May	May	Jun
Total Vehicle $\bar{X}$	608	515	532	502	527	482	484	454	443	442	437	405	448	419	402	416	382	373	390
s	146	193	153	107	151	131	85	97	80	91	93	92	115	94	102	108	85	94	81
Visual $\bar{X}$	438	369	397	363	405	362	380	329	323	327	353	318	300	328	298	309	299	283	291
Functional $\bar{X}$	170	146	135	139	122	120	104	125	120	115	84	86	147	91	104	107	82	90	99

Fig. 9.26 Present method of management reporting using $(\bar{X}, s)$ chart

Fig. 9.26 shows how the present method is being applied, for a specific product, to the results shown in Fig. 9.25. The chart illustrates not only a steady improvement in the product throughout the year but also, from the standard deviation plot, how car-to-car variation improved.

The company believes this method provides a good statistically valid management report, concentrating on stability and rates of improvement rather than on actual numbers versus objectives. Calendarised UPAS objectives have been abandoned, because there was evidence that they inhibited improvement in the better plants.

Procedures have been developed for the extension of this concept at plant level. This enables plants to sort and batch the data and establish charts for significant manufacturing areas or model derivates, concentrating in the same way on reducing variation and achieving improvement.

Extending the application of statistical charting into administrative functions is in line with the Ford Continuous Improvement model shown in Fig. 9.27. This model can apply to any process. With UPAS, customer feedback determines the concern criteria. Statistical methods feedback is obtained from the chart and is integrated into the Deming cycle (plan/do/check/act). This repeated use of the cycle forms part of the never-ending programme to improve customer satisfaction in all respects.

This case study has illustrated the use of the ($\bar{X}$, s) chart in improving process performance within the Ford Continuous Improvement model. Its use as part of management reporting also shows the relevance of charting techniques to areas outside manufacturing.

9.5 Dealing with drifting processes

Some processes exhibit a pronounced drift over time. For example, changing constituents in a chemical solution may cause a drift in the properties of that solution, or tool wear in machining operations may cause a component dimension gradually to increase in size. Techniques are available which can eliminate some of these problems. For example, tool wear can be automatically compensated for over time.

For drifting processes (and drifting is usually only evident on the mean chart, not the range chart), continuous improvement has a new meaning. The SPC manual

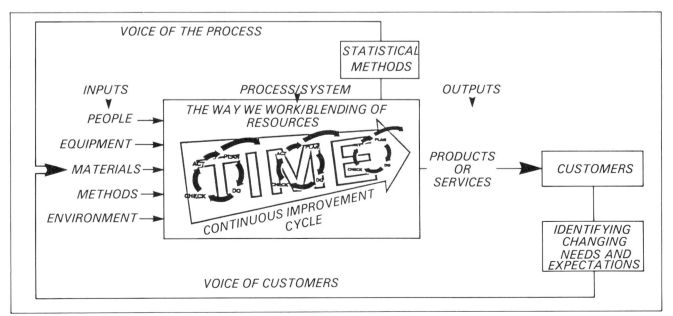

Fig. 9.27 Ford continuous improvement model

produced by Ford of Europe provides clear guidance. A brief summary will suffice here.

Fig. 9.28 shows a typical process with a drifting mean and a stable within-sample variability. The corresponding control chart shows a pattern somewhat similar to that in Fig. 9.29. The first point to be recognised is that calculating control limits for $\bar{X}$ in the usual way is no help in this situation The presence of the drift renders the conventional horizontal lines no longer appropriate. However, there is no reason why the control limits should not be inclined along the line of the drift. These lines will enable special causes to be determined which may operate over and above the permanent drift factor.

The limits are set at A_2 outwards from the line of drift. With the set of values given, $\bar{R}$ is 4.6. Hence

$$A_2\bar{R} = 0.577 \times 4.6$$

$$= 2.65$$

The line of drift and control limits are now shown in Fig. 9.30. Here the line of drift has been drawn in as a visual estimate. There is an analytical method available for producing the best straight line.

With a drifting mean there is a requirement constantly to set and reset the process. Setting and resetting decisions will have been based traditionally

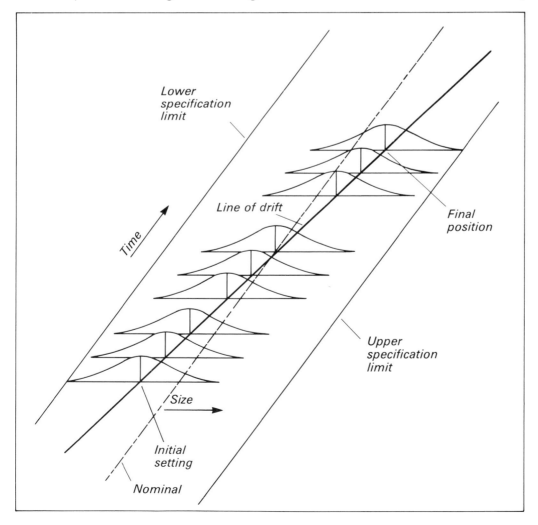

Fig. 9.28 Stable process with drifting mean

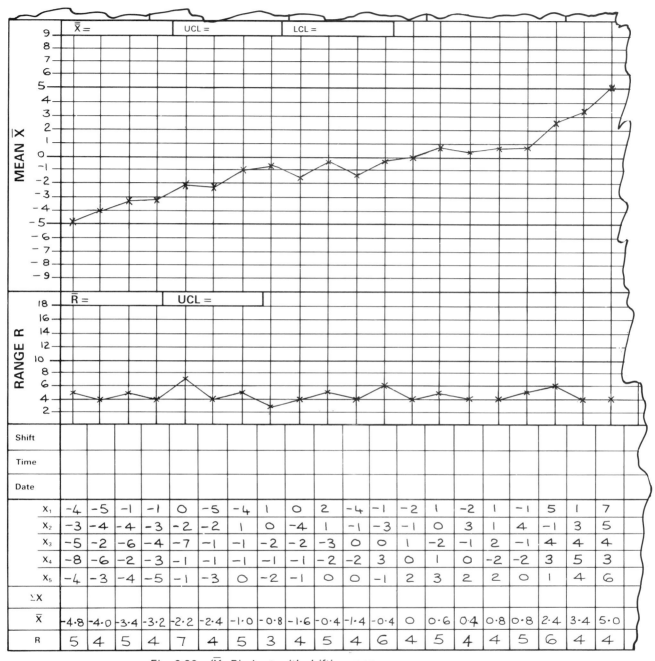

X_1	-4	-5	-1	-1	0	-5	-4	1	0	2	-4	-1	-2	1	-2	1	-1	5	1	7
X_2	-3	-4	-4	-3	-2	-2	1	0	-4	1	-1	-3	-1	0	3	1	4	-1	3	5
X_3	-5	-2	-6	-4	-7	-1	-1	-2	-2	-3	0	0	1	-2	-1	2	-1	4	4	4
X_4	-8	-6	-2	-3	-1	-1	-1	-1	-1	-2	-2	3	0	1	0	-2	-2	3	5	3
X_5	-4	-3	-4	-5	-1	-3	0	-2	-1	0	0	-1	2	3	2	2	0	1	4	6
ΣX																				
$\bar{X}$	-4.8	-4.0	-3.4	-3.2	-2.2	-2.4	-1.0	-0.8	-1.6	-0.4	-1.4	-0.4	0	0.6	0.4	0.8	0.8	2.4	3.4	5.0
R	5	4	5	4	7	4	5	3	4	5	4	6	4	5	4	4	5	6	4	4

Fig. 9.29 ($\bar{X}$, R) chart with drifting mean

on the need to satisfy the customer's specifications. Continuous improvement requires moving away from procedures based on specification limits.

If the results of, say, 10 setting and resetting steps are available, then a straight line corresponding to the average movement of the mean over a period of time is obtained. Performance-based control limits would then be set at the usual distance of $A_2\bar{R}$ outwards from the extreme points of the line, as in Fig. 9.31. Continuous improvement implies, though, that with the line pivoted centrally the angle of inclination to the horizontal is gradually reduced. Eventually the line becomes horizontal and the usual format for the $\bar{X}$ chart with control limits set outwards from $\bar{X}$ at distances of $A_2\bar{R}$ results (Fig. 9.32).

Here, continuous improvement brings in aspects such as improved tooling technology, higher performance chemicals, etc. In addition to reducing the angle of the line, attempts would also be continuing to reduce the within-sample variability.

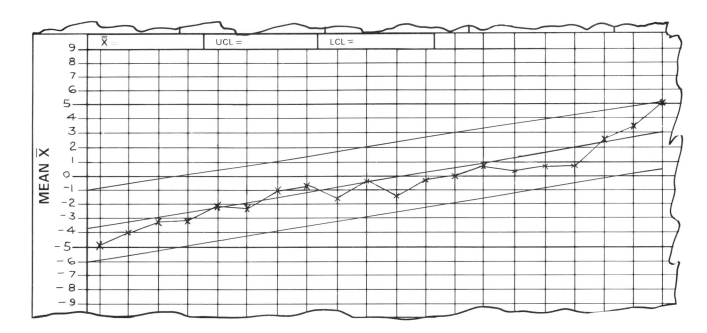

Fig. 9.30   Control limits for drifting mean

A final look at the information contained in Fig. 9.29 brings out further features. The basic data is the same as that used in generating the ($\overline{X}$, R) chart in Fig. 7.3. In other words, the data boxes in Fig. 9.27 and Fig. 7.3 represent different arrangements of a set of 100 readings. If the 100 readings in the data box in Fig. 9.29 were represented in tally mark form, they would generate the pattern shown in Fig. 9.33. It can be seen that this pattern is in fact identical to that shown in Fig. 7.5. The limitations of the histogram were discussed in some depth in Section 7.6. This example further proves the point.

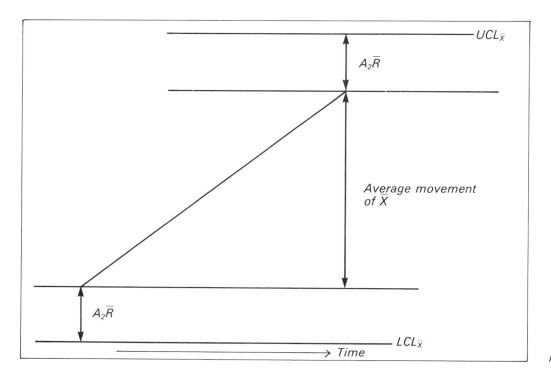

Fig. 9.31
Performance-based
limits for drifting mean

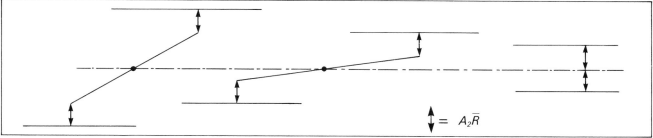

Fig. 9.32 Continuous improvement with a drifting mean

Processes with a drifting mean are often inherently capable because the natural variability of the process at a particular time is small in comparison with the tolerance. The key lies in the R chart. For the same set of data there are two different R charts depending on whether the process is drifting or not. These are compared in Fig. 9.34. $\bar{R}$ for the drifting process is now 4.6 compared with 6.85 for the non-drifting process. This greater consistency is reflected in improved C_p and C_{pk} values.

It is not intended to include details of every chart-based technique in this text. The references in Appendix I provide excellent material if required. Some detail on short production runs will prove useful, however.

There seems to have been a tendency in some organisations to seek out the reasons for not introducing SPC rather than pursuing the opportunities it provides. Short run analysis is a case in point. Management concerned has argued that they cannot introduce charts: 'Our industry is different', 'We don't have enough items', 'It won't work here', 'It's alright for those in mass production, but . . .' These views are understandable but based on attitudes more than technical reasons. Problem-solving techniques, use of charts in administrative areas and introducing SPC in the supply base are possibilities even in companies making aircraft carriers, oil rigs or pressure vessels.

A variation of the standard ($\bar{X}$, R) chart is available to assist in monitoring the quality levels of small batches.

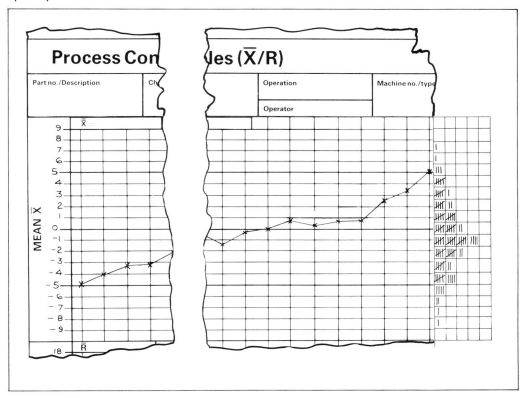

Fig. 9.33 $\bar{X}$ plot and tally chart

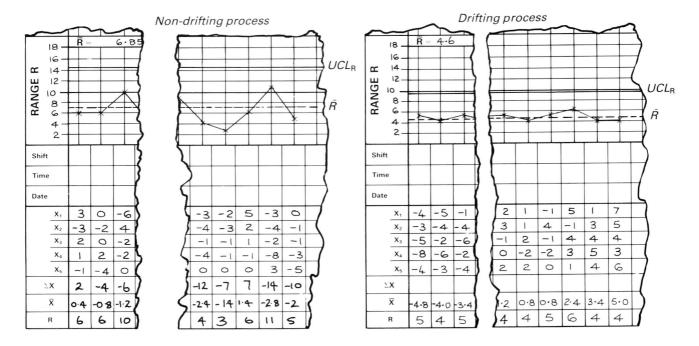

Fig. 9.34   R chart for drifting and non-drifting $\bar{X}$

9.6   **Dealing with small batches**

Small batch analysis emphasises a key aspect of SPC, which is the process that must be monitored, not the product.

Small batches arise in one of two ways: either there are a large number of similar items but with varing nominal values, or the required time scale to produce the items is a long one.

9.6.1   **Varying nominal values**

Instances of varying nominal values typically require repeated process adjustments to allow for different specifications, although the items produced are similar.

If the natural variability, irrespective of setting, is constant, then the R chart is plotted in the usual way. UCL_R is calculated, as usual, as $D_4\bar{R}$. For the $\bar{X}$ chart, $\bar{X} - \phi$ is plotted, where ϕ is the relevant process nominal for each batch. The control limits are then set at $\pm A_2\bar{R}$ about the zero value. The interpretation of the chart and the search for special causes follow in the usual way.

If the natural variability does change with process setting, then this can be accommodated. Details are provided in one of the references in Appendix I.

9.6.2   **Emphasis on the process**

When considering processes which produce few items over a long time scale, the approach adopted in the preceding section is not appropriate. There now has to be a heavy emphasis on the process parameters, so that they can be controlled and improved. For example, if the tooling can be improved then the product must be improved. Similarly factors such as machine speeds or process temperatures are seen to have a critical influence on the final product, and so effort is directed towards controlling these rather than the many varying product characteristics. In addition, techniques relating to the design of experiments, and problem-solving techniques such as brainstorming and fishbone diagrams come into their own as cost-effective methods for quality improvement.

SPC programmes are based on continuous improvement, and this feature applies as much to the techniques themselves as it does to the product or process to which they are directed. It is clear that the development, application and further development of the techniques of SPC is an ongoing process. Standard techniques are available to deal with most situations but the programme is an evolving one. In some cases, though, it will still be a case of applying a technique which has been known to operate in similar, but not identical situations.

One technique that has been available for many years, yet is still underused, is the multi-vari , and this chapter will be concluded with a discussion on this method.

9.7 **Multi-vari analysis**

Reference has been made to the idea of variability between samples and variability within samples. The multi-vari chart provides a visual representation of the different types of variability.

A good introduction to the technique is to look again at a typical median chart. This is the chart used in Fig. 9.5, as shown again in Fig. 9.35. For each sample a line can be drawn spanning the least and greatest value, i.e. equivalent to the range. For visual impact the other lines are removed and the same information can now be shown in a different form (Fig. 9.36). This form of chart makes it easier to detect changes in the pattern of variability.

Other features can be illustrated. For example Fig. 9.37 shows the presence of excessive variability between samples. Fig. 9.38, on the other hand, shows how variability can change with time. In many ways the multi-vari chart provides support in interpreting chart patterns as discussed in Chapter 8.

It is open to question how far the multi-vari chart would be utilised at operational level. It is probable that it will be applied by technical staff as an extension of the basic chart. Whatever the use, its visual effect is dramatic.

The following case study shows how it has been applied in one organisation.

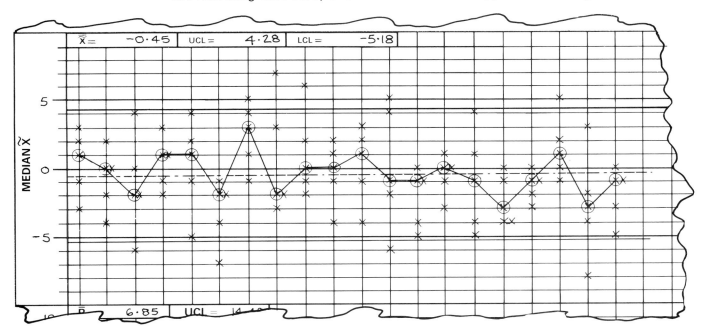

Fig. 9.35 Median chart with control limits

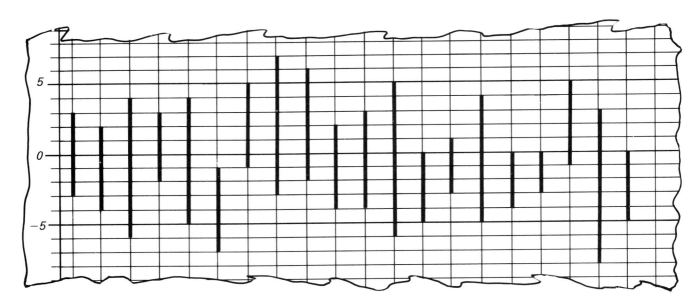

Fig. 9.36   Basic multi-vari chart

Case Study 6: Sauer Sundstrand, Swindon

With a wealth of practical experience of internal SPC charting behind them, Sauer Sundstrand turned their attention to improving the quality of bought-in product. SPC would be a requirement for future business, but before implementing a supplier awareness/training programme, the company carried out a statistical analysis on components supplied by prospective suppliers.

This particular case study relates to a cast iron cover which had been causing continuous problems for some months. When machining the face it was found that two cuts were required when milling in order to remove hard skin and maintain a flat surface. It was decided to carry out a hardness check on a sample of 50 components. The actual component is as shown in Fig. 9.39. Hardness readings were taken at the positions indicated and the results, given in Fig. 9.40, show the corresponding Brinell readings.

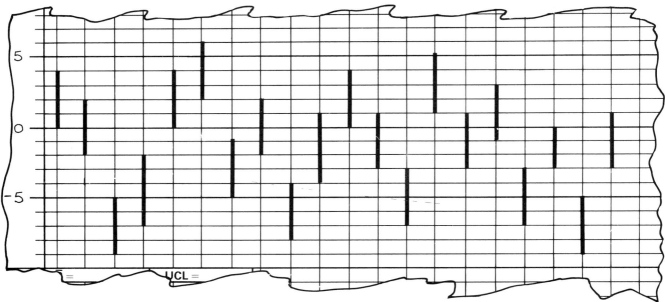

Fig. 9.37   Variability between samples

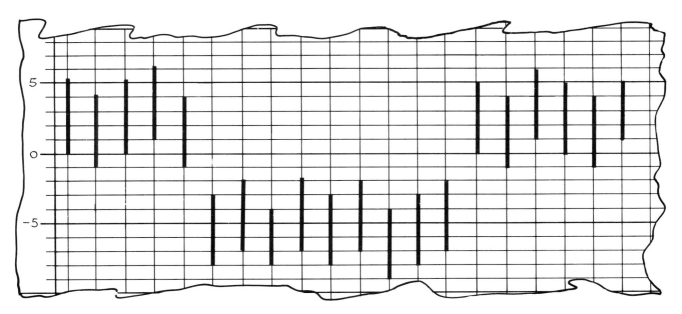

Fig. 9.38 Effect of jumps in setting

It is impossible to detect any patterns or trends from this table of figures. It is only when a multi-vari chart is drawn, as in Fig. 9.41, that the nature of the variation becomes clearer.

In addition to the range of all nine readings, the critical reading at the centre of the component was also indicated, together with relevant specification limits. The information indicated an excessive number of items outside the supplier's specification limits. More than that, there was large piece-to-piece variability as well as within-sample variability. Copies of the table and the multi-vari chart were sent to the supplier, with a request for improvement.

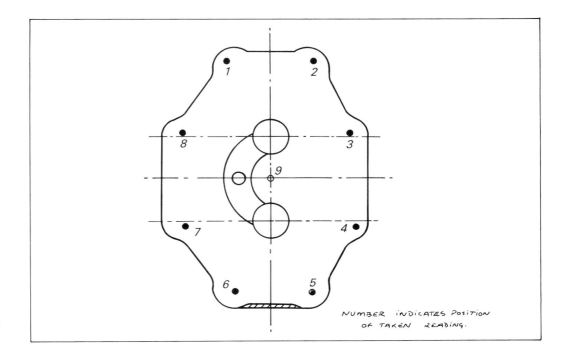

NUMBER INDICATES POSITION
OF TAKEN READING.

Fig. 9.39 Cast-iron cover housing

SAMPLE N°	HARDNESS READING AT POSITIONS INDICATED								
	1	2	3	4	5	6	7	8	9
1	214	175	172	179	207	194	206	205	148
2	171	165	185	209	286	232	220	198	179
3	193	157	158	178	189	149	154	186	133
4	216	213	210	207	200	199	193	189	179
5	157	172	206	195	178	172	189	184	164
6	185	186	199	212	225	205	183	192	162
7	198	228	226	200	206	214	207	204	162
8	239	212	216	180	180	184	207	223	177
9	215	197	197	214	220	201	204	192	163
10	210	227	227	216	196	170	198	209	160
11	182	181	213	203	181	176	177	178	165
12	202	204	207	218	185	167	207	219	170
13	214	167	191	211	217	215	197	213	160
14	237	195	195	210	206	204	197	209	190
15	196	183	186	190	192	174	174	188	160
16	210	191	189	212	231	187	206	221	157
17	185	194	210	216	169	166	206	209	161
18	224	224	210	204	215	201	221	235	183
19	231	237	238	207	209	212	200	190	165
20	224	201	210	195	198	194	195	231	163
21	166	199	224	202	168	155	170	163	148
22	217	208	204	192	171	206	224	217	147
23	226	193	205	213	223	190	216	216	170
24	224	213	199	162	171	171	211	212	150
25	144	165	216	207	189	146	159	155	153
26	201	217	217	225	210	221	219	213	183
27	166	178	186	174	148	149	153	156	149
28	198	185	159	166	164	172	174	167	147
29	208	232	219	228	216	203	196	203	198
30	215	197	164	160	172	191	190	201	152
31	226	191	165	160	193	182	189	208	162
32	201	208	224	243	241	230	204	216	162
33	186	210	216	206	179	141	168	192	161
34	217	194	188	184	191	206	223	220	174
35	166	194	206	191	171	191	213	190	163
36	200	198	220	230	219	187	206	228	173
37	192	188	196	208	202	173	177	187	161
38	173	215	168	160	162	191	214	197	156
39	205	193	166	162	193	187	188	198	153
40	180	164	199	180	154	164	153	153	140
41	224	202	175	174	198	212	229	236	177
42	217	189	192	200	237	223	223	243	181
43	196	184	224	223	219	225	227	199	178
44	170	131	211	214	195	160	155	173	159
45	204	205	205	189	206	196	177	187	156
46	158	168	185	189	188	200	176	167	146
47	195	200	187	157	152	163	180	200	148
48	181	167	151	152	154	173	204	203	148
49	210	191	152	151	153	172	178	200	149
50	174	155	151	170	198	207	188	193	158

Fig. 9.40   Multi-vari chart of hardness readings

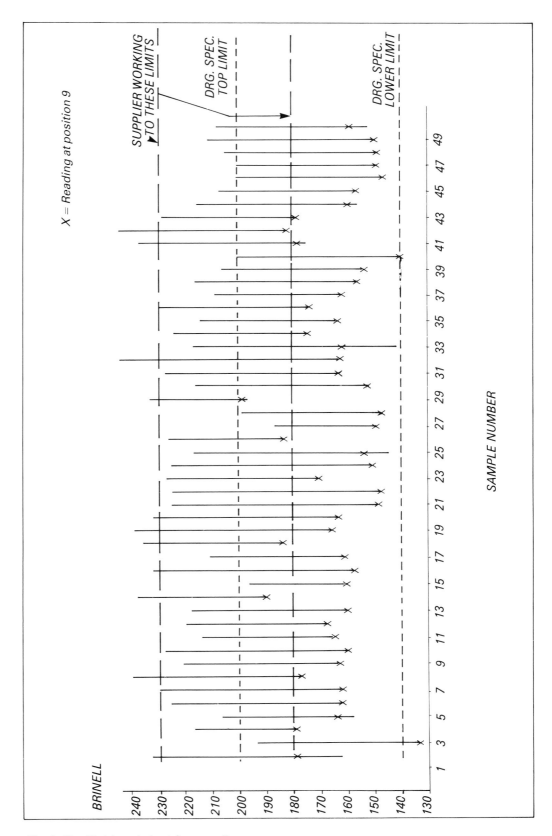

Fig. 9.41   Multi-vari chart for supplier

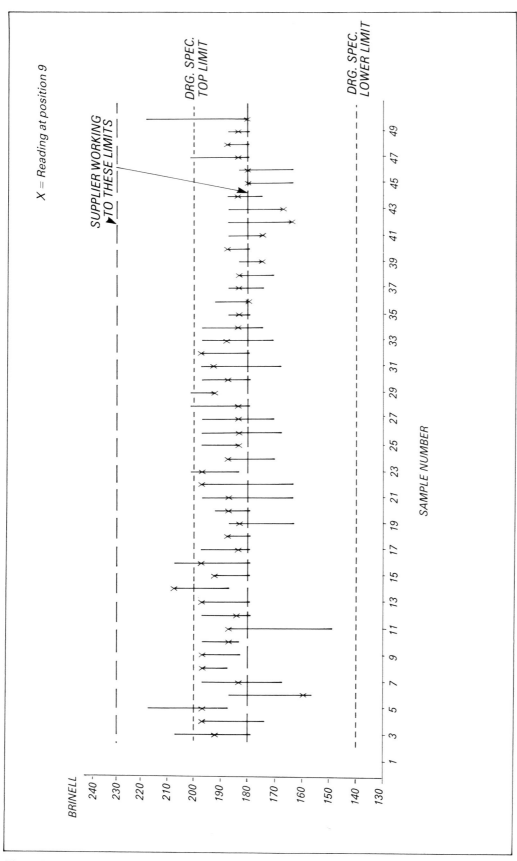

Fig. 9.42 Multi-vari chart for new supplier

Over the next 12 months, the supplier's performance showed little improvement in quality level and Sundstrand decided to involve a new supplier. Fig. 9.42 shows how the results are now considerably improved, so that a much reduced proportion of product is outside the specification limits.

Without this graphical approach the customer would not have been able to detect the pattern of results and take action in improving the quality of bought-in components.

Summary

This chapter has covered several techniques for charting variables, though other specialist approaches may be required. The current expanding interest in SPC is resulting in different applications of charting methods and the development of techniques specific to one industry or process. It is unwise to assume that for any process a statistically based system can be drawn out of the hat, almost on demand. The reverse does seem to be true, however, in that some organisations tend to hide behind the uniqueness of their process or product, stating that SPC will not work in their industry. This has far more to do with attitude than statistical limitations.

For those requiring specific methods for unique situations, there is a growing database of references. Internal company documents also provide valuable information. SPC is resulting in collaboration and the gradual removal of barriers, as departments talk to one another and organisations open their doors to others.

- Powerful as it is, the $(\overline{X}, R)$ chart does not provide all the answers.
- The $\widetilde{X}$ (median) chart has the advantage of simplicity.
- Templates are an added feature of the $\widetilde{X}$ chart.
- Control limits for $\widetilde{X}$ charts are wider apart than control limits for $\overline{X}$ (mean) charts.
- Moving mean/range charts are used when only single readings are possible.
- Readings on the moving mean/range chart are not independent.
- A mask can help the operator.
- Use the s chart when backup computing facilities are available.
- Drifting processes need special attention.
- The conventional $(\overline{X}, R)$ approach is not valid when monitoring a drifting mean.
- Continuous improvement now means concentrating on improving the technology.
- The histogram cannot distinguish between different time series.
- Multi-vari charts emphasise different variability patterns.

Having concentrated on variables so far in the book, attributes will now be considered. Attribute charts provide scope for extensive progress in quality improvement, particularly in the non-manufacturing areas. This aspect will be emphasised in the following chapters, which consider charting methods with real potential for monitoring and improving reject levels in defective units and for reducing the defects within those units.

Chapter 10 Control charts for attributes

10.1 Introduction

It is disappointing, yet at the same time not surprising, that attribute charts have played a minor role in SPC programmes. The companies who have introduced charts over the years have restricted them, for a variety of reasons, to the familiar $(\bar{X}, R)$ variety. This has resulted in a negative approach to SPC implementation. If the $(\bar{X}, R)$ chart cannot be applied, then the conclusion has followed that SPC will not work.

As well as indicating a lack of understanding of the wider implications of SPC, the approach will result in missed opportunities, particularly in administrative activities. By using control charts and problem-solving techniques in reducing errors in the service departments of organisations, customer satisfaction, in an external sense, must improve. There is little point in producing high-quality products if they are delivered at the wrong time, in the wrong quantity or to the wrong address.

A medium-sized company in the chemical processing industry surveyed its customers regarding the quality of its product and service. The response showed considerable satisfaction with the product itself but the packaging fared less well. The biggest complaint, however, was with invoicing. With its company-wide approach to SPC training, invoicing errors could now be tackled and reduced.

Similarly, an organisation with some 1200 personnel involved in a processing industry recognised that its main problem was not the product quality but on-time delivery. It was known that deliveries were late because the various processes were clearly not under control. The process obviously involved non-manufacturing areas, hence a need for attribute analysis.

For those organisations involved entirely in direct service activities, the gains could be even more impressive. Insurance companies, banks and finance houses, services for gas, electricity, water or telephone – the scope for improvement programmes based on charting appropriate data is enormous. The vast majority of SPC applications in these services are attribute-based.

Three chapters here are devoted to attribute charting. The present chapter covers the general aspects of the subject while the following two deal with the specific types of attribute charts.

10.2 What are attributes?

Previous chapters have considered the analysis of data relating to measurable items. Often, however, the data refers to quantities which cannot be measured, but instead satisfy a yes/no definition. The following are examples:

- Scratches on a car panel.
- Incorrect invoices.
- Telephone calls received.
- Visitors to the company.
- Faulty deliveries.
- Packaging errors.

For all these items, and others, the feature is either present or it is not. Continuous improvement means reducing the level of these faults to zero.

If the data is not already available in one form or another, then a check sheet is required for collecting it. Data collection was discussed in Chapters 2 and 3, and Fig. 3.3 showed a typical data collection sheet for attributes. The basic information displayed there has been condensed to provide the simplified check sheet shown in Fig. 10.1.

The data available from the check sheet can be used in two ways. Firstly, the totals on the right can be used to generate a Pareto diagram, as illustrated in Fig. 3.5. Secondly, the total number defective each day, irrespective of type, can be used to generate an attribute chart. Fig. 10.2 summarises the way in which the check sheet provides the basis for deciding priorities for action and at the same time monitoring performance.

This simple check sheet in turn provides the basis for the multiple characteristics chart. This is a readily understood, easy to use chart with a wide range of possible applications and much potential as a quality improvement tool. It will be seen in Chapter 12 how the chart is used and the range of environments in which it can operate.

It is unfortunate that simple check sheets of this nature have not been made use of over the years. Other sources of attribute data have been neglected in a similar fashion. The introduction of SPC programmes results sooner or later in a consideration of attribute charts after early forays into charting variables. The same experiences are shared by many: inspection reports providing a vast amount of detail which was documented in tabular form but never represented as a graph, least of all plotted as a performance-based chart; audit analyses containing considerable

CHECKING CHARACTERISTIC	3/1	4/1	5/1	6/1	7/1	10/1	11/1	12/1	13/1	14/1	17/1	18/1	19/1	20/1	21/1	24/1	25/1	26/1	27/1	28/1	TOTAL
NO PERSONAL DATE STAMP				I																	1
WRONG NAMEPLATE	II	II	III	I	II	II		III		IIII	I	II		III	III	I	II	I	III		35
DAMAGED FLANGE				I							I			I							3
WRONG FLANGE		I		II		III	I		I		II		I		I		II	I			15
BODY SCORED										I											1
TOTAL	2	3	3	4	3	5	0	4	0	6	0	4	2	1	3	5	1	4	2	3	55

Fig. 10.1 Simple check sheet

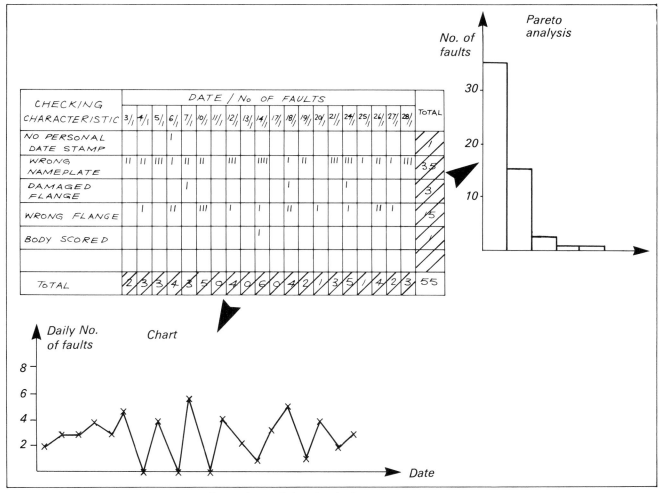

Fig. 10.2 Extensions of the check sheet

information which could have been put to better use had they been presented in a more easily interpreted form; and management summaries comprising tables of figures which could have pointed the way to more effective action had they been represented graphically.

As has been mentioned elsewhere the advent of the computer has often tended to hinder decision-making rather than assisting it. With computer print-outs involving detail at the level indicated in Fig. 10.3 it is virtually impossible to track the pattern of performance. Decision-making is not possible unless clear-cut indicators of performance are displayed. Amongst the many spin-offs resulting from SPC programmes is a recognition that far too many existing computer programs are unhelpful as a guide to monitoring progress. As a result, programs are being redesigned to provide the attribute and variable data in a form that assists in monitoring the various process performances.

10.3 Definition of a fault

Because it is based more on personal interpretation attribute data is not so clear-cut as data on variables. For example, there cannot be much doubt about the weight of a container of paint; provided it is measured correctly with adequate equipment, then the resulting measurement is absolute. This is not so with attributes. Because decisions are made on personal judgement then there can be different views on, for example, whether the print on a container label is satisfactory or not.

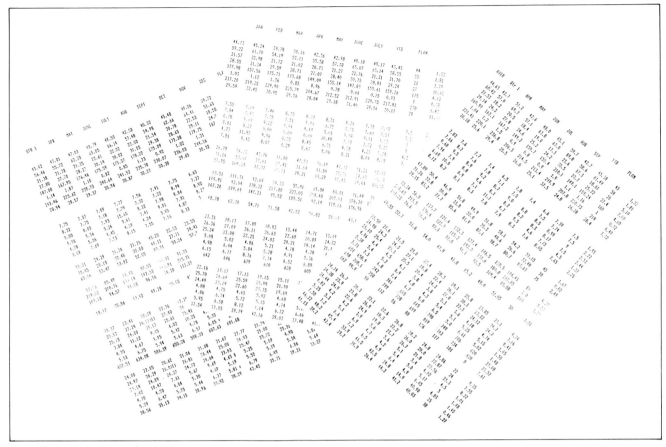

Fig. 10.3   Selection of print-outs

A standard definition of a fault is therefore required. There are various options. A display board could be provided comprising known types of faults so that personnel can see at a glance the nature and degree of the fault. Layout boards can be drawn up, with specific indicators of what is acceptable and what is not. Fig. 10 provides an indication of the approach adopted by one company.

Case study 7: Lucas Electrical

The Lucas organisation has several Divisions in the UK associated one way or another with the automotive/defence industry. Each Division was left to choose its own programme for quality improvement and how it would be implemented. In some Divisions there was a concentration on systems and assessment whereas in others SPC, and the implications of company-wide involvement, were recognised as being more appropriate.

The Starter and Alternator Division of Lucas Electrical undertook a programme of implementing SPC as one part of a complete restructuring plan that was essential to their policy of continuous improvement. There was naturally a heavy emphasis on the use of ($\overline{X}$, R) charts, but it became clear that attribute charts also had a very important part to play in the SPC programme. Analysis of the finished armature proved to be a case in point.

Fig. 10.4 illustrates the attribute faults which are looked for, both at final testing and at the earlier stages in the operation. The company makes use of well-designed visual defect display boards at each stage of production where visual faults need detecting. Clear guidelines are provided for the operators so that, as far as possible, errors in the interpretation of a fault are reduced to zero. The visual faults are plotted

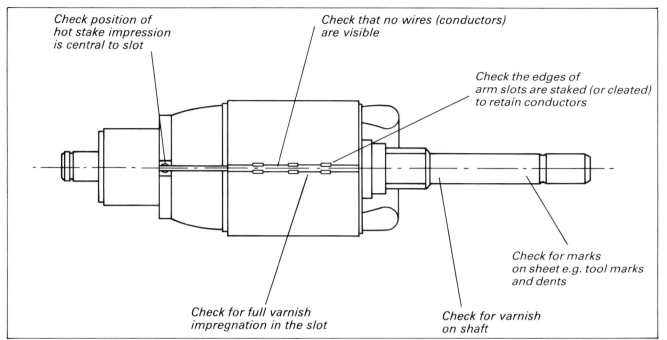

Check position of hot stake impression is central to slot

Check that no wires (conductors) are visible

Check the edges of arm slots are staked (or cleated) to retain conductors

Check for marks on sheet e.g. tool marks and dents

Check for full varnish impregnation in the slot

Check for varnish on shaft

Fig. 10.4 Finished armature

on an appropriate attribute chart and a programme of continuous improvement established so that the level of these faults is reduced to zero over time.

Detecting the presence of unacceptable items in a group of acceptable ones, or picking out a scratch, or an error in a form, means that the personnel concerned must have the appropriate faculties. An organisation cannot expect foolproof results in checking on colour faults if the staff have not been tested for colour blindness, or associated failings. There are related issues here which call into play factors probably outside the mainstream of SPC. Nevertheless, SPC programmes are known to carry with them wide-ranging organisational implications. Hence at the earliest possible stage in the programme staff must be assessed as to their ability to detect faults, trained in detecting these faults, and provided with the right physical environment in which to detect them (adequate lighting, ventilation etc.).

Charts for variables tend to be based on a fixed sample size – typically 5. It would be rare for the sample size to change for a given process, and if it does then limits have to be recalculated. In attribute work a change in sample size does not mean just a recalculation: it also results in the use of a different type of attribute chart altogether.

10.4 Size of sample

Attribute sample sizes can vary considerably. It is no longer a case of, say, 5 items every hour. The task may be to look at 15 sub-assemblies put together each week and to monitor the number of those which are satisfactory. Deliveries of a product could average 75 a month, with actual figures ranging from 68 to 79, for example. This data provides a basis for controlling and reducing the proportions of those deliveries which are unacceptable in one form or another.

It may be necessary, because of high production rates, to take a representative sample from a larger batch produced per hour. The sample size chosen should be sufficiently large for several items per subgroup to appear on average. If a sequence of zeros is recorded then either the level of rejects is low enough already or, more likely, the sample size chosen is not sufficiently large to detect the presence of rejects. An appropriately increased sample size should then be chosen. For example, in a process running at a 1% defective level, a sample size of 100 is required before there is a mean of 1 defective per sample. Similar calculations can provide guidance on the appropriate sample size to take for recording purposes.

This means that the sample of 5 used in analysis of variables should not also be used for simultaneous attribute analysis. Two distinct charts are therefore required in such a case: one for monitoring the variable and one for the attribute.

The critical point is that if it is not the entire output for a given time period that is being checked, then the sample chosen must be representative of the group. In addition costs associated with taking the sample and acquiring the related test equipment must be taken into consideration.

Because in the majority of cases samples are being dealt with which are much larger than those in analysis of variables, then much more time is spent inspecting the items. This means there is a need for visual aids and detection skills.

10.5 Types of attribute charts

Different options are available when plotting attribute data. In practice it is necessary to be quite specific as to whether numbers or proportions or being plotted. Equally it needs to be clear whether defective units or defects are being considered. Some explanation of these statements is required.

A defective unit, sometimes called a non-conforming unit, is a complete item which is unacceptable for one reason or another. An item in this case could be a car door, a starter motor, a coil of steel, an invoice or a computer, for example. These items are classified in the sense that they are either acceptable or unacceptable. It is a clear go/no-go situation for the unit as a whole.

However, each of these items may possess a number of defects, sometimes called nonconformities. The car door may be scratched or dented. Several types of error could appear on the invoice. The coil of steel may be stained. Depending on the circumstances involved, the presence of these nonconformities does not necessarily mean that the unit as a whole has to be rejected. Obviously the aim is to have no nonconformities and no non-conforming units, and in trying to achieve this the two types of fault are dealt with in slightly different ways. As discussed above, sample size is also a factor; Fig. 10.5 summarises the alternatives. As a result, four different types of attribute charts are available, as represented in Fig. 10.6.

A clear distinction must be made regarding the choice of chart to use. In practice, the choice is not always an easy one and some examples will be provided. But before looking at the specific charts the next section will provide some background on the attribute chart in general.

10.6 The attribute chart

10.6.1 Introduction

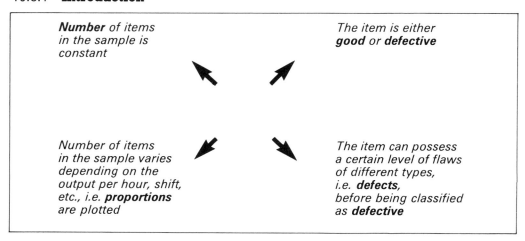

Number of items
in the sample is
constant

The item is either
good or *defective*

Number of items
in the sample varies
depending on the
output per hour, shift,
etc., i.e. *proportions*
are plotted

The item can possess
a certain level of flaws
of different types,
i.e. *defects*,
before being classified
as *defective*

Fig. 10.5 Alternatives when charting attributes

Fig. 10.6 Types of attribute charts

	Defective	*Defects*
Number	*np*	*c*
Proportion	*p*	*u*

Unlike charts for variables, where two charts are required for monitoring the process (one for mean/median and one for range/standard deviation, for example), attribute charts are based on plotting a single characteristic. This means that the plotting section of the chart carries one grid only. Apart from that, the basic format of the attribute chart follows that of the variable chart. Hence two other zones are necessary: one for recording the data and calculations and one for recording the administrative information. Further reference to the latter is made when the specific charts are discussed later.

A typical attribute chart appears in Appendix H and the same blank chart is used for plotting all four variations of attribute chart. The reference box in the top right-hand corner provides an indication of the type of chart being plotted.

10.6.2 **Collecting, recording and plotting**

To see how the chart is constructed, use will be made of a sequence of readings which have been recorded in the data box in the chart shown in Fig. 10.7. The readings record the number of defective items out of 100 units being inspected at the end of a given time period. The sample size is therefore constant at 100.

It is clear that the 100 units could come from many different situations – manufacturing and non-manufacturing alike. Rather than restrict the analysis to a particular operation, the example will be used in a general sense to illustrate the basic principles of the attribute chart. Specific applications will be used in generating the p, np, u and c charts covered in the two following chapters.

Twenty samples have been taken, in line with the convention referred to in the discussion of variable charting. Some organisations, though, may typically use 25 or 30 samples as a basic requirement in their situation. As with variables, these samples will have been taken over a sufficiently long period of time for the natural variation to express itself.

The 20 readings are plotted in the same manner as for variables. A suitable scale is chosen and the readings graphed, with the result shown in Fig. 10.8. As a guide the vertical scale should be chosen to span a range from 0 to 1.5 times the largest value recorded. This will allow the control limits, when calculated, to register on the scale.

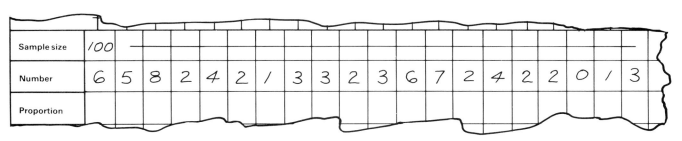

Sample size	*100*																			
Number	*6*	*5*	*8*	*2*	*4*	*2*	*1*	*3*	*3*	*2*	*3*	*6*	*7*	*2*	*4*	*2*	*2*	*0*	*1*	*3*
Proportion																				

Fig. 10.7 Data block from attribute chart

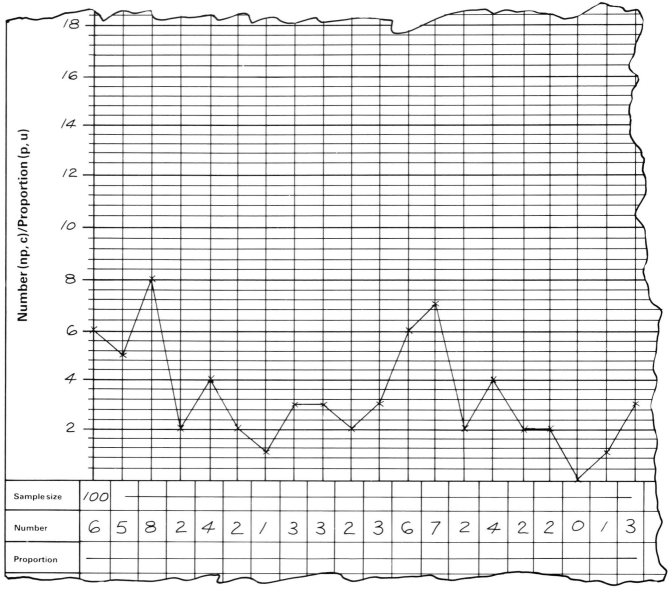

Fig. 10.8   Plot of attributes

10.6.3   **Obtaining the control limits**

As with all the charts discussed to date, a central value is required. This corresponds to the mean of the readings, i.e.

$$(6 + 5 + \ldots + 1 + 3)/20 = 3.30$$

A line is drawn on the chart corresponding to this value. (Fig 10.9)

The difficulty comes in proceeding to the next step where the control limits are required. The first problem is that, unlike the $\bar{X}$ chart for variables, the readings giving rise to the attribute chart do not generate a symmetrical pattern. This can be seen in Fig. 10.10. Here the 20 sample readings which have been plotted on a time sequence are recorded as crosses on the vertical axis, so that a tally diagram is generated. This shows that even with just 20 readings there is evident lack of symmetry.

If more and more readings are made available from the same process over a much longer time period, then the initial pattern shown in Fig. 10.10 will stabilise on a form which more closely represents the performance of the process. Fig. 10.11 shows what happens if 250 readings are recorded. A skew distribution is now

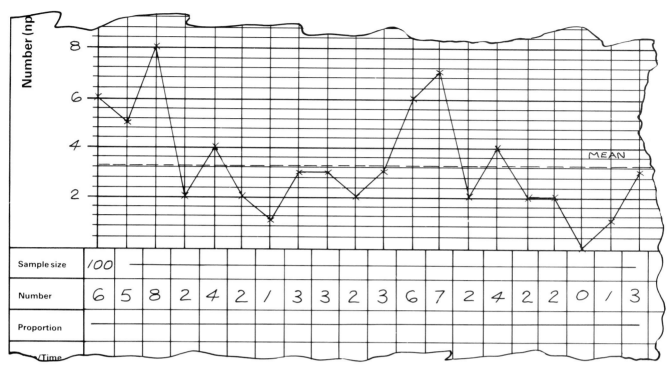

Fig. 10.9   Plot and central line

clearly illustrated. Any control limits must therefore be obtained using the properties of the theoretical distribution which matches this pattern.

The information in Fig. 10.11 can also be represented in the form of a bar chart.
Even though the diagram represents a skew pattern it is nevertheless still possible
to fit the symmetrical normal curve to the result. Above the axis of symmetry of the
normal curve there is a quite acceptable level of agreement of the curve with the
observed readings, as seen in Fig. 10.13. Below the axis of symmetry the agreement
is not so good, but in practice this does not cause any great problems. It will be
appreciated by this stage that an essential aspect of SPC is to keep things simple,

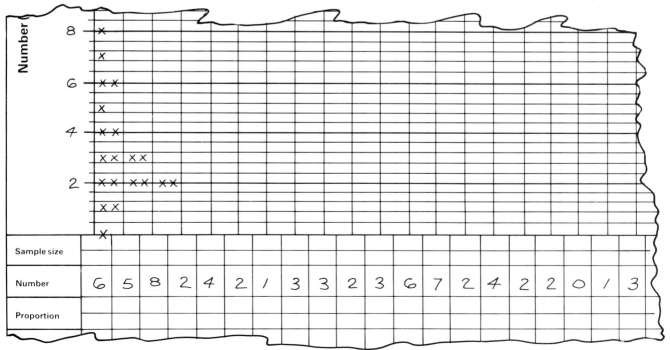

Fig. 10.10   Asymmetrical pattern

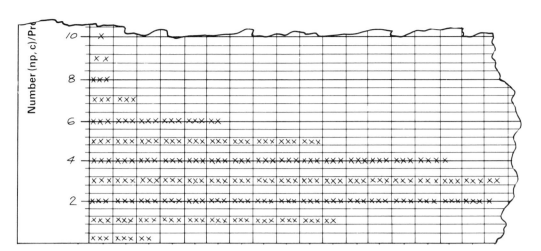

Fig. 10.11   Pattern with 250 readings

and the fitting of the normal curve to attribute data is a very good example of this. This approach has been the basis of the appropriate American National Standards Institute (ANSI) documentation on which many of the current UK industrial practices have been based.

Now that there is a normal distribution, its properties can be made use of, in particular the fact that 99.97% of the values must fall within plus or minus 3 standard deviations from the mean. Hence, as for the charts for variables, upper and lower control limits (UCL and LCL) for all attribute charts will be given by:

UCL = mean + 3 standard deviations

LCL = mean − 3 standard deviations

One problem remains. With variables use is made of constants such as A_2 to obtain the control limits. With attributes this approach is not possible. Instead the appropriate expressions for the standard deviation have to be used, depending on which

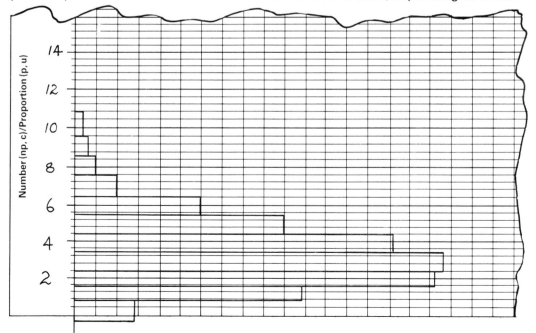

Fig. 10.12   Bar chart for readings

Fig. 10.13 Normal approximation

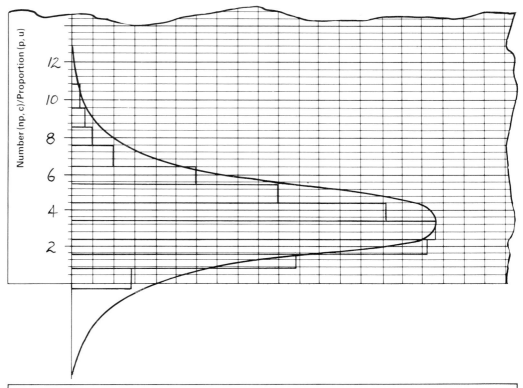

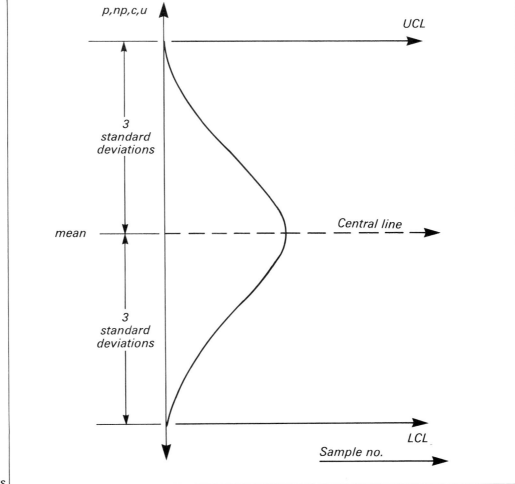

Fig. 10.14 Basis of limits on attribute charts

type of attribute chart is being plotted. The results required are given below:

Type of chart   *Control limits*

p $\bar{p} \pm 3 \sqrt{\dfrac{\bar{p}(1 - \bar{p})}{\bar{n}}}$

np $\overline{np} \pm 3 \sqrt{\overline{np}\left(1 - \dfrac{\overline{np}}{n}\right)}$

u $\bar{u} \pm 3 \sqrt{\dfrac{\bar{u}}{\bar{n}}}$

c $\bar{c} \pm 3 \sqrt{\bar{c}}$

Note that for all charts there is a common format for calculating control limits whereby the standard deviation is denoted by the expression involving the square root. Each calculation also involves the relevant mean. And for the p and u chart, the mean sample size $\bar{n}$ is needed.

The various expressions are verified in standard statistical references. The results can be used with confidence, in the knowledge that even though they are based on the normal approximation, they provide a perfectly adequate working approach to attribute charting.

10.6.4 **Probability levels**

Practical usage in many areas of work indicates that even though the control limits for attribute charts are based on an approximation to a normal distribution, this causes no problems. It is quite legitimate to fit the normal curve as long as any resulting provisos are accepted. In particular, this means that when evaluating UCL on the basis of the expressions defined in Section 10.6.3, the associated probability levels are actually nearer to 1 in 200 rather than 1 in 1000. This does not cause any practical difficulty.

Changing the probability level from 1 in 1000 to nearer 1 in 200 does mean there is an increased risk of a chart suggesting that the process setting needs changing when in fact is does not. Set against that is the advantage of having a ready-made set of expressions which enable the quick calculation of the appropriate values for UCL and LCL. These results can be evaluated using only an understanding of the properties of the normal curve; there is no need to introduce statistical distributions which are more likely to confuse than to clarify.

10.6.5 **Calculating the control limits**

Which formula should be used for calculating the control limits in the current example? It is known that non-conforming units as opposed to nonconformities are being dealt with, and also that the sample size is constant at 100. An $\overline{np}$ chart must therefore be used, and this is indicated by ticking the box at the top of the chart.

The mean, np, has already been calculated as 3.3. Hence:

$$\text{UCL}_{np} = \overline{np} \pm 3 \sqrt{\overline{np}\left(1 - \dfrac{\overline{np}}{n}\right)}$$

$$= 3.3 + 3 \sqrt{3.3\left(1 - \dfrac{3.3}{100}\right)}$$

$$= 3.3 + 5.36$$

$$= 8.66$$

Similarly:

$$LCL_{np} = \overline{np} - 3\sqrt{\overline{np}\left(1 - \frac{\overline{np}}{n}\right)}$$

$$= 3.3 - 5.36$$

$$= -2.06$$

Obviously UCL_{np} cannot have a negative value because there cannot be a negative number of rejects. If the calculations do result in a negative quantity then the value is treated as zero. The minus appears as a result of fitting a normal curve and is due to the fact that the lower tail of the curve in Fig. 10.13 extends below the horizontal (time) axis.

The central value, $\overline{np}$, and UCL_{np} can now be recorded in the appropriate data boxes and lines drawn on the chart.

The same procedures apply as for variables in that charts are set up on the basis of an initial study and the limits recalculated if necessary after excluding special causes. The examples in Chapters 11 and 12 will illustrate any specific features relating to particular types of chart.

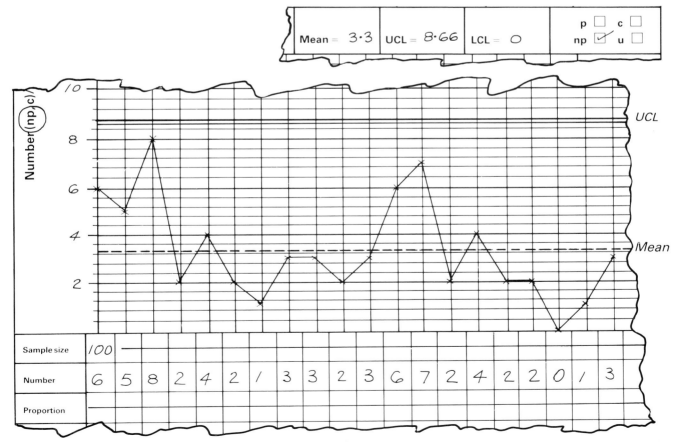

Fig. 10.15 Completed attribute chart

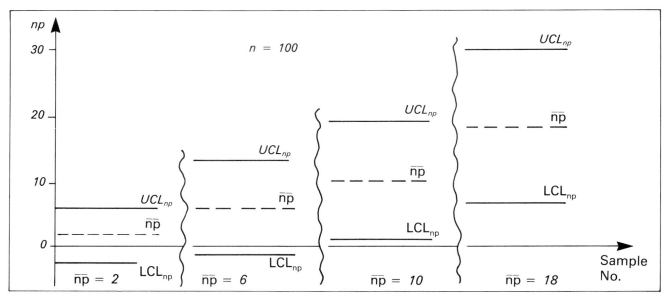

Fig. 10.16 Effect of increasing mean

10.6.6 The lower control limit

The value of LCL is not always negative. As the mean value increases, whatever the type of attribute chart, then a stage will be reached where the lower limit, based on the normal approximation, moves from a negative value to a positive value.

For the current example, with the sample size kept constant at 100, the following table shows the effect of varying the mean on the value of the control line:

$\overline{np}$	UCL_{np}	LCL_{np}
2	6.2	−2.2
6	13.12	−1.12
8	16.14	−0.14
9	17.59	0.41
10	19	1
14	24.41	3.59
18	29.53	6.47

Thus as $\overline{np}$ increases from 8 to 9, the value of LCL_{np} becomes positive.

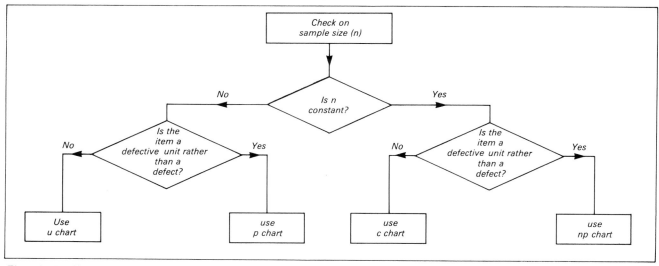

Fig. 10.17 Flow chart for deciding which attribute chart to use

Fig. 10.18 Examples of chart use

ITEM	SAMPLE SIZE	TYPE OF CHART
INCORRECT INVOICES	250-300	P
SCRATCH ON GLASS	325 - 400	μ
DEFECT CABINETS	70	nP
MARKS ON PRINTED CIRCUIT BOARDS	400	C
UNITS FAILED ON TEST	225-275	P
ERRORS IN DOCUMENTS	100 -110	μ
REJECT BOOK BINDINGS	400	nP
COLOUR FAULTS ON PACKAGING	150	C
DEFECTIVE ROLLS OF CLOTH	50-60	P
ERRORS IN CONTROL CHARTS	75 -80	μ
FAULTS IN COMPUTERS	100-110	μ
CAR PARK SPACES AVAILABLE	500	nP
GRIT IN PAINT FINISH	30	C

Fig. 10.16 shows graphically the effect of increasing the mean for a sample size 100. It can be seen that as the mean increases:

- The LCL becomes positive.
- The distribution becomes more symmetrical.
- The distance between the control limits increases.

These features hold for all attribute charts, not just the np chart.

It may be asked why a lower control limit is necessary at all. The answer is that it is required as part of the improvement process. In the same way as a point falling below LCL_R on the R chart means the presence of a special cause which indicates improvement, then similarly a point falling below LCL on the attribute chart is an indication of reduced rejects and consequently improved performance. The importance of the lower limit on the attribute chart cannot be overstated. The upper control limit admittedly plays an important role in detecting problems, but it is a controlling role rather than an improving one. The lower limit, on the other hand, plays a vital part in instigating improvement – and then further improvement.

10.7 Interpreting attribute charts

Attribute charts are interpreted in exactly the same way as charts for variables in that the usual rules for determining the presence of special causes apply. But caution is necessary in applying rule 4, the middle third rule, to a case where the mean is not symmetrical with respect to the control limits (such as when UCL is negative). Because of the use of a normal approximation, care should be taken in not placing too much credence on using this rule for the zone of the chart below the central line.

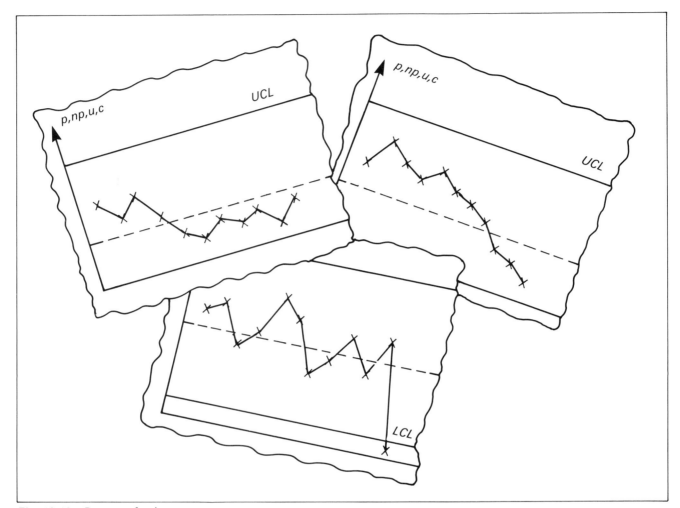

Fig. 10.19   Patterns for improvement

10.8 Which chart to use

Making the right choice of attribute chart is not easy. A useful guide is to check on the sample size; if that is constant then it restricts the choice to either an np chart or a c chart. If it is the number of defective units that is to be assessed then the np chart not the c chart is used. A simple flow chart is useful as in Fig. 10.17. The table in Fig. 10.18 also gives guidance on the correct chart to use.

10.9 Capability

With attribute charts there is no tolerance and hence capability, as understood in dealing with variables, does not have any real meaning.

A simplistic measure of capability (applicable only in the case of non-conforming units, not nonconformities) is given by the mean value, i.e. $\bar{p}$. The aim is to make this figure zero. An alternative is to define capability as $(1-\bar{p})$ and to aim to make this equal to 1. Each alternative provides the same end result in that a process with no reject items is, by definition, producing 100% acceptable items. Both forms are commonly used in the automotive industry.

10.10 Continuous improvement

With variables, the target is to reduce the variability around a nominal value to zero.

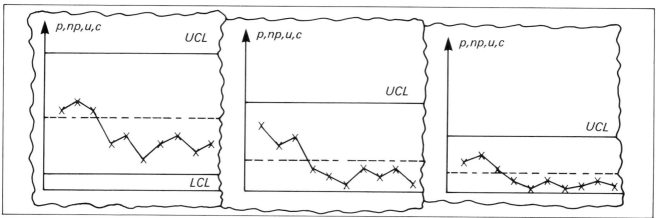

Fig. 10.20 Continuous improvement for attributes

With attributes the target is zero defects. Continuous improvement means first looking for the presence of special causes. While causes that result in out-of-control situations associated with deterioration of the process must be acted on, the special causes of real interest in this context are those which suggest an improvement in the process. Such a cause might be indicated by a sequence of seven points below the central line on the attribute chart, a downward trend, or a point below the LCL (provided the mean level is sufficiently high for this lower control limit to be present).

Once a special cause indicating improvement has been incorporated in the process a new mean level of performance is calculated, which means that a new UCL and LCL need establishing. The cycle is then repeated: after a further sign of improvement a new mean and control limits are established, and so forth. The repeating cycle is illustrated in Fig. 10.20.

Continuous improvement means a continuous commitment to reducing the level of both non-conforming units and nonconformities.

Summary

- Attribute charts can be used extensively in non-manufacturing functions.
- Attribute charting enables SPC programmes to be genuinely company-wide.
- Attribute analysis is based on a yes/no definition.
- Use attributes when the item cannot be measured.
- Attribute charts and Pareto analysis are related.
- Attribute plotting enables more sensible decisions to be made from data.
- A clear definition of a fault is required.
- Sample sizes can vary greatly.
- Defective units and defects need clear definitions.
- Different attribute charts are used depending on whether the item is a defective unit or a defect and whether the sample size is variable or constant.
- Control limits for attribute charts can be based on the normal curve.
- Use of the LCL provides one of the keys to continuous improvement.
- Capability, $(1 - \bar{p})$, is not as important in attribute analysis as in analysis of variables.
- Continuous improvement as regards attributes means working towards reducing the level of rejects to zero.

The two chapters that follow deal with the four types of attribute charts in detail, starting with the p and np charts for assessing non-conforming units.

Chapter 11   Attribute charts for defective units

11.1   Introduction

Chapter 10 provided an introduction to attribute charts in general, and it was seen how the normal distribution provided a perfectly acceptable model on which to base calculations of defect levels. What is now required is detail regarding the attribute chart chosen for a particular situation. This chapter will concentrate on the analysis of non-conforming units (defectives). The two related charts (p and np) will be studied and guidance given on setting them up and interpreting them.

The definition of a defective unit is already familiar. Is the disc-drive acceptable or not? How many error-free invoices were sent through for processing? What proportion of the units in production were subject to a concession? These questions bring with them discussions on sample size. Are a constant number of disc-drives manufactured per week? Does the number of invoices to be analysed remain constant or is there variation over time? Are the same number of units made in each time period analysed?

Ideally the aim in any process should be constant throughput. In other words, if the cumulative number of items processed per week or per month is plotted against time, then the result should be a straight line the angle of which varies depending on the projected output. Any deviation from constant throughput is reflected in a zigzag line, as shown in Fig. 11.1. All organisations aim for linearity, with constant production rates and hence constant sample sizes. In practice the output per shift, day or week tends to vary, and hence the sample size varies. For this reason, and others, the p chart will be considered first.

11.2   The p chart

11.2.1   Completing the administrative details

It is as well to complete the required administrative details on the chart early in the analysis. Some features are obvious: the box indicating a p chart is ticked and the descriptive items on the vertical scale which do not apply are deleted. Fig. 11.2 shows the completed boxes for this analysis. It will be seen that not all headings are appropriate for this project. As with any chart, there is a balance to be achieved in designing the chart to make it neither too general nor too specific.

The target sample size of 70 was the ideal sample size to match the existing schedules.

11.2.2   Collecting, recording and plotting the data

The general rules regarding collecting and recording apply to the p chart just as they do to any other chart.

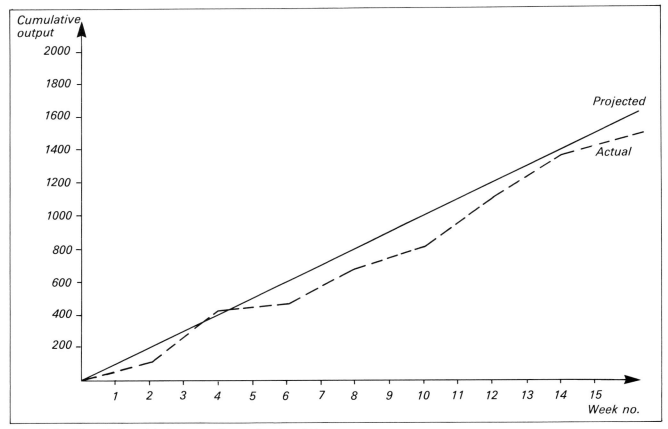

Fig. 11.1   The relationship between throughput and time

At least 20 samples are required. The sample size in each case is recorded in the appropriate data box, as shown in Fig. 11.3.

The company in question was concerned about its delivery performance. There was an ongoing quality improvement programme and it had been recognised early on that restricting this to manufacturing areas was not sufficient to fulfil the broad objective of customer satisfaction. Many other issues were involved. For example, deficiencies in invoicing, labelling, packaging, delivery times and delivery quantities would all be irritating to a customer demanding quality of service as well as quality of product.

The company was aware of a lack of consistency in the performance of its non-

Process Control Chart – Attributes							Division		Department	Week no.
							A		CUSTOMER LIASON	*14* →
Part no./Description	Operation DELIVERY PERFORMANCE	Machine no./Type	Sampling frequency WEEKLY	Target sample size 70	Mean =	UCL =	LCL =		p ☑ c ☐ np ☐ u ☐	

Fig. 11.2   Administrative data for p chart

Sample size	65	77	74	50	45	84	75	70	63	59	57	88	62	59	69	63	75	44	65	63
Number	3	7	4	3	4	4	6	5	9	4	3	5	3	1	2	3	5	3	2	4

Fig. 11.3   Data box

manufacturing and support departments. A first step was to obtain information on delivery faults, irrespective of type. A further analysis of the type of fault would produce more detail which could then be assessed using a multiple characteristic chart as described in Chapter 12. An initial analysis based on 20 weeks' deliveries showed the number of those deliveries which were unacceptable because of incorrect invoice, short delivery, etc. (Fig. 11.3). For each sample the proportion of unsatisfactory deliveries in each case can be calculated and recorded with the results shown in Fig. 11.4. Taking into account the usual guidelines regarding choice of scale, the proportions can be plotted on the chart to give the pattern shown in Fig. 11.5.

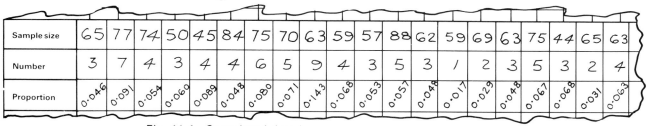

Sample size	65	77	74	50	45	84	75	70	63	59	57	88	62	59	69	63	75	44	65	63
Number	3	7	4	3	4	4	6	5	9	4	3	5	3	1	2	3	5	3	2	4
Proportion	0.046	0.091	0.054	0.060	0.089	0.048	0.080	0.071	0.143	0.068	0.053	0.057	0.048	0.017	0.029	0.048	0.067	0.068	0.031	0.063

Fig. 11.4   Completed data box

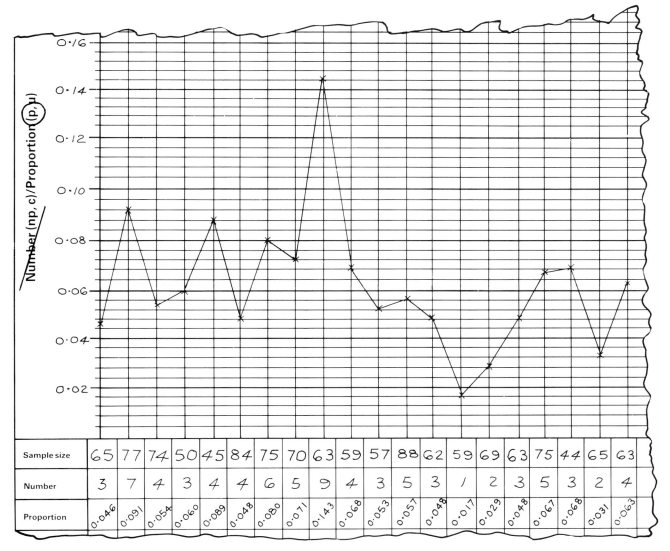

Sample size	65	77	74	50	45	84	75	70	63	59	57	88	62	59	69	63	75	44	65	63
Number	3	7	4	3	4	4	6	5	9	4	3	5	3	1	2	3	5	3	2	4
Proportion	0.046	0.091	0.054	0.060	0.089	0.048	0.080	0.071	0.143	0.068	0.053	0.057	0.048	0.017	0.029	0.048	0.067	0.068	0.031	0.063

Fig. 11.5   Plot for p values

11.2.3 **Obtaining the control limits**

The first step in calculating control limits is to obtain the central value, $\bar{p}$. There are two ways of obtaining this. The first is to calculate the separate totals for the number of deliveries and the number of those which were unsatisfactory. Thus:

    Total number of deliveries = 1307
    Number of unsatisfactory   =   80
    deliveries
Therefore

$$\bar{p} = \frac{80}{1307}$$

$$= 0.061(2)$$

Alternatively, since the 20 sample proportions are already available, the mean of these can be calculated, as 0.061(6). There is a slight difference in the two results due to rounding off error in the individual sample proportions.

It is preferable to use the first approach – and not just because it is more accurate. A benefit in obtaining the separate totals is that one of them will be needed in any case to evaluate the mean sample size, $\bar{n}$. In addition, there will be occasions when out-of-control points are obtained in the first 20 readings. Limits may then need to be recalculated, and it is easier to do this if the two separate totals referred to are already available.

The value for $\bar{p}$ is therefore 0.061(2), which is used to determine the control limits. For the p chart, control limits are obtained from

$$\bar{p} \pm 3 \sqrt{\frac{\bar{p}(1-\bar{p})}{\bar{n}}}$$

The value of $\bar{n}$ is given by 1307/20 = 65.4. Therefore for the upper control limit

$$\mathrm{UCL}_p = \bar{p} + 3 \sqrt{\frac{\bar{p}(1-\bar{p})}{\bar{n}}}$$

$$= 0.061 + 3 \sqrt{\frac{0.061(1-0.061)}{65}}$$

$$= 0.061 + 0.089$$

$$= 0.15$$

For the lower control limit

$$\mathrm{LCL}_p = 0.061 - 0.089, \text{ which is treated as } 0$$

The mean, UCL and LCL boxes can now be completed on the chart and lines drawn on the pattern of figures to monitor the process (Fig. 11.6).

11.2.4 **Interpreting the chart**

Applying the usual four rules for detecting the presence of special causes, the process is seen to be under statistical control. However, a further investigation is required to see whether the sample size has changed significantly. If it has this will

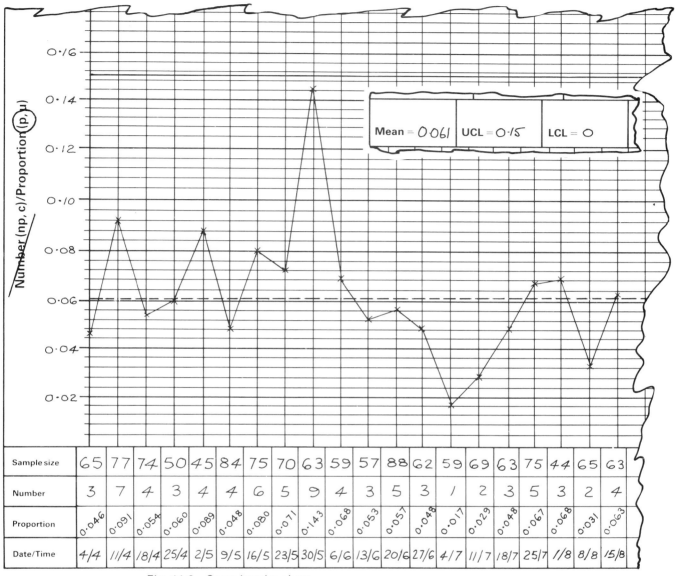

Fig. 11.6 Completed p chart

affect the control limits. The reason for this is tied in with the fact that the control limits make use of the mean sample size $\bar{n}$.

Strictly speaking, UCL_p and LCL_p should be evaluated using the expression

$$\bar{p} \pm 3 \sqrt{\frac{\bar{p}(1-\bar{p})}{n}}$$

The actual sample size, n is used not $\bar{n}$, and this means that the limits will vary in position as the value of n changes. Fig. 11.7 shows the effect of recalculating the limits for each n value.

It will be helpful to indicate how the position of the upper control limit has been obtained for one of the 20 samples. A value for n will be chosen which is sufficiently different from $\bar{n}$ to bring out any difference in the value of the corresponding

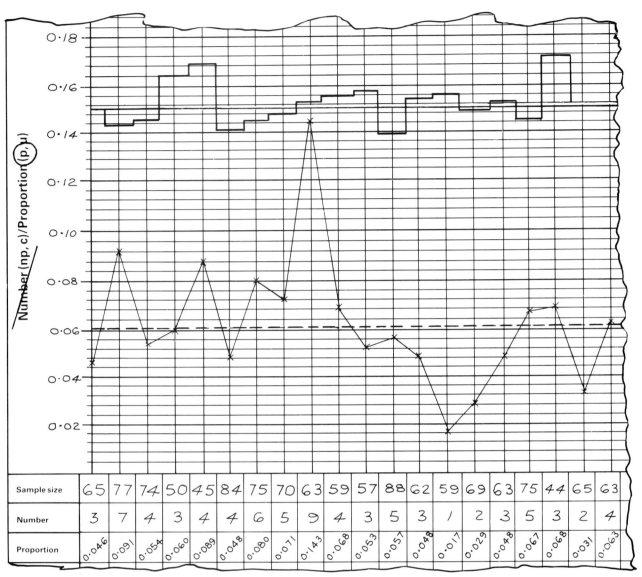

Sample size	65	77	74	50	45	84	75	70	63	59	57	88	62	59	69	63	75	44	65	63
Number	3	7	4	3	4	4	6	5	9	4	3	5	3	1	2	3	5	3	2	4
Proportion	0.046	0.091	0.054	0.060	0.089	0.048	0.080	0.071	0.143	0.068	0.053	0.057	0.048	0.017	0.029	0.048	0.067	0.068	0.031	0.063

Fig. 11.7   *p chart showing exact control limits*

control limit. The fourth sample, n = 50, is suitable in this respect. Hence

$$UCL_p = \bar{p} + 3 \sqrt{\frac{\bar{p}\,(1 - \bar{p})}{n}}$$

$$= 0.061 + 3 \sqrt{\frac{0.061\,(1 - 0.061)}{50}}$$

$$= 0.061 + 0.102$$

$$= 0.163$$

In a similar way the values of UCL_p for the remaining 19 samples can be determined. These produce the fluctuating line shown on the control chart.

It is clearly time-consuming to calculate the control limit this way each time. In any case it is not worth the trouble in many instances because the difference between the two lines is negligible. At what stage should it be considered that the sample size has changed sufficiently to warrant using n rather than n̄? A generally accepted guideline is to use the figure of a 25% change in n̄: if the sample size is outside the limit of n̄ ± 0.25 n̄, then a recalculation of the control limits is necessary.

In the case being considered 25% of 65 is 16 and hence the critical values for n are 81 and 49. Control limits should therefore be recalculated for those values of n which are either greater than 81 or less than 49. There are four sample readings outside this range: n = 45, n = 84, n = 88 and n = 44. Recalculating may therefore be necessary for these.

In this particular example, however, recalculation is not going to help. There is really only a need to recalculate if in doing so the status of the sample is affected, i.e. if there is a change from an in-control situation to an out-of-control situation or vice versa.

Fig. 11.9 shows the effect of changing n. With n in the denominator, it means that as n increases the control limits are drawn closer to p̄. As n decreases the control limits move away from the central line p̄. With this in mind, Fig. 11.7 suggests that recalculating the limits is a waste of time. With n = 45 or 44, i.e. less than 65, the process is under control. Hence recalculating, and therefore widening, the limits will not influence the status of the points. For n = 84 and n = 88, recalculating the lower control limit will similarly not make any difference to the result. The lower limit is used here because both sample readings are closer to LCL_p than they are to UCL_p.

The situation is therefore that the process is under statistical control with one limit at UCL = 0.15, and a capability given by $(1 - \bar{p}) \times 100 = (1 - 0.061) \times 100 = 94\%$.

Continuous improvement means working to improve this figure to nearer 100%. A level of 6% unsatisfactory deliveries does not reflect well on the company and action is required to improve the system.

11.2.5 Improving the process

An analysis of the system showed that several factors were operating which affected delivery performance. Data had already been obtained, and a Pareto diagram indicated the major items contributing to poor results. Plotting the chart is only a start; problem-solving techniques are needed to provide guidance on improvement measures.

Fig. 11.10 shows the results at a later stage following an improvement programme in the invoicing and delivery departments. The effect of the change is evident. After the initial period during which the process settles, the level of performance has improved considerably. There is a run of seven points below the central line. Once the specified 20 have been achieved these would then be used to determine a new mean and new control limit. The process is then worked on further

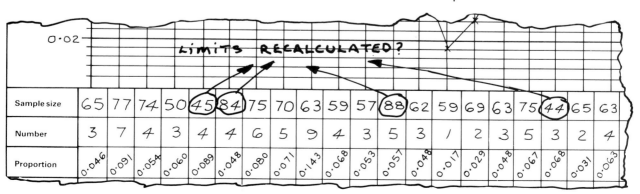

Fig. 11.8 Data box indicating specific n values

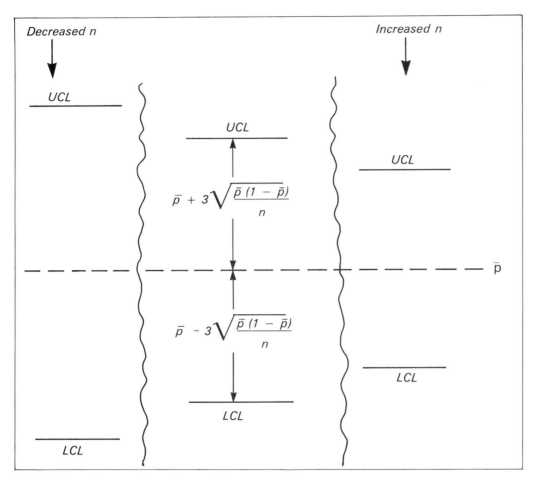

Fig. 11.9   Effect of variations in n on control limits

to reduce yet again the level of unacceptable items. This means tackling all categories of delivery performance, including, for example, early delivery.

11.2.6   Limitations of the p chart

The p chart assumes that there is a constant probability of obtaining a defective. This requires each sample to be independent of the previous one and the next one in succession. If this is not so then the majority of the plotted points tend to fall outside the control limits. In such cases the limits are no longer appropriate. That does not mean that a chart should not be used. It is still beneficial to plot the proportions because they provide a picture of the process, but control limits are not drawn on the chart.

The following case study shows how a simple chart results in useful information which can be used to reduce the level of rejects whilst at the same time stabilising the process. This will eventually enable performance-based limits to be introduced.

Case study 8: Hewlett-Packard

The Hewlett-Packard plant in Bristol is part of the Computer Peripherals Division. SPC was gradually introduced across the plant over a period of months and at the same time the Purchasing Department began placing greater emphasis on the use of SPC within the company's supply base. This case study refers to a supplier of an item used in the disc drive produced at Bristol.

Fig. 11.11 shows the situation in the early stages of introducing charting techniques for monitoring test performance. The product was being 100% tested in the materials laboratory and the results relate to a time before discussions with the supplier on quality improvement programmes. There was an unacceptably high

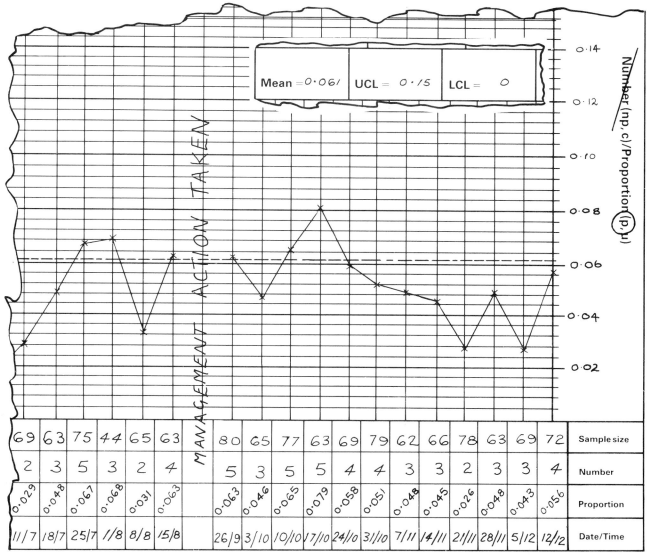

Fig. 11.10   On-going p chart

level of reject items and the supplier was asked to indicate the intended programme for improvement.

The central line and control limits were calculated in the usual way, based on the first 20 readings. The mean sample size was 95 and the 25% change limits 119 and 71. Of the seven points outside the control limits, six had sample sizes beyond the allowable range. Recalculating the limits for the six points in question would not radically change the situation, though, as the process was showing excessive, unnatural variability. This was because the probability of getting a defective item was not constant, due to some factor affecting the supplier's production processes. Hence control limits were not valid in this case and so were not projected forward in the usual way. However, plotting the proportions themselves provided much useful information.

Fig. 11.12 provides comments on features that were evident in the testing programme at particular times. It was clear that quite apart from the supplier's performance, the chart was revealing factors relating to the testing process itself.

Further readings were plotted as shown in Fig. 11.13, with comments referenced to points on the chart.

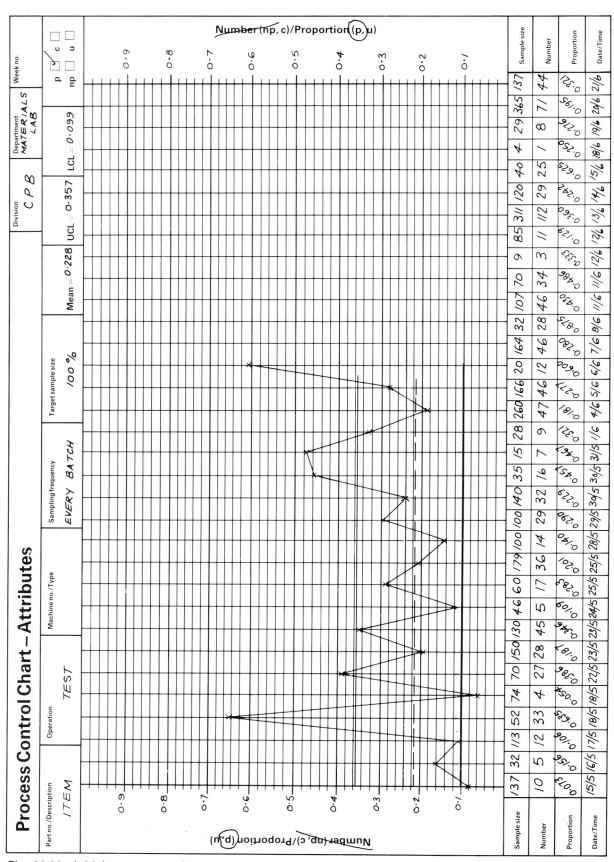

Fig. 11.11   Initial process results

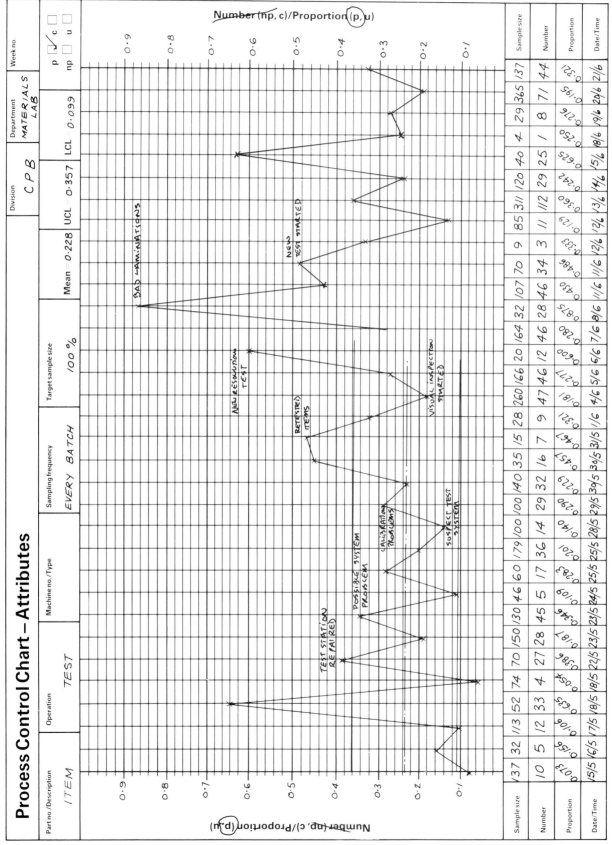

Fig. 11.12 Record of events

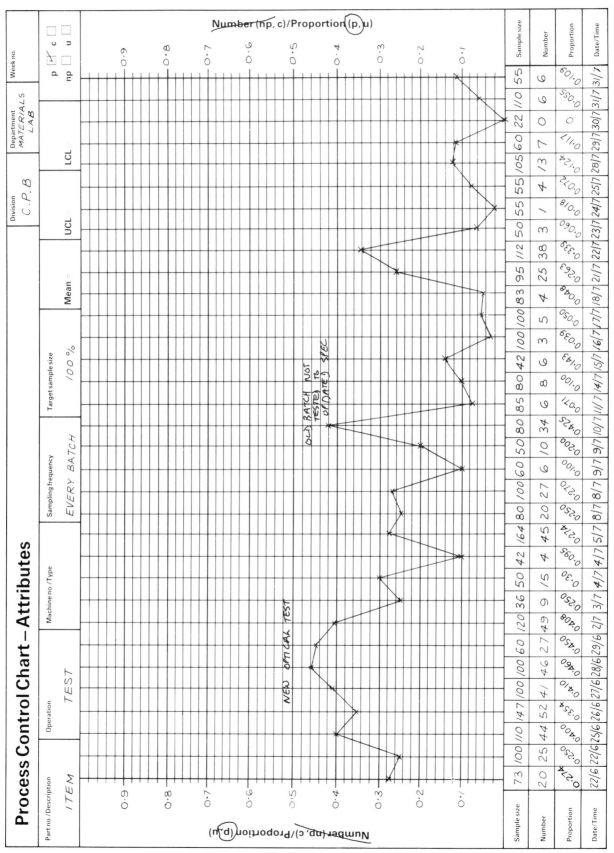

Fig. 11.13   Further results

Fig. 11.14 shows the next control chart sheet in time sequence. Apart from three points, the cause for which had not been traced, it is evident that the process is running at a level considerably below the original mean of 0.228 for the period 11/7 onwards. Not only that, there are no longer excessive fluctuations and therefore there would be justification in calculating a new mean and new control limits. In order to conform to the convention of using 20 points for setting up the system, the last 20 values on the sheet were used to generate the control limit values indicated in the data boxes. New lines were then drawn on the chart as indicated to cover the last 20 points in question.

As expected, the process is now under much better statistical control. However, the drive to reduce the reject level continued. Further progress was achieved, brought about by Hewlett-Packard's commitment to improved supplier quality.

Charting proportions is only one aspect of monitoring non-conforming units. Equally as important is the charting of the actual numbers of units.

11.3 **The np chart**

11.3.1 **Introduction**

The np chart was introduced as an example of an attribute chart in Section 10.6.2. It tends to be used in two typical situations:

- Where the sample size is fixed by the very nature of the process: for example, a printed circuit board with 10 soldered connections, a housing with 12 welds, an injection moulding with 30 cavities, a department with 15 employees, cigarettes packed in boxes of 20. In each case the sample size is constant, and if it does change, as a result of different housing, moulding or package size being used, for example, then that new sample size is also fixed.
- Where the sample size is constant by choice as part of a programme of improved throughput, or simplified visual inspection procedures. For example, 20 boxes of a product packed in boxes of 10 may always be chosen to provide the 200 items which give the right level of rejects for charting purposes.

The following example covers the essential features of the np chart. The data relate to a safety critical item which requires a functional test at a certain stage in the production process. The product is tested in batches of 250.

11.3.2 **Completing the administrative details**

The boxes in the administrative section of the chart should be completed as shown in Fig. 11.15.

11.3.3 **Collecting, recording and plotting the data**

Twenty samples were taken, one sample per shift, and the results recorded and then plotted on the chart in the usual way as indicated in Fig. 11.16.

11.3.4 **Obtaining the control limits**

The mean, $\overline{np}$, is given by

$$(3 + 2 + \ldots + 1 + 2)/20 = 59/20$$

$$= 2.95$$

The sample size, n, is 250. Therefore for the upper control limit

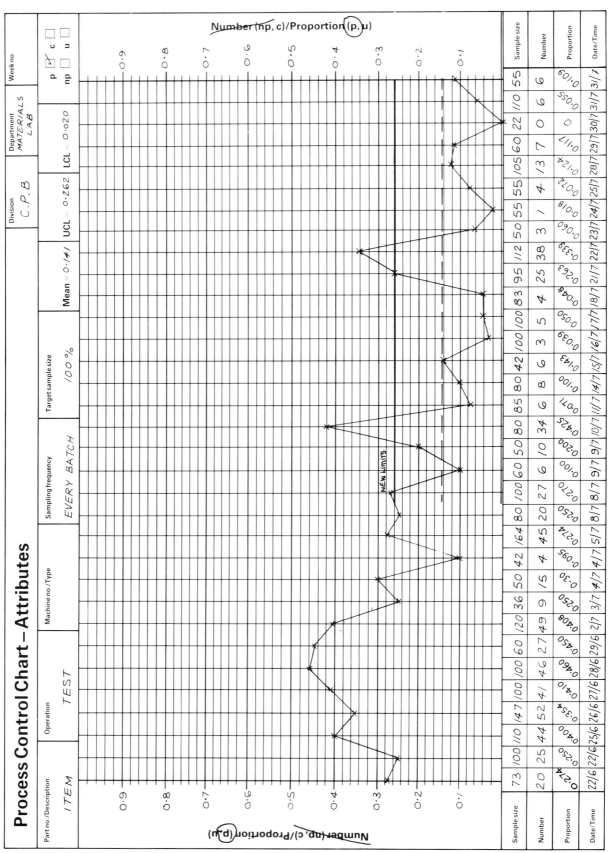

Fig. 11.14   Improved performance

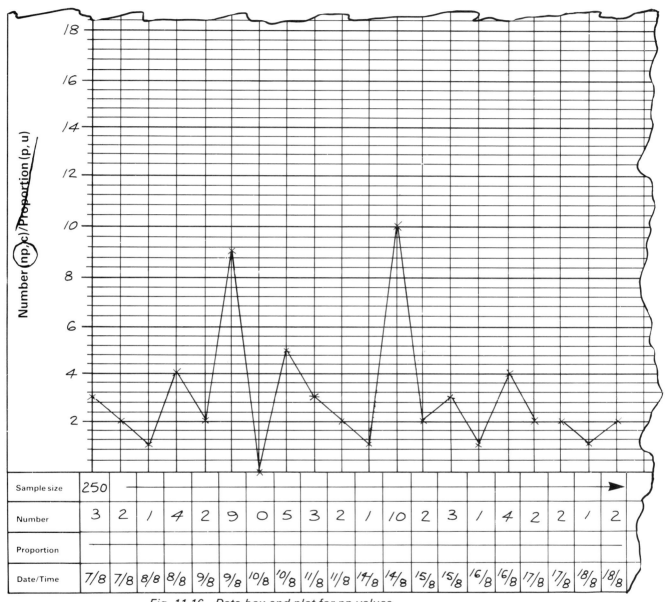

Fig. 11.15   Administrative data for np chart

Fig. 11.16   Data box and plot for np values

$$UCL_{np} = \overline{np} + 3\sqrt{\overline{np}\left(1 - \frac{\overline{np}}{n}\right)}$$

$$= 2.95 + 3\sqrt{2.95\left(1 - \frac{2.95}{250}\right)}$$

$$= 2.95 + 5.12$$

$$= 8.07$$

For the lower control limit

$$LCL_{np} = 0$$

The $\overline{np}$ and UCL_{np} values can now be recorded in the data boxes and the corresponding lines drawn on the chart. The result is as shown in Fig. 11.17.

11.3.5 Interpreting the chart

The chart shows two points outside UCL_{np} and therefore there is sufficient statistical evidence to suggest that a special cause (or causes) is present. In this case the presence of a special cause is unduly influencing the position of the central line and control limits and the reason or reasons for this must be found and eliminated from the system.

Finding the cause, however, is often not easy. If, for one reason or another, it cannot be tracked down in the short term, then the problem has to be accepted, and it must be assumed for the interim that the points belong to the system. But this would only be a short-term approach to get the control chart operating. Continuing effort is required to eliminate these features which increase the variability; at the same time the process needs to be improved by noting features which influence it for the better. These are then included as a permanent feature.

In this instance it was found that the test rig was the source of the problem in those samples corresponding to out-of-control signals. The rigs were checked and it was verified that the problem related only to those samples and no others. The

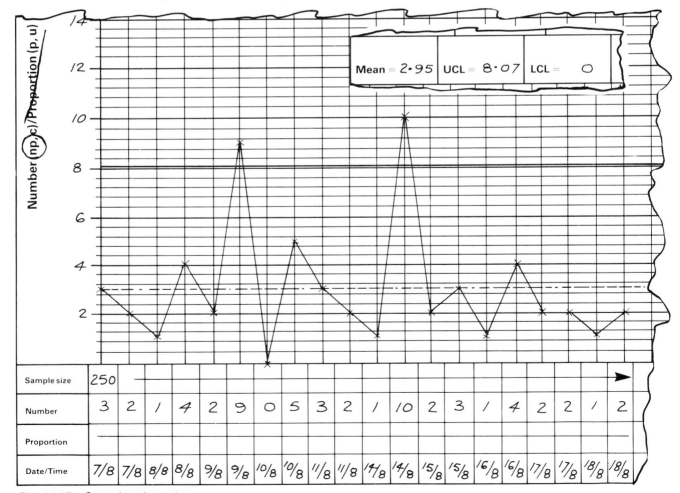

Fig. 11.17 Completed np chart

readings of 9 and 10 were then deleted from the first 20 samples and new calculations performed based on the remaining 18 readings. As a result $\overline{np}$ was reduced to 2.22 and UCL_{np} became 6.67. The more representative situation appears in Fig. 11.18.

The capability of the process is now

$$(1 - \bar{p}) \times 100 = \left(1 - \frac{40}{250 \times 18} \right) \times 100$$

$$= 99.1$$

The process is continuously monitored and progress made, as with any attribute chart, by looking for those special causes associated with improvement. This can then mean a reduction in $\overline{np}$, and a new system of operation which can provide the basis for further improvements.

Case study 9: British Alcan (Rolled Products)

Tracking customer response is essential for any organisation committed, as British Alcan is, to a programme of continuous improvement.

Following a training course for the senior managers, projects were undertaken by each participant. The purpose was twofold. First, such a project would provide much-needed practice in using the various techniques covered. Attending a 3-day course is not in itself sufficient to develop the necessary confidence to handle data and plot charts. By setting up control charts on features which were directly involved with their own activities, the managers gained first-hand experience of

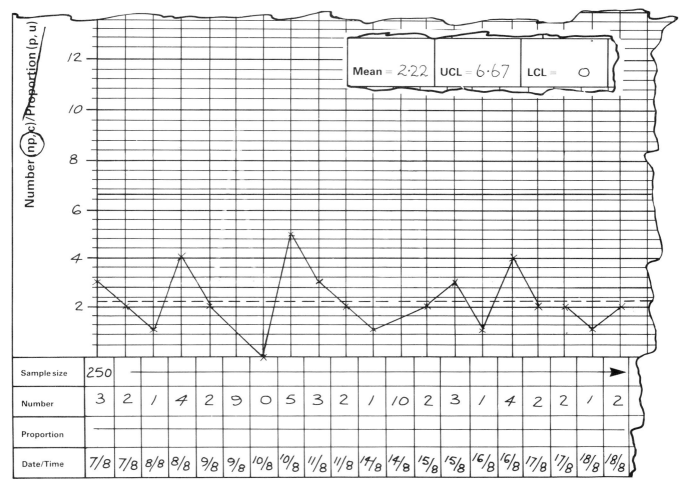

Fig. 11.18   np chart with recalculated limits

how to monitor performance in a more powerful, effective manner. Secondly, and perhaps even more important, is the element of leadership. In displaying charts for their own processes, the managers were indicating that they had confidence in charting and were using it. Supervisory and operational staff would not then be in a position to criticise their colleagues for lack of example.

The General Manager chose, for his project, an analysis of late deliveries reported by customers. Data was collected over a 32-week period. As there was a constant sample size (corresponding to the number of customers) an np chart was chosen and control limits drawn based, according to the usual convention, on the first 20 readings. Whilst the process appeared to be under statistical control in that there were no points outside the limits, and no runs of seven, there was an unusual pattern in that regular peaks were occurring. Further investigation showed that the peaks represented an internal company issue rather than an external feature and action was taken to remedy the problem and at the same time reduce the level of late deliveries to zero. Without charting, the true picture of customer complaint level, and the fluctuation of that level, would almost certainly not have been detailed.

Progress was maintained in dealing with customer complaints of this nature. A further analysis was carried out which classified complaints by their nature. This meant, for example, that delivery times could be monitored using an $(\bar{X}, \text{moving } R)$ chart.

Summary

- p charts are used for monitoring non-conforming units when the sample size varies.
- It is important with a p chart, as with any other chart, to complete the administrative section.
- The variability in the sample size determines whether n or $\bar{n}$ is used in calculating control limits.
- Recalculation of the control limits is only necessary when the status of the point in question could change as a result of using n instead of $\bar{n}$.
- Excessive variability in the p values indicates that control limits cannot be statistically justified.
- Plotting the p values when control limits cannot be used is still advised because of the advantage gained in presenting the information more graphically.
- np charts are used for monitoring non-conforming units when the sample size is constant.

Whilst p and np charts tend to have priority because they refer to units rather than defects, this should not imply that charts for defects are not of real value. Checking and eliminating the defects will obviously improve the throughput of acceptable units.

The c and u charts for defects, detailed in Chapter 12, operate in very much the same way as the charts discussed in this chapter. The multiple characteristics chart, considered in the final section of Chapter 12, makes use of the Pareto principle when monitoring several defects at the same time.

Chapter 12 Attribute charts for defects

12.1 Introduction

A continuing theme of SPC is that of controlling the process at a very early stage so that minor difficulties do not grow over time into major obstacles. Thus non-conforming units may be unacceptable because of the high level of minor nonconformities they contain. A small particle of dust included at the paint-spraying stage of a high-cost luxury limousine will appear as a real blemish to a customer who is expecting a mirror finish. Similarly aircraft navigation equipment will fail if a greasy thumb mark is registered in a critical item of the component at the assembly stage. Controlling and reducing the level of nonconformities is therefore just as important as monitoring the rejection rate for non-conforming units. The two charts used are the u chart and the c chart.

These two charts have essentially the same areas for potential application. Marks on glass, imperfections in cloth, scratches in surface finish, breakdowns in insulation, injuries recorded in the works surgery and errors in paperwork analysis all lend themselves to analysis by u or c charts. It is the sample that indicates which chart to use.

If the sample size varies, then the u chart is used. In the same way as the p chart allows for varying sample sizes when analysing non-conforming units, then similarly the u chart allows analysis of nonconformities when the sample size varies.

12.2 The u chart

12.2.1 Completing the administrative section

In the example to be considered here, visual faults are being monitored by means of samples of varying size that are analysed daily as part of an audit programme. The appropriate data is as shown in Fig. 12.1.

12.2.2 Collecting, recording and plotting the data

The same rules regarding collecting and recording apply to the u chart as to all other charts considered so far. At least 20 samples are required. Fig. 12.2 shows the data for visual faults.

The category of fault is not specified. There will be data available in the company which provides a breakdown on the type of fault and the multiple characteristics chart (Section 12.4) would be useful in further analysis. Note that it is the number of faults within a sample that is being monitored, and not the number of items (in this case coils) which have been rejected. In the latter case, a p chart would be the appropriate chart to use, as discussed in Chapter 11.

Process Control Chart – Attributes					Division C	Department QUALITY	Week no 36
Part no./Description COIL	Operation VISUAL FAULTS ANALYSIS	Machine no./Type	Sampling frequency DAILY	Target sample size 80	Mean =	UCL =	LCL =	p ☐ c ☐ np ☐ u ☑

Fig. 12.1 Administrative data for u chart

Sample size	83	76	95	101	96	65	68	92	98	66	77	90	75	78	93	68	98	65	68	82
Number	6	4	7	4	8	4	4	5	4	3	4	1	4	5	5	2	5	2	3	2
Proportion																				
Date/Time	1/9	2/9	3/9	4/9	5/9	8/9	9/9	10/9	11/9	12/9	15/9	16/9	17/9	18/9	19/9	22/9	23/9	24/9	25/9	26/9

Fig. 12.2 Data box

One further point is worth mentioning here. The data could relate typically to coils of steel or aluminium, rolls of paper, containers of chemicals. The reject levels would probably not be the same in all cases and the categories of faults will differ to a greater or lesser extent. However, the principle of charting, as indicated here, is the same whatever the particular product being analysed.

For each sample the proportion of faults is calculated and the results recorded as shown in Fig. 12.3. The 20 values for u are then plotted on the chart with the result shown in Fig. 12.4.

12.2.3 **Obtaining the control limits**

The procedure for calculating the control limits is the same as for all previous charts in that first the mean value, $\bar{u}$, must be determined. The total number of items sampled is 1634 and the total number of defects is 82. Therefore

$$\bar{u} = \frac{82}{1634}$$

$$= 0.050$$

It will be found that in order to determine the control limits the mean sample size, $\bar{n}$, is also required. This is given by

$$\bar{n} = \frac{1634}{20}$$

$$= 81.7$$

Sample size	83	76	95	101	96	65	68	92	98	66	77	90	75	78	93	68	98	65	68	82
Number	6	4	7	4	8	4	4	5	4	3	4	1	4	5	5	2	5	2	3	2
Proportion	0.072	0.053	0.074	0.040	0.083	0.062	0.059	0.054	0.041	0.045	0.052	0.011	0.053	0.064	0.054	0.029	0.051	0.031	0.044	0.024
Date/Time	1/9	2/9	3/9	4/9	5/9	8/9	9/9	10/9	11/9	12/9	15/9	16/9	17/9	18/9	19/9	22/9	23/9	24/9	25/9	26/9

Fig. 12.3 Completed data box

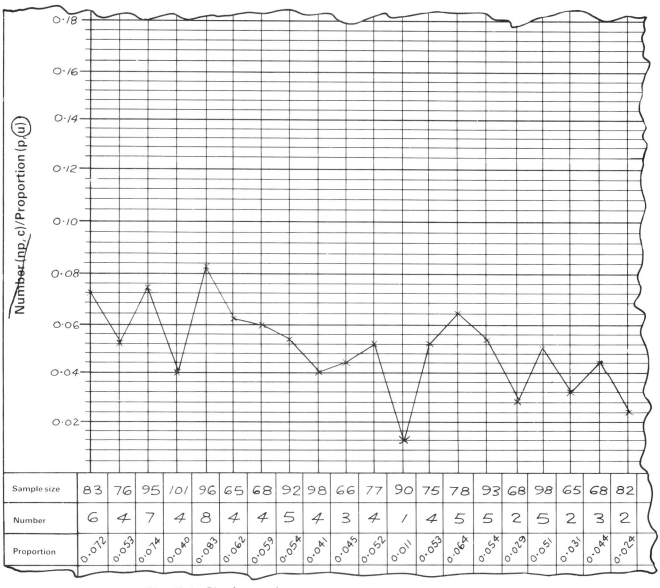

Sample size	83	76	95	101	96	65	68	92	98	66	77	90	75	78	93	68	98	65	68	82
Number	6	4	7	4	8	4	4	5	4	3	4	1	4	5	5	2	5	2	3	2
Proportion	0.072	0.053	0.074	0.040	0.083	0.062	0.059	0.054	0.041	0.045	0.052	0.011	0.053	0.064	0.054	0.029	0.051	0.031	0.044	0.024

Fig. 12.4   Plot for u values

For the u chart, the upper control limit is given by

$$UCL_u = \bar{u} + 3 \sqrt{\frac{\bar{u}}{n}}$$

$$= 0.05 + 3 \sqrt{\frac{0.05}{82}}$$

$$= 0.05 + 0.074$$

$$= 0.124$$

The lower limit is given by

$$LCL_u = 0.05 - 0.074, \text{ which is treated as 0}$$

The mean, UCL and LCL boxes can now be completed. Lines corresponding to these values are drawn on the chart and the process is analysed (Fig. 12.5).

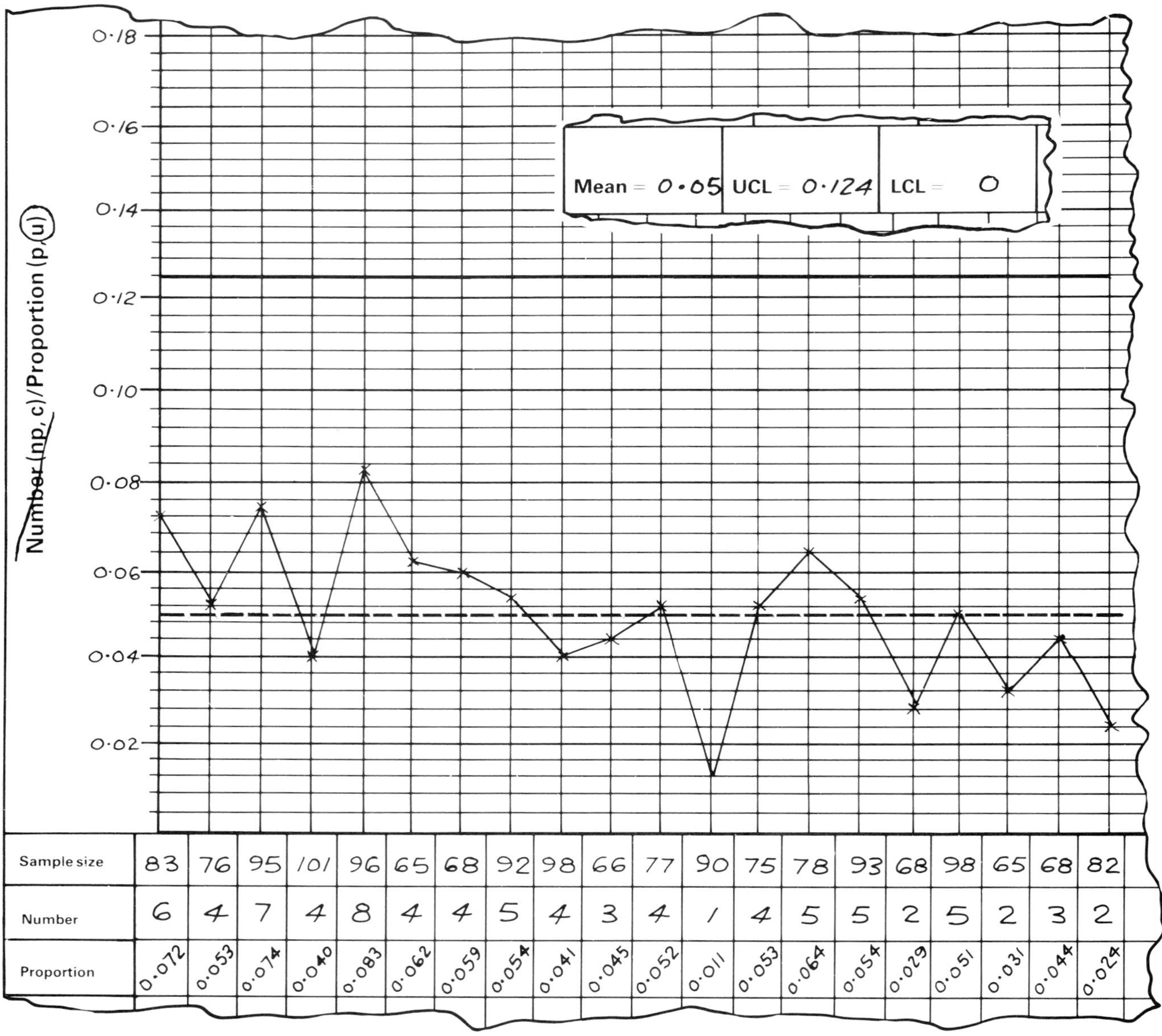

Fig. 12.5   *Completed u chart*

12.2.4   **Interpreting the chart**

From the chart the process would seem to be under statistical control. One further calculation is required, though, before the presence of special causes can be excluded: the variability in sample size needs to be checked using the same 25% rule as for the p chart. The mean sample size is 82 and 25% of this is 21. Therefore the 25% change limits are 103 and 61. In this case all the sample sizes are within the limits.

In the previous examples dealing with p and np charts, a capability value was determined. For u and c charts, capability cannot be measured as such. This does not mean, though, that achieving zero defects is not the aim.

12.2.5 **Improving the process**

Given that the level of nonconformities is under statistical control, the next step is to improve the process. Whatever the item being monitored, a 5% level of nonconformities gives plenty of scope for achieving improved performance.

The chart is now monitored on an ongoing basis as more readings become available. The data should provide the basis for action in specific areas of the process. Which is the major category of fault? Is this fault a major one in financial terms? How can the fault be eliminated? Can the process relating to that fault be improved? Do the control charts for the various faults show any common pattern?

This u chart has prompted more questions than it has answered. That is not a criticism, but it does illustrate the need to keep a tight control on charts and projects as SPC programmes develop. A common fault is to hurry the programme, get the charts up and working, smother the shop floor with control sheets. It is no wonder that programmes fail. A planned schedule for introducing charts is vital, and even then it is advisable to delay the introduction of further charts at any stage if problems arising from existing ones are not resolved.

12.2.6 **Choosing a standard unit**

The u chart applies to those cases where the sample size varies, as in the previous example where the number of coils examined per batch is not constant from day to day. But u charts can also operate in a different situation, where the length, area or volume involved varies with each reading. For example, the length of wire used when recording insulation breaks may not always be the same, or the area of cloth being checked for flaws may vary each time. This aspect of u charting demands an understanding of the need for a standard unit of length, area or volume. All other lengths, areas or volumes are then calculated relative to this reference value. Fig. 12.6 provides some examples.

In each case one sample is taken for charting purposes. The size of this sample relative to the standard unit is the basis of fixing the sample size n. Thus if the standard unit has been defined as 1 square metre of fabric, then 2.4 square metres of fabric would correspond to n = 2.4. The number of defects in that particular area is then c, and the u value for each case is given by c/n. Therefore if the number of defects recorded in each case is known, the u values corresponding to the n values in Fig. 12.6 are given in the table in Fig. 12.7.

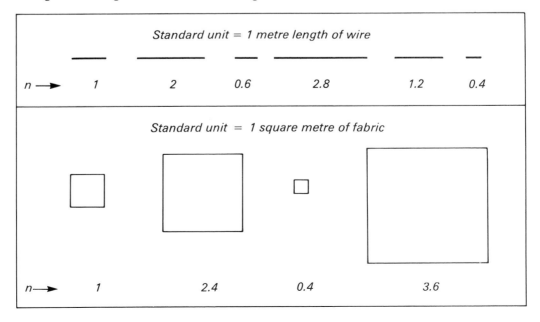

Fig. 12.6 Standard units

DEFECTS IN WIRE						
NUMBERS OF DEFECTS (c)	3	7	2	9	7	3
n	1	2	0·6	2·8	1·2	0·4
u = c/n	3	3·5	3·3	3·2	5·8	7·5

(a)

(b)

FLAWS IN FABRIC						
NUMBERS OF DEFECTS (c)	8	10	3	7	9	6
n	1	2·4	0·4	3·6	1·7	0·8
u = c/n	8	4·2	7·5	1·9	5·3	7·5

Fig. 12.7   Table of u values

Taking flaws in fabric as the example, in practice 20 samples would be taken and the figures in Fig. 12.7(b) would form the first few readings of a typical data set, as shown in Fig. 12.8. The total number of defects is 176 and the total number inspected (which could now involve decimal quantities) is 37.6. Hence

$$\bar{u} = \frac{176}{37.6}$$

$$= 4.68$$

A useful tip when generating u values of the type described in this section is to choose a subgroup size which will provide values of u of the order of 1–3. Suppose, for example, the standard unit is a 0.5 metre length. The data gathered for a series of 20 samples then appears as in Fig. 12.9. Here the level of defects is so low that not enough are detected to allow analysis. Increasing the standard unit to a 2.5 metre length produces results shown in Fig. 12.10. The value of n is now such that the u values are running at a more appropriate level for recording and analysis purposes. Standard unit size will be referred to again when discussing the c chart.

12.3 The c chart

12.3.1 Introduction

If the number of defects in a standard length of wire is being counted, or the number of blemishes in a specified area of painted surface, the number of particles in a specified volume of liquid, then these figures can be plotted directly to give a c chart. In each case the sample size is known and fixed. From clerical errors in forms to dust particles on painted surfaces, from computer failures to surgery visits, the potential applications of the c chart abound in almost any organisation.

12.3.2 Completing the administrative section

The example discussed here relates to invoicing. Several hundred invoices are issued by the company concerned each week. The number of errors in the invoices,

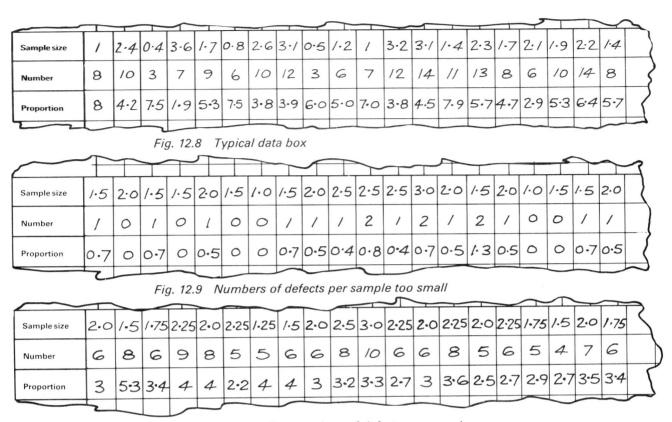

Sample size	1	2·4	0·4	3·6	1·7	0·8	2·6	3·1	0·5	1·2	1	3·2	3·1	1·4	2·3	1·7	2·1	1·9	2·2	1·4
Number	8	10	3	7	9	6	10	12	3	6	7	12	14	11	13	8	6	10	14	8
Proportion	8	4·2	7·5	1·9	5·3	7·5	3·8	3·9	6·0	5·0	7·0	3·8	4·5	7·9	5·7	4·7	2·9	5·3	6·4	5·7

Fig. 12.8   Typical data box

Sample size	1·5	2·0	1·5	1·5	2·0	1·5	1·0	1·5	2·0	2·5	2·5	2·5	3·0	2·0	1·5	2·0	1·0	1·5	1·5	2·0
Number	1	0	1	0	1	0	0	1	1	1	2	1	2	1	2	1	0	0	1	1
Proportion	0·7	0	0·7	0	0·5	0	0	0·7	0·5	0·4	0·8	0·4	0·7	0·5	1·3	0·5	0	0	0·7	0·5

Fig. 12.9   Numbers of defects per sample too small

Sample size	2·0	1·5	1·75	2·25	2·0	2·25	1·25	1·5	2·0	2·5	3·0	2·25	2·0	2·25	2·0	2·25	1·75	1·5	2·0	1·75
Number	6	8	6	9	8	5	5	6	6	8	10	6	6	8	5	6	5	4	7	6
Proportion	3	5·3	3·4	4	4	2·2	4	4	3	3·2	3·3	2·7	3	3·6	2·5	2·7	2·9	2·7	3·5	3·4

Fig. 12.10   Appropriate numbers of defects per sample

irrespective of type, is to be analysed using a fixed sample of 100 invoices a week. Fig. 12.11 shows the relevant administrative data.

12.3.3   Collecting, recording and plotting the data

Previous estimates by the company have already indicated that the weekly sample of 100 invoices is of the right size to provide a level of rejects appropriate for charting purposes. Fig. 12.12 shows the results for the first 10 weeks.

The usual requirement for an initial process study is that at least 20 samples are taken. This does cause difficulties at times. In this case, for example, there was a real need to assess the process performance in as short a time as possible, and there-

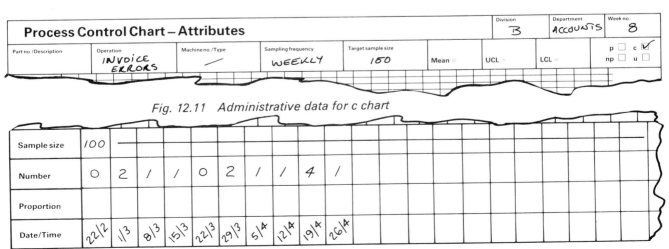

Fig. 12.11   Administrative data for c chart

Sample size	100									
Number	0	2	1	1	0	2	1	1	4	1
Proportion										
Date/Time	22/2	1/3	8/3	15/3	22/3	29/3	5/4	12/4	19/4	26/4

Fig. 12.12   Data box for 10 readings only

fore the first 10 weeks' results were used to provide interim control limits. These could then be updated over time as the number of results increased.

The first 10 values of c were plotted on the chart with the result shown in Fig. 12.13.

12.3.4 **Obtaining the control limits**

As usual the first step in obtaining the control limits is to calculate the mean value. The total number of errors is 13 and the number of samples is 10. Therefore

$$\bar{c} = \frac{13}{10}$$

$$= 1.3$$

For the c chart, the upper control limit is obtained from

$$UCL_c = \bar{c} + 3 \sqrt{\bar{c}}$$

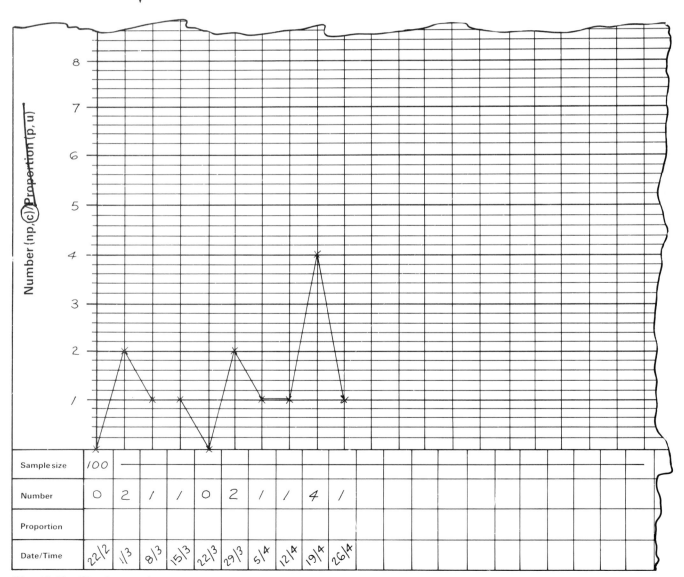

Sample size	100									
Number	0	2	1	1	0	2	1	1	4	1
Proportion										
Date/Time	22/2	1/3	8/3	15/3	22/3	29/3	5/4	12/4	19/4	26/4

Fig. 12.13   Plot for c values

$$= 1.3 + 3 \sqrt{1.3}$$

$$= 1.3 + 3.42$$

$$= 4.72$$

The lower limit is given by

$$LCL_c = 1.3 - 3.42$$

which is treated as 0

The mean, UCL and LCL boxes can now be completed. Lines are drawn on the chart at the corresponding levels and an initial analysis carried out. (Fig. 12.14).

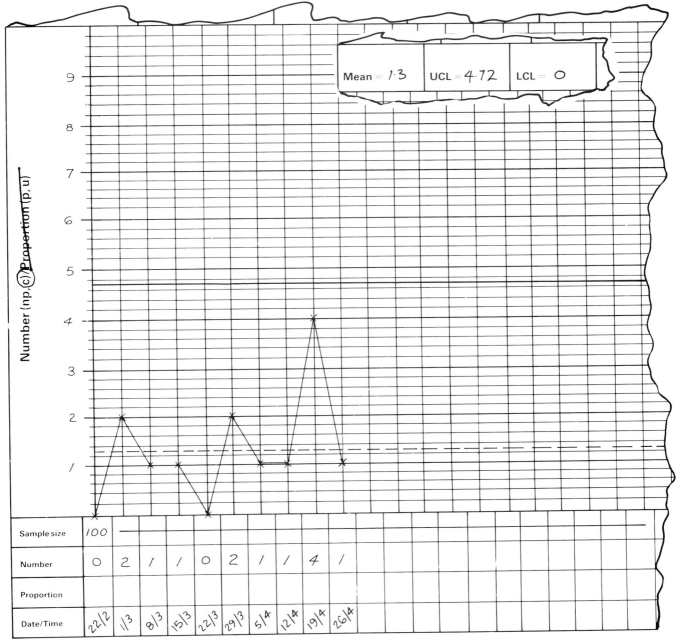

Sample size	100																	
Number	0	2	1	1	0	2	1	1	4	1								
Proportion																		
Date/Time	22/2	1/3	8/3	15/3	22/3	29/3	5/4	12/4	19/4	26/4								

Fig. 12.14 Completed c chart

12.3.5 Interpreting the chart

Rules for identifying special causes from this chart would need to be interpreted with caution. With only 10 points available, it would be unwise to jump to conclusions. However, even with 10 points a value of, for example, 9 would be sufficiently out of line with the bulk of the readings to warrant suspicion. The points on the chart at least indicate the pattern of results to be expected.

Further readings continued to be taken over the next 10 weeks under the same conditions. The full 20 readings were then used to provide a new mean and new control limits as shown in Fig. 12.15.

The chart shows the effect of basing results on 10 rather than 20 readings. Even now there would appear to be slight evidence of a drift upwards, so that the mean based on 20 readings has stabilised at a higher level than before. In addition there is possibly a cyclical effect present. If no special causes can be found then it must be

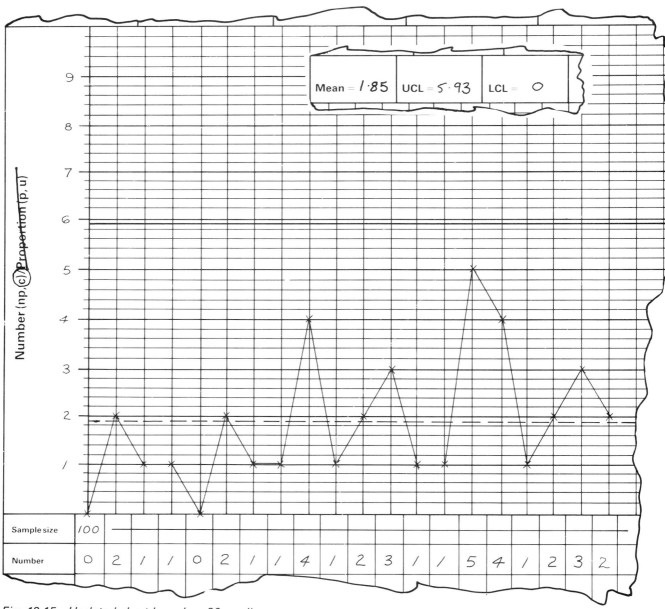

Fig. 12.15   Updated chart based on 20 readings

assumed that these 20 points represent the performance. The objective is then to work on the invoicing system to bring the level of defects down to zero over as short a time scale as possible.

12.3.6 Choosing a standard unit

In some instances it may be preferable to relate the mean value to a corresponding expected mean value based on a different target sample size.

For example, the target sample size in this case was 100. This figure was convenient for a variety of reasons: it provided an adequate picture of the process and it could be sustained by the resources available. In another division of the company (A) the sample size chosen was 250. For the same week, the level of errors in this division was running at 3.2. How can the two performances be compared? Comparison is achieved by using a standard unit, typically defects per hundred units (DHU). Hence with 3.2 defects on average for 250 units, the DHU for division A will be 100/250 × 3.2, i.e. 1.28. This compares with DHU figure of 1.85 for division B

Case study 10: PPG Industries (UK) Ltd

PPG Industries is a major supplier to leading organisations involved with resins and coatings. Its headquarters are in Pittsburgh, USA, and it has plants in many countries. The UK plant in Birmingham has contracts with a variety of customers, which include Ford, Nissan, Rover Group, Vauxhall, Reliant, Renault Trucks and many component companies.

The demands laid down by the Ford Motor Co. in particular have resulted in a positive response by PPG. The company has recognised the need for SPC programmes and committed itself to the philosophy of continuous improvement. In addition to the Ford approach, there is a drive from the US corporate group to instigate quality improvement programmes based on the teachings of Crosby.

Training in SPC around a management philosophy based on Deming's 14 points (as proposed by Ford) was undertaken on a top-down basis. The senior executive and management courses were followed by 1-day workshops where various personal projects were presented and discussed. It was evident that there was a major commitment to the SPC programme on the part of the senior executives. This was shown in a practical way by a positive response to supporting in financial and material terms the projects by the personnel who had attended the earlier training courses. Various projects were undertaken, which involved different applications of the SPC techniques. In most cases a moving mean/moving range chart was used because of the routine of the industry; however, attribute charting was a major feature in some projects.

This case study relates to the investigation of 'dirt count' on car electropaint primer at one of the major car manufacturing plants in the UK. Each day, at approximately the same time, six car bodies are monitored in sequence on the production line. This takes place after seam sealing but before any preparation. The body has therefore been through the electrodeposition tank but has not been sanded. The number of inclusions (or defects) present in the electropaint paint film is then counted.

The method uses a specially designed 1 square foot frame that can be placed in exactly the same position on each car depending on the panel being examined. Particular areas examined are the hood, roof, and left-hand and right-hand front doors. Counting is both visual and by touch. The touch method requires running a fine-gloved finger slowly over the area from side to side; all inclusions that can be felt are counted. The definition of a defect is therefore severe and the results which follow show that the paintwork is very clean with a low mean defect value. This mean value is an important measure because it gives an indication of the cleanliness of the electropaint tank, the oven and the surrounding environment where the paint film is wet. Perhaps more importantly, the mean level of inclusions gives a guide to the amount of sanding preparation required before the next stage of the painting operation. The car manufacturer concerned is understandably interested in this. The whole of the paint process – time, manpower and resources – is geared to minimum preparation of bodies between coats of paint and therefore a reduction in the dirt count has considerable implications.

The c chart in Fig. 12.16 is based on 20 samples from the hood, each sample reflecting the total dirt count in six cars. A sample size of 6 was chosen to give a satisfactory level of defects for recording purposes.

The process is seen to be under statistical control, and progress was being made in looking for special causes which would result in a reduced level of inclusions. As the process is improved the lower mean level of defects will indicate that sample size will need to be increased.

Similar charts for the roof and front doors enable further analysis to be carried out. This could result in detecting a change in the pattern of defects dependent on the position of the frame on the car, or a relationship between the patterns. Attention to detail regarding what may appear to be a minor feature will justify itself many times over as the applications of SPC begin to show results.

This first stage of process analysis will ultimately lead to concentrating resources on reducing and then eliminating the level of dirt count at source.

12.4 **Multiple characteristics chart**

12.4.1. **Introduction**

Several references have been made to the multiple characteristics chart. This chart has many advantages, not least the fact that it can provide a means of assessing priorities in charting. In addition, its application is limitless: it can be used to monitor faults in the paint shop just as easily as the reasons for visits to the works surgery; telephone enquiries can be categorised by this method in the same way as can errors in paperwork. It is an extremely effective technique, considerably under-used and with much potential for improving product and service quality in a wide range of areas.

12.4.2 **Design of the chart**

The chart differs somewhat from the typical control chart considered to date. Fig. 12.17 illustrates the basic format of the multiple characteristics chart and indicates how the ideas expressed in Fig. 10.2 are formally represented in a chart. It can be seen that the chart has five main sections.

- An administrative section.
- A main block for recording the frequency of occurrence of a particular feature.
- A recording/calculation block.
- A graph section where a control chart can be plotted. Various options are available: p, np, c or u.
- An analysis section which gives guidance on priorities for action.

12.4.3 **Completing the administrative section**

In the example to be discussed the data to be recorded and analysed relates to a packing audit. The information is the same as that already expressed on the simple check sheet shown in Fig. 10.1. The audit is being carried out daily and the appropriate administrative section of the multiple characteristics chart appears in Fig. 12.18.

12.4.4 **Collecting, recording and plotting the data**

The relevant fault characterists are shown in Fig. 12.19. In order to proceed with the analysis, some further information is required. The packing audit is to be based on a

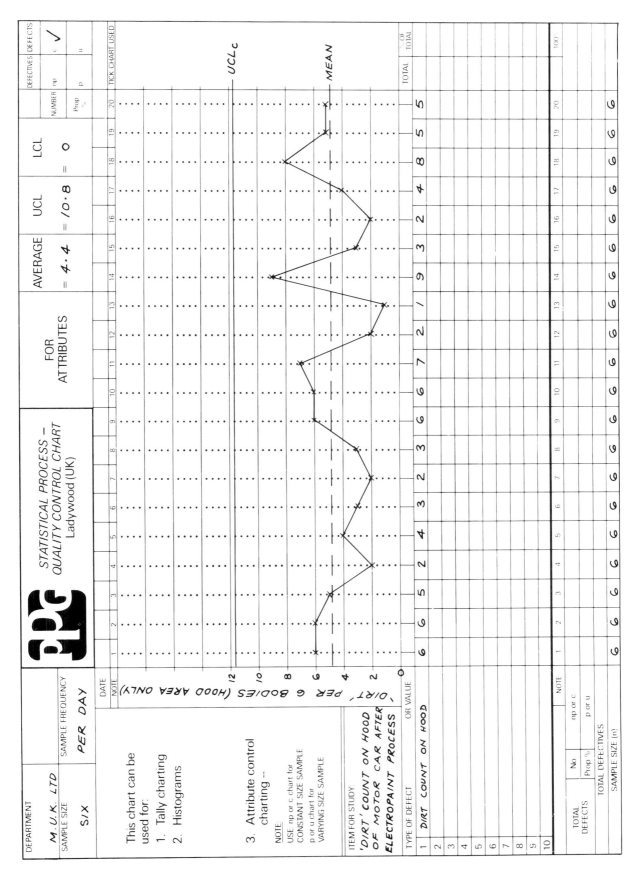

Fig. 12.16   c chart for dirt count

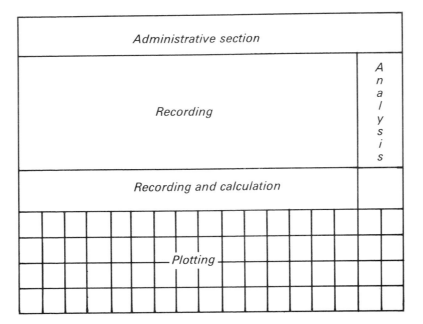

Fig. 12.17   Multiple characteristics chart: schematic diagram

Process Control Chart – Multiple Characteristics						Division	Department	Week no.
						E	*PACKING/ DESPATCH*	*1 – 4*

Part no./Description	Operation	Machine no./Type	Sampling frequency	Target sample size		Mean	UCL 0·047	LCL	O	p ☐ c ☐
—	*PACKING AUDIT*	*—*	*DAILY*	*100 %*						np ☐ u ☐
No.	Characteristic			Defective/Defect frequency						£ %

Fig. 12.18   *Administrative data for multiple characteristics chart*

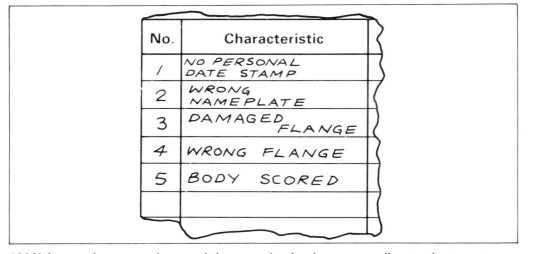

Fig. 12.19   Fault categories

100% inspection procedure and the sample size (corresponding to the target production rate per day) is 200. This target sample size of 200 is not achieved in practice in this case, and the actual sample sizes relating to the 20 samples required for the initial process study are as shown in the data boxes in Fig. 12.20. The number of defects corresponding to a particular fault category is also shown.

Corresponding totals can now be obtained and the proportions of unacceptable items determined and recorded in the appropriate boxes as indicated in Fig. 12.21.

Various possibilities seem to present themselves when it comes to plotting the results on the grid section of the chart. In fact only one option is available. It has already been suggested that the sample size provides the key to choosing the right chart to plot. The sample size in this case is varying, and therefore the chart has to be either a u chart or a p chart. The faults analysis indicates that it is different types

No.	Characteristic	Defective/Defect frequency																			
1	NO PERSONAL DATE STAMP				/																
2	WRONG NAMEPLATE	//	//	///	/	//	//		///		////		/	//		///	///	/	//	/	///
3	DAMAGED FLANGE					/							/				/				
4	WRONG FLANGE		/		//		///		/		/		//		/		/		//	/	
5	BODY SCORED										/										
	Total Defective/Defect	2	3	3	4	3	5	0	4	0	6	0	4	2	1	3	5	1	4	2	3
	Sample size	230	178	183	215	205	203	172	176	195	210	221	203	172	198	190	203	220	215	193	200
	Proportion																				

Fig. 12.20   Record of fault frequencies

Process Control Chart – Multiple Characteristics

Part no./Description	Operation	Machine no./Type	Sampling frequency	Target sample size	
	PACKING AUDIT	—	DAILY	100 %	Mea

No.	Characteristic	Defective/Defect frequency																			
1	NO PERSONAL DATE STAMP				/																
2	WRONG NAMEPLATE	//	//	///	/	//	//		///		////		/	//		///	///	/	//	/	///
3	DAMAGED FLANGE					/							/				/				
4	WRONG FLANGE		/		//		///		/		/		//		/		/		//	/	
5	BODY SCORED										/										
	Total Defective/Defect	2	3	3	4	3	5	0	4	0	6	0	4	2	1	3	5	1	4	2	3
	Sample size	230	178	183	215	205	203	172	176	195	210	221	203	172	198	190	203	220	215	193	200
	Proportion	0.009	0.017	0.016	0.019	0.015	0.025	0	0.023	0	0.029	0	0.020	0.012	0.005	0.016	0.025	0.005	0.019	0.010	0.015

Fig. 12.21   Completed data

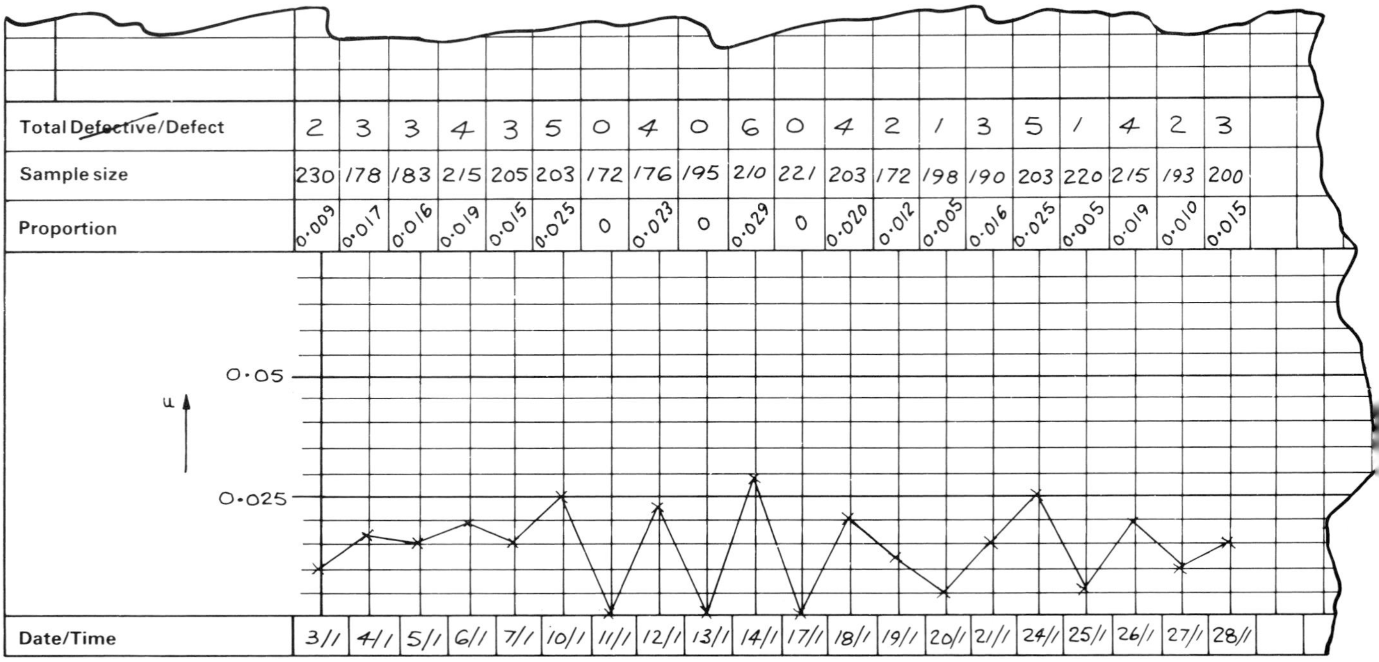

Total Defective/Defect	2	3	3	4	3	5	0	4	0	6	0	4	2	1	3	5	1	4	2	3
Sample size	230	178	183	215	205	203	172	176	195	210	221	203	172	198	190	203	220	215	193	200
Proportion	0·009	0·017	0·016	0·019	0·015	0·025	0	0·023	0	0·029	0	0·020	0·012	0·005	0·016	0·025	0·005	0·019	0·010	0·015

| Date/Time | 3/1 | 4/1 | 5/1 | 6/1 | 7/1 | 10/1 | 11/1 | 12/1 | 13/1 | 14/1 | 17/1 | 18/1 | 19/1 | 20/1 | 21/1 | 24/1 | 25/1 | 26/1 | 27/1 | 28/1 |

Fig. 12.22   Plot for u values

of faults within a unit that are being considered, rather than how many units failed. Hence the appropriate chart to use is a u chart.

Following the usual guidelines for choosing the scale, the proportions are plotted as shown in Fig. 12.22.

12.4.5 Obtaining the control limits

The procedure for obtaining control limits is the same as that used for previous charts. First $\bar{u}$ is evaluated using the formula.

$$\bar{u} = \frac{\text{Total number of defects}}{\text{Total number inspected}}$$

Therefore

$$\bar{u} = \frac{55}{3982}$$

$$= 0.014$$

The mean sample size $\bar{n}$ is the total number inspected/number of samples, i.e.

$$\bar{n} = \frac{3982}{20}$$

$$= 199$$

Thus for the upper control limit

$$\text{UCL}_u = \bar{u} + 3\sqrt{\frac{\bar{u}}{\bar{n}}}$$

$$= 0.014 + 3\sqrt{\frac{0.014}{199}}$$

$$= 0.014 + 0.025$$

$$= 0.039$$

For the lower control limit

$$LCL_u = 0$$

The mean, UCL and LCL boxes can now be completed and the corresponding lines drawn on the u chart(Fig.12.23).The sample size remains within the $\bar{n} \pm 0.25\bar{n}$ limits of 249 and 149.

12.4.6 Interpreting the chart

The chart indicates that the level of faults is under control.

12.4.7 Improving the process

Although the process is under control the level of rejects is unacceptably high and improvement is called for. A guide to priorities is provided by looking at the number of occurrences of a particular fault. In this case the totals are based on the results of the first 20 readings only. In practice they could equally be obtained using the total number of samples available on the chart (30 in the case of the version shown in Appendix H). The right-hand column in Fig. 12.24 shows that 'wrong nameplate' contributes 64% of the faults and therefore should be the major feature to work on

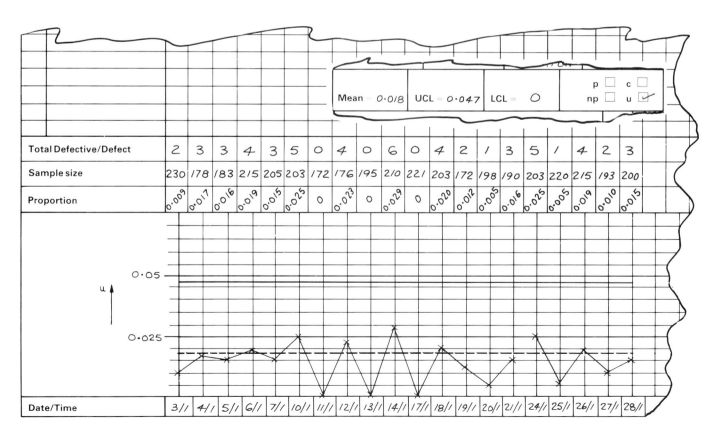

Fig. 12.23 Completed u chart

as part of a quality improvement programme. In the same way as the total number of defects have been plotted in Fig. 12.22, then a control chart, and corresponding limits, can be set up for plotting the number of 'wrong nameplates' against time.

However, as was made clear in the earlier discussion on Pareto analysis, taking into account only the numbers of occurrences of a fault is not necessarily the best approach. A further data collection exercise is required to ascertain the costs associated with the different fault categories. This may indicate that 'wrong flange', for example, could be the major cost carrier and the feature to be concentrated on initially.

12.4.8 **Varying the characteristics**

Of all the charts available for use in SPC, the multiple characteristics chart is one of the most powerful in its ability to assign priorities and allow for various options when plotting. The current example has looked at data from a packing audit, but the range of applications is almost endless. Fig. 12.25 shows a selection of multiple characteristics charts. Not only are there many options regarding the characteristic to plot. The nature of the chart also means that depending on the characteristic, any one of the four alternatives of p, np, c or u can be used.

Case study 11: British Alcan (Rolled Products)

British Alcan (Rolled Products) has applied SPC charts in a variety of areas. Extensive use has been made of the individual/moving mean and moving range chart as it is far more applicable to the industry than the conventional ($\bar{X}$, R) chart. British Alcan has also recognised the value of the attribute chart in non-manufacturing areas, and this case study indicates how the multiple characteristic chart has been used in analysing surgery visits.

Fig. 12.26 shows a completed multiple characteristics chart based on results spanning a period of 20 months. There is an out-of-control signal on the first point and the reason for this was investigated. Failure to determine the reason meant that in the short term this point had to be considered part of the system and the control limits, as calculated, projected ahead to monitor, and then improve, the system.

The associated Pareto analysis showed that manual handling was the major feature to be concentrated on and a subsequent c chart was generated for monitoring this item. Fig. 12.27 shows the chart. This illustrates that towards the latter period of the analysis there was indication of an improvement in the process; further readings would then either substantiate this or not.

Process Control Chart—Multip

sion: E | Department: PACKING/ DESPATCH | Week no.: 1—4

Part no./Description | Operation: PACKING AUDIT | Machine no./Type

= 0·047 | LCL = 0 | p ☐ c ☐ np ☐ u ☑

No.	Characteristic							f	%
1	NO PERSONAL DATE STAMP				/			1	2
2	WRONG NAMEPLATE	//	//	///	/	//	//	35	64
3	DAMAGED FLANGE					/		3	5
4	WRONG FLANGE		/		//		///	15	27
5	BODY SCORED							1	2

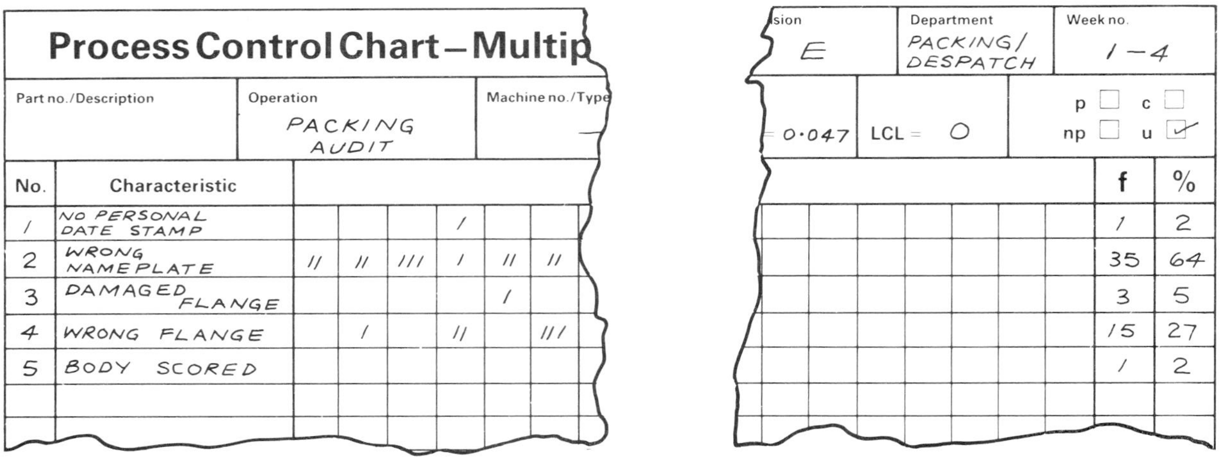

Fig. 12.24 Pareto analysis

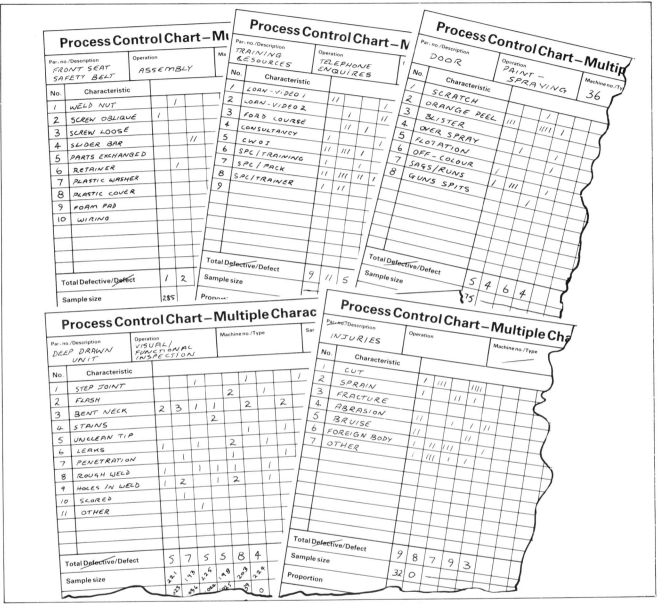

Fig. 12.25   Examples of multiple characteristic charts

This analysis of surgery visits meant that other related charts were required and the company also analysed the associated factors such as types of injuries, time and place of injury and particularly cost. Some of the reasons for visits, such as machine in motion, may not be major in incidence but could have considerable implications in terms of the extent of injury, financial compensation, sickness benefits, industrial injuries liability, and so forth.

The multiple characteristics chart has enabled Alcan to monitor non-manufacturing processes in a much more effective way. More than that, it has provided a common forum for analysing data and making decisions.

Summary

- Reducing the number of nonconformities will bring about a reduction in the number of non-conforming units.
- u charts are used for monitoring non-conformities when the sample size varies.
- When using u charts check on the limits of $\bar{n}$.

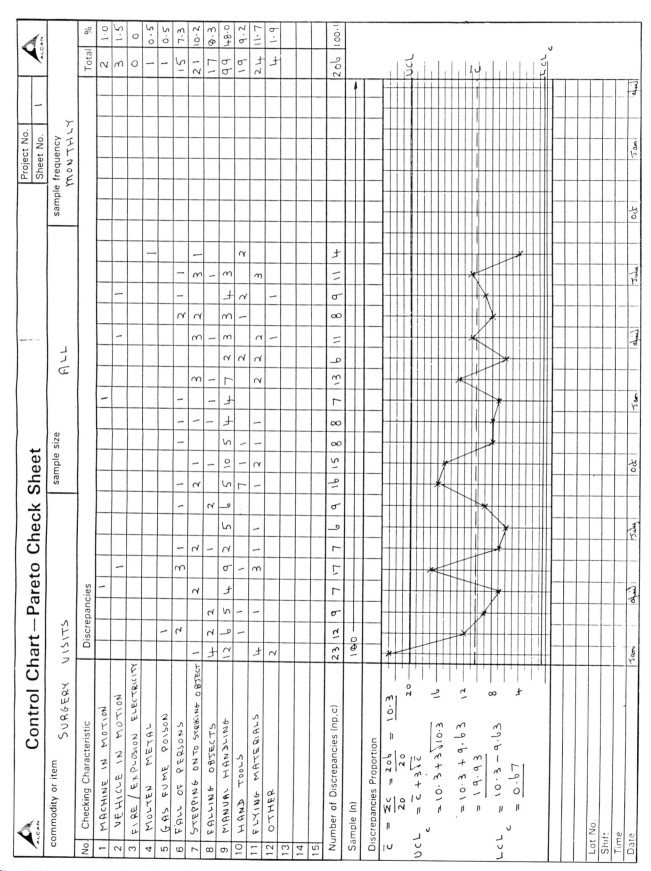

Fig. 12.26 Analysis of surgery visits

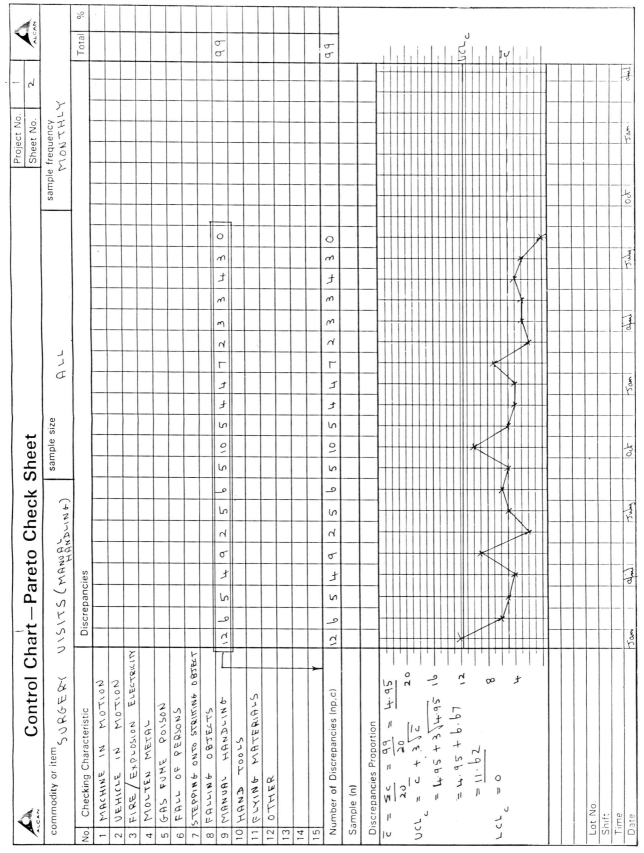

Fig. 12.27   Analysis for manual handling

- Choose a standard unit large enough to provide an appropriate level of readings for recording and analysis purposes.
- c charts are used for monitoring nonconformities when the sample size is constant.
- The administrative sections of the chart are important and must be completed.
- In some cases, trial limits can be used based on less than 20 readings.
- Use of a standard unit provides a basis for direct comparison of different-sized samples.
- Multiple characteristics charts enable many features to be monitored at the same time.
- Priorities for action can be quickly determined from multiple characteristics charts.
- The multiple characteristics chart allows various options for plotting.
- Multiple characteristics charts have a wide range of applications.

This chapter completes the section of the book allocated to attribute charts. It also completes the study of the traditional chart: the Shewhart chart. In Chapter 2 the distinction was made between variables and attributes; subsequent chapters then discussed one set of charts for analysing variables and another for attributes. The following chapter considers a technique which can be applied equally to variables or attributes: the Cusum technique.

Chapter 13   The Cusum technique

13.1 **Introduction**

In the analysis of both variables and attributes as described in the previous chapters, the readings were plotted in time sequence and out-of-control situations determined by using the four rules for special causes. Two of these rules (the rule of seven and the middle third rule) did take into account the position of points previous to the one being plotted at a particular stage. However, no allowance was made for the numerical values of those previous points. The moving mean, moving range chart was an improvement in this respect, but even so it was not always suitable.

Other limitations of conventional Shewhart charts were evident when the time taken to detect changes in process level was considered. Particularly when changes in level of performance are small, the Shewhart chart is seen to be at some disadvantage. The Cusum technique provides an answer. It has the advantage of enabling decisions to be made much earlier than with the Shewhart chart. However, it also has one major limitation which is that it can become rather technical, and it is no doubt because of this that the benefits of the technique have not always been fully exploited. The statistical and mathematical aspects need not be off-putting, though. There are many examples of the technique being used effectively in a range of industries and in quite a simple, yet powerful, manner. This chapter may assist those looking for a tool to deal with problems somewhat different from those discussed so far.

13.2 **What is the Cusum technique?**

The origin of the word Cusum can best be seen by considering a simple example. Take a sequence of numbers as follows

12, 9, 12, 11, 7, 8

that relate to a target of 10. It is immaterial at this stage what the figures actually refer to. They could be sales figures, rejects, decimal units of weight – almost any application is permissible. Some specific applications will be developed later in the chapter, but for now the letter X will be used to denote the feature.

A table can then be generated as shown in Fig. 13.1. The second column is obtained by subtracting the target 10, often denoted in a general sense by T, from each value of X in turn. Finally these differences are accumulated, one by one, giving the cumulative sum. 'Cumulative sum' is a cumbersome expression and so it is shortened to 'Cusum'. The next thing to consider is how a Cusum table can be used.

13.3   **From Shewhart to Cusum**

13.3.1   **Introduction**

The data in the attribute chart in Fig. 13.2 relates to the same situation as discussed in Chapter 11, i.e. units in samples of 250 which fail a functional test. This will provide a suitable example but it must be stressed that the principle holds for other situations.

Fig. 11.18 showed the situation based on the first 20 samples (reduced to 18 samples after the special causes have been removed). These gave a mean of 2.22 and a UCL of 6.67. In Fig. 13.2 further points have been obtained and monitored against the central line and the UCL. The process continues to be under statistical control. The chart is not a very good indicator of process change, however. It is not possible to determine whether the mean has changed from a level of 2.22 for the points plotted after the completion of the initial process study. The ability to detect such a change by simple visual means would be of immense value, and the Cusum technique provides the answer.

	Target, T = 10	
X	X − 10	CUmulative SUM
12	2	2
9	− 1	1
12	2	3
11	1	4
7	− 3	1
8	− 2	− 1

Fig. 13.1   Cusum table

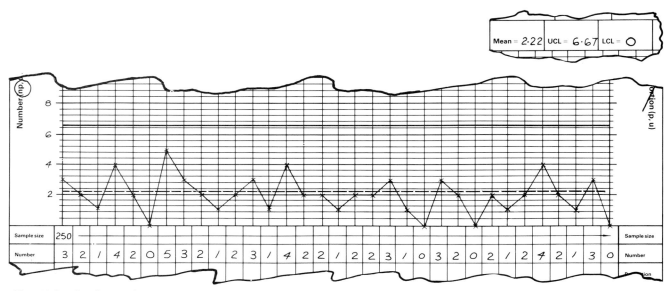

Fig. 13.2   Attribute chart

13.3.2 **The slope protractor**

To see how the technique operates the data in Fig. 13.2 first need to be expressed in a slightly different form.

Fig. 13.3 shows part of a table of cumulative defective units. The number of these in each sample is added cumulatively. If these cumulative values are plotted against the corresponding sample number, then a graph is produced as shown in Fig. 13.4. Now the slope of this graph over a given section corresponds to the mean level of performance. For example, a line corresponding to the first 18 readings (Fig. 13.5) will have a slope of 40/18 which is 2.22, the value already known. It follows that the level of performance over other time intervals can be determined. For example, the slope over the section from sample 18 to sample 33 is given by 26/15 = 1.73 (Fig. 13.6).

Fig. 13.3 Table of cumulative defects

Sample no.	No. of defective units (X)	Cumulative no. of defective units (ΣX)
1	3	3
2	2	5
3	1	6
4	4	10
15	2	35
16	2	37
17	1	38
18	2	40

Sample no.	No. of defective units (X)	Cumulative no. of defective units (ΣX)
19	2	42
20	3	45
21	1	46
22	0	46
30	2	62
31	1	63
32	3	66
33	0	66

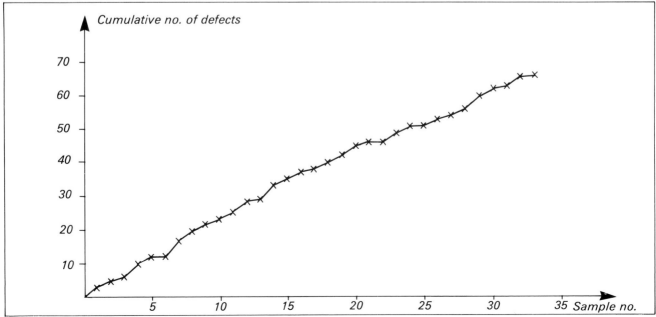

Fig. 13.4 Graph of cumulative defects

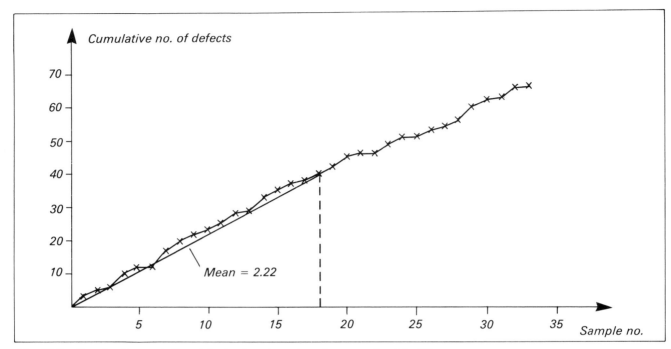

Fig. 13.5   Graph of cumulative defects showing slope for samples 1 to 18

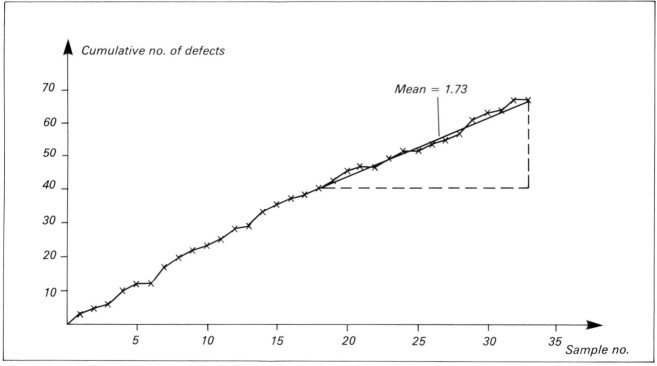

Fig. 13.6   Graph of cumulative defects showing slope for samples 18 to 33

This relationship between the slope of the line and the mean level of performance is of assistance in determining the point which corresponds to a change in the process mean. For example, if a change in slope in Fig. 13.7 can be detected from sample 18 onwards, then this will indicate that the process mean has changed in the region of that point.

Such a procedure is not visually easy. The problem is highlighted by using a sheet of paper to cover over all the points in Fig. 13.7 to the right of sample 18. As the

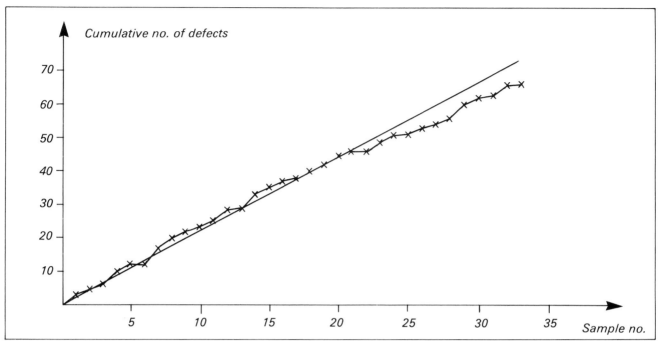

Fig. 13.7 Detecting the process change

paper is moved to the right, gradually revealing more and more points, where is the situation reached where there are enough points displayed to be confident that the slope has indeed changed? Such a visual approach suggests that sample 25, at least, needs to be reached before there is enough evidence to indicate a real change in the process.

A difficulty has arisen because the task is to detect a change in slope in a line that is already inclined. This difficulty could be minimised if the reference line were horizontal. Fig. 13.8 shows how much easier it is to detect a change in slope in this situation. The angle between the lines is constant in both cases, but the change in slope is more apparent when referenced against a horizontal line.

A Cusum graph provides the means of transforming an inclined reference line to a horizontal one. In Fig. 13.3 a table for X was generated. Now a table for $(X - T)$ is generated, where T is the target value. It is known that the mean value over the first 18 points is 2.22. It is sufficiently accurate in this example to round up to 1 decimal place and fix 2.2 as the target. Fig. 13.9 shows part of the Cusum table which results. The Cusum graph follows in Fig. 13.10, the Cusum value being plotted vertically against the corresponding sample number horizontally.

Two features are immediately apparent:

• The graph shows two distinct sections. For the first half of the graph the line

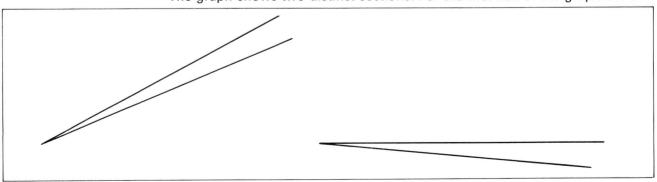

Fig. 13.8 Improving the visual interpretation

Sample no.	X	X − 2.2	Σ (X − 2.2)
1	3	0.8	0.8
2	2	−0.2	0.6
3	1	−1.2	−0.6
4	4	1.8	1.2
15	2	−0.2	2.0
16	2	−0.2	1.8
17	1	−1.2	0.6
18	2	−0.2	0.4

Sample no.	X	X − 2.2	Σ (X − 2.2)
19	2	−0.2	0.2
20	3	0.8	1.0
21	1	−1.2	−0.2
22	0	−2.2	−2.4
30	2	−0.2	−4.0
31	1	−1.2	−5.2
32	3	0.8	−4.4
33	0	−2.2	−6.6

Fig. 13.9 Cusum table

oscillates about a horizontal level. For the remainder of the graph the mean line is inclined downwards. The picture is clarified by drawing in two lines corresponding to the mean levels of performance (from samples 1 to 18 and 18 to 33 respectively), as shown in Fig. 13.11. The change in level that takes place around sample 20 is much more emphatic here than it was in Fig. 13.4.

- Overall trends, rather than the random fluctuation, are sought. It is therefore necessary to choose a scale for plotting Cusum values which will be sensitive enough to indicate genuine changes in the process yet not so sensitive that it responds to the natural random changes. Guidelines on the choice of scale are given in Section 13.4.3.

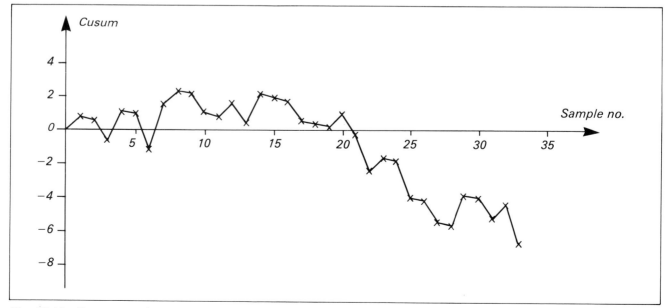

Fig. 13.10   Cusum graph

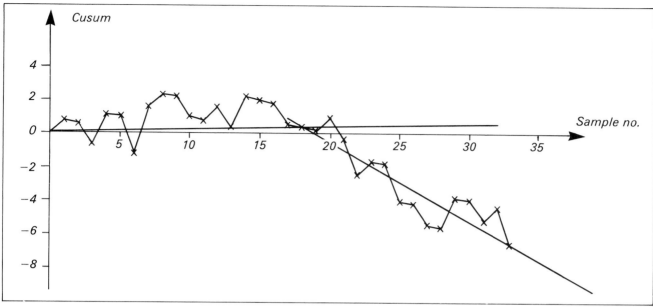

Fig. 13.11 Cusum graph showing slopes for samples 1 to 18 and 19 to 33

The Cusum technique, therefore, provides a method of detecting changes in process level from the target, or nominal value. A method of actually determining the process level at any point using the Cusum graph is still needed.

13.3.3 Measuring the slope

The basis of the Cusum chart is that when the process is running on target, then the Cusum line is horizontal. Changes from the target are then measured by measuring the slope of the graph, whether positive or negative. This numerical value is added to, or subtracted from, the target to give the value of the actual process level.

An example will be given using the figures shown in Fig. 13.10, though initially only the scale and axes will be considered, as represented again in Fig. 13.12(a).

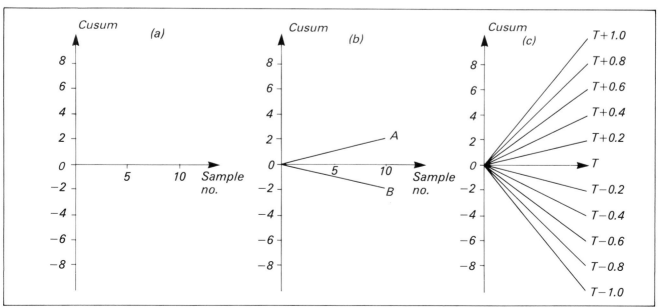

Fig. 13.12 Development of slope protractor

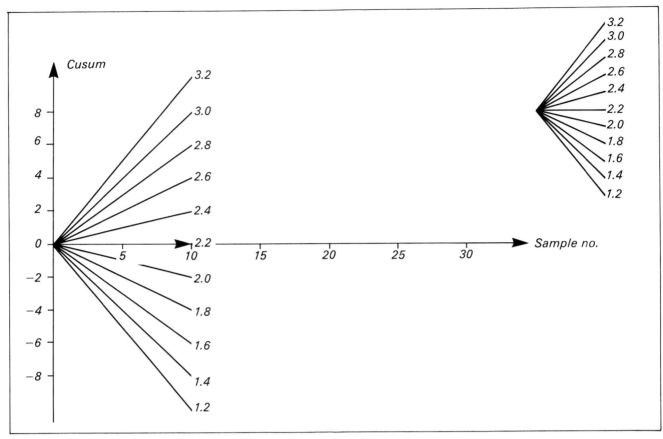

Fig. 13.13   Location of protractor

Lines are drawn fanning outwards from a reference point on the sample number axis. For the present, it does not matter where this point is and therefore the origin, 0, will be used. The first pairs of lines are shown in Fig. 13.12(b). A constant horizontal distance of 10 sample intervals has been used whilst the vertical distance is 2 units of Cusum. The line 0A has a slope of 2/10, i.e. 0.2, and the line 0B a slope of −2/10, i.e. −0.2. Other lines can be drawn in proportion and are shown in Fig. 13.11(c).

Hence, moving upwards, the respective slopes are 0.2, 0.4, 0.6, etc.; moving downwards the slopes are −0.2, −0.4, −0.6, etc. But these are all changes from the nominal, T. Hence the actual process levels are T + 0.2, T + 0.4, T + 0.6, etc., upwards and T − 0.2, T − 0.4, T − 0.6, etc., downwards.

Since a Cusum graph based on a target of 2.2 has been generated, then in order to apply these results to the graph in Fig. 13.11, T is replaced by 2.2. This slope guide, or protractor, is generally reduced in size. As in Fig. 13.13 it would not be located on an axis but placed in an area of the graph which does not interfere with the actual plotting of the results. The current level of performance can then be readily determined from the Cusum graph by comparing the slope at any point with the corresponding slope on the protractor. Fig. 13.14 shows the protractor on the Cusum plot for the level of non-conforming units.

At the start of this analysis reject levels were plotted in the usual way on an attribute chart, as shown in Fig. 13.2. It is impossible to distinguish any changes in process level. One cannot decide whether the process is running at a level of 1.4, 2.0 or 2.6. Equally it is not possible, with any degree of confidence, to indicate where, if anywhere, the process mean has changed to a new level. It has now been demonstrated that use of the Cusum method makes it much more likely that decisions such as these can be made. By drawing lines on Fig. 13.14 (as in Fig. 13.11) it is seen that

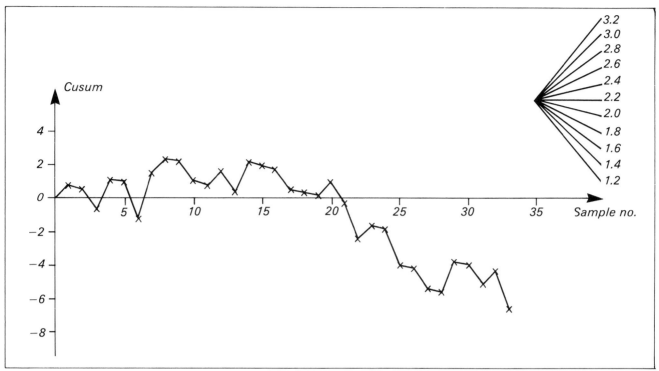

Fig. 13.14   Cusum graph plus protractor

for the first 18 points the process level is 2.25 approx. and for the remaining points the process mean is 1.7 approx.

In practice it is not always easy to trace the reason for the change. For example, there may be a time delay between the presence of a special cause and its influence on the process. Problem-solving groups are necessary whatever the type of chart being used.

This introductory example has highlighted the ability in most situations of the Cusum method to detect changes more efficiently than the Shewhart method. There are still some situations, though, where the Shewhart method is preferable (see Section 13.6.5).

13.4   Interpreting the Cusum chart

13.4.1   The vertical scale

The vertical scale records the Cusum value. It is used only to assist in generating the protractor so that the slopes can then be determined. Otherwise the Cusum axis is not used, i.e. the level of performance is *not* indicated for any sample point by reading off the value on the Cusum scale. With the Cusum graph it is the *slope* that is an indicator of process level (Fig. 13.15).

13.4.2   Choice of target

To a certain extent the choice of target is immaterial: the same sort of pattern will appear irrespective of the target value. For example, Fig. 13.16 is based on the same data as was used to produce the Cusum graph in Fig. 13.10. This time, however, instead of using a target of 2.2 a target of 1.4 was chosen. The effect of changing the target is twofold:

● The process will be satisfactory when the slope is horizontal. Thus the protractor values will be set at intervals of 0.2 about 1.4, the new target. Note that since

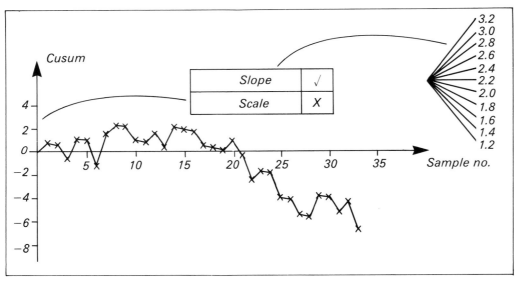

Fig. 13.15   Slope not scale indicates process level

exactly the same scales are used for the Cusum graphs in Fig. 13.10 and Fig. 13.16, then the protractor scale intervals are equal.

- Changing the target to 1.4 means that the Cusum graph in Fig. 13.16 has steeper lines than those in Fig. 13.10. However, over the two intervals 0–18 and 18–33, the slopes are numerically identical. By relating the respective slopes to the protractor in Fig.13.16 it is seen that the process levels are again 2.25 and 1.7.

Even though a change in the target does not affect the result, it is always advisable to choose a target which will correspond to the ideal, e.g. the nominal aimed for in

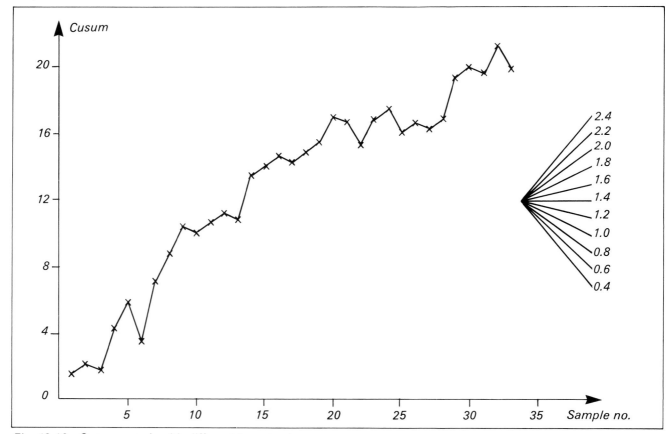

Fig. 13.16   Cusum graph with different target values

analysis of variables. This will then result in a Cusum graph where the slope is horizontal when the process is running satisfactorily.

13.4.3 Choice of Cusum scale

The scale chosen for the Cusum axis in Fig. 13.10 allowed genuine changes to be detected but did not exaggerate the random changes within the process. Fig. 13.17(a) and (b) shows the same graph but with different scales. In Fig. 13.17(a) the scale is too small. In smoothing out the natural irregularities there is a danger of not being able to detect a genuine change in the process. Fig. 13.17(b) shows that too large a scale highlights the natural random variation to an extent that could signal false alarms and suggest that action should be taken on causes which are common rather than special in nature.

Clearly a balance is needed in choosing the scale, and guidelines are available. For example, if the readings are individual ones from a normal distribution with a standard deviation of s, then a suitable choice of scale is as shown in Fig. 13.18. It is arranged for one interval of Cusum to equal one sampling interval, and for both to

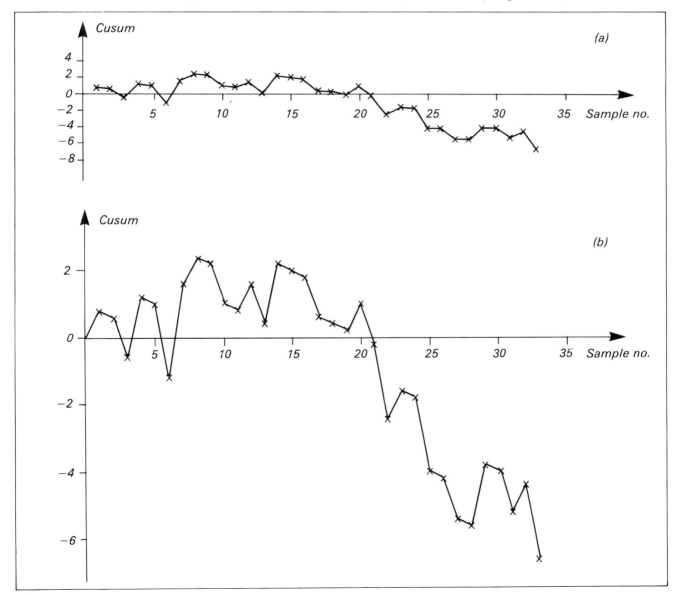

Fig. 13.17 Effect of changing Cusum scale

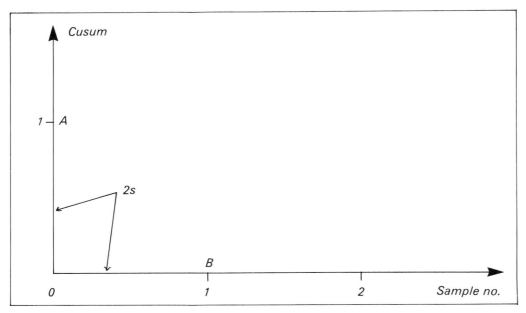

Fig. 13.18 Choice of Cusum scale

be 2s, i.e. 0A = 0B = 2s. This relationship only holds true for individual readings from a normal distribution; the relationship varies for other distributions. For those who require the detail, there is appropriate material in BS 5703 (1), the first part of the British Standards Institute document on the Cusum technique.

When detailed use of the Cusum as a monitoring agent is envisaged, then a correct scale must be used. For the majority of practical Cusum graphs, however, a purely visual choice of scale is adequate. Experience will quickly ensure that a sensible choice is made between the extremes of Fig. 13.17(a) and (b).

13.5 The Cusum technique as a retrospective tool

The ability of the Cusum technique to detect changes in process level means that management has at its disposal a powerful method for process improvement.

Fig. 13.14 has shown that there appeared to be the first indication of a change in slope at about sample 20. Use of the protractor gives a process level of about 1.6 units as opposed to 2.2. What happened at about sample 20 therefore needs to be determined. This means that records and administrative details need to be kept. Fig. 13.19 shows the data, graph and protractor on a standard cusum chart .

The Cusum chart has the same basic features as all SPC charts. There is an administrative section which needs completing, a section for recording data and performing calculations, and a graphical section for recording Cusum values against sample number. The protractor is superimposed on the graphical section in a position which does not interfere with the graph yet makes it possible easily to read off the slope at any point.

Fig. 13.20 provides a direct comparison of the Shewhart approach and the Cusum method for the same data. Whereas it is virtually impossible to distinguish any change in process setting on the conventional attribute chart, the corresponding Cusum graph indicates the change in setting that occurs at about sample 20.

The major emphasis so far in this chapter has been on the need to understand the significance of slope and the fact that a change in slope corresponds to a change in process performance. This is using the Cusum method in a retrospective sense, which has been seen to be very effective. However, the method can also operate to monitor performance in real time and detect changes as soon as they occur.

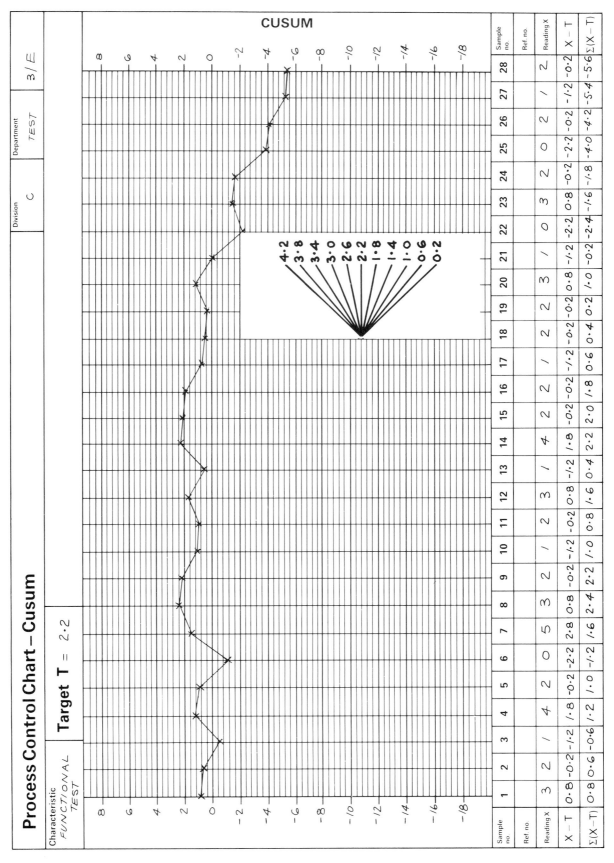

Fig. 13.19 Cusum graph with protractor

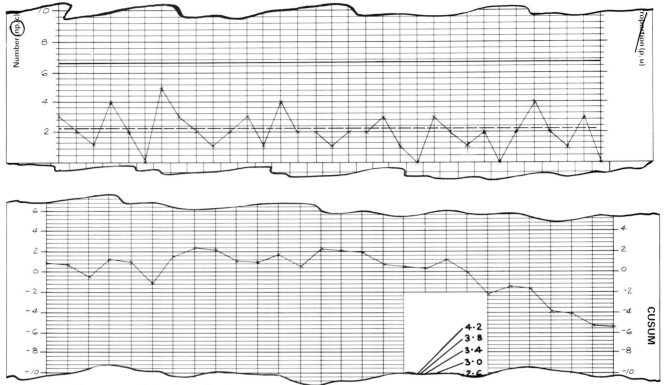

Fig. 13.20   Cusum and Shewhart charts

13.6 **Detecting process changes**

13.6.1 **Introduction**

On a Shewhart chart, changes may be detected by relating the position of a point to the control lines. This approach cannot be used on the Cusum graph because slope is used as an indicator of performance. It does suggest, however, that relating the slope of the Cusum graph to a known standard may be an appropriate basis for decision-making. This is in fact the basis of the V mask – a graphical technique that enables a significant change in slope to be determined.

13.6.2 **The V mask**

The mask, or template, used to monitor performance on a Cusum chart is essentially a truncated V as shown in Fig. 13.21. The mask is placed on the appropriate Cusum chart with its axis horizontal and is located with the notch on the last reading in time sequence on the Cusum graph. The decision as to whether or not the process is under control then rests on the pattern of the previous points and their relationship to the arms of the mask. If any of the previous points fall outside the arms of the V then the process is out of statistical control because the slope of the points exceeds that corresponding to the particular arm.

Fig. 13.22 shows a process which is under control. As the nose of the mask is moved from sample point to sample point in sequence, none of the previous points fall outside the lines of the arms.

As the slope of the Cusum graph increases or decreases, then so the mask signals that the process level has changed significantly. Fig. 13.23(a) and 23(b) show processes which are out of control due to an increase in level of performance and a decrease in level of performance respectively.

Fig. 13.21 The truncated V mask

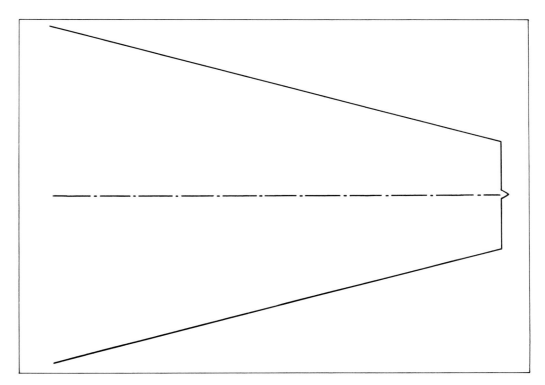

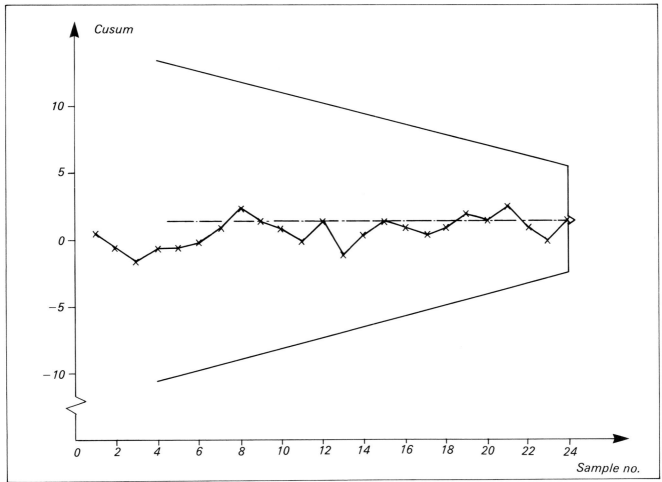

Fig. 13.22 V mask showing in-control situation

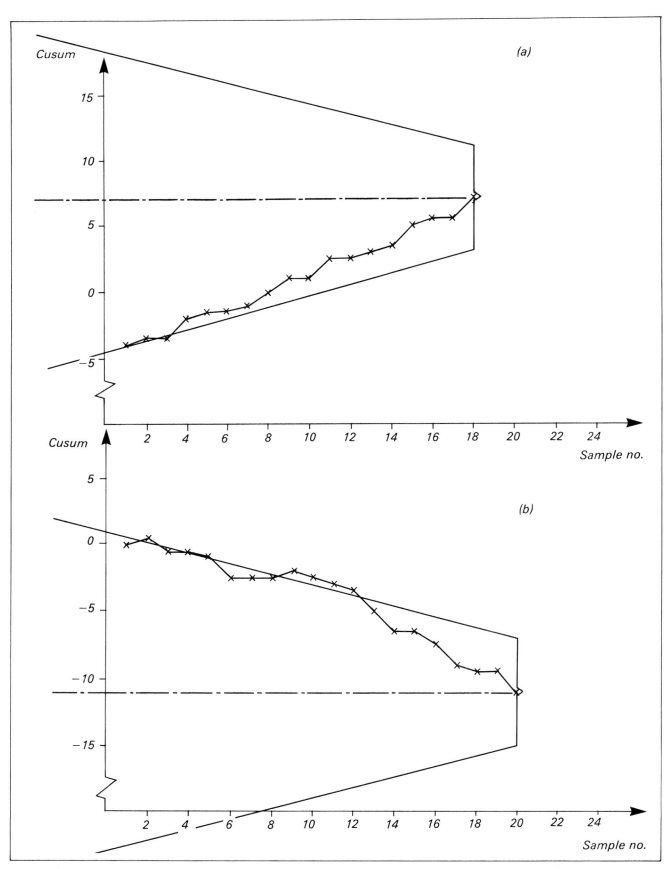

Fig. 13.23   V mask showing out-of-control situations

The properties of the V mask are such that the point falling outside the arm is not necessarily the point at which the process went out of control. This may seem contradictory, but the explanation lies in the notion of average run length, which is explained in the following brief and elementary description of the design of the V mask.

13.6.3 **Average run length and V mask design**

The average run length (ARL) can best be explained by reference to the basis of the control limit as used on the Shewhart chart. The control chart for $\bar{X}$ had an upper control limit ($UCL_{\bar{x}}$) such that the chance of obtaining a point outside $UCL_{\bar{x}}$ was about 1 in 1000 (to be specific 0.0013). The reciprocal of this number, 1/0.0013, i.e. 770, is then the ARL. In other words, the position of the control limit has been so chosen that for every 770 points, one, on average, will fall by chance beyond the limit. More generally, the ARL is the number of points required, on average, to detect a change in the process.

This idea of the ARL enables the V mask to be designed to suit the requirements in hand. The mask itself can be produced with different parameters: for example the arms can vary in steepness, as can the width of the nose of the mask. Fig. 13.24 illustrates some typical options.

The shape is fundamentally determined by the two parameters of slope and width. For convenience, and to be consistent with the current BSI notation, the slope – called the decision line – can be denoted by F, and half the width – called the decision interval – denoted by H. F and H can vary depending on the requirements. The less steep the arm, i.e. the smaller the value of F, the more sensitive is the mask in detecting genuine small changes in process level. This sensitivity is also influenced by H, in that the smaller the value of H, the greater the sensitivity, and vice versa.

Values of F and H are determined on the basis of the ARL required. Tables are available which provide F and H values in line with the sensitivity demanded.

If the ARL is to be 10, for example, then specified values of F and H can be looked

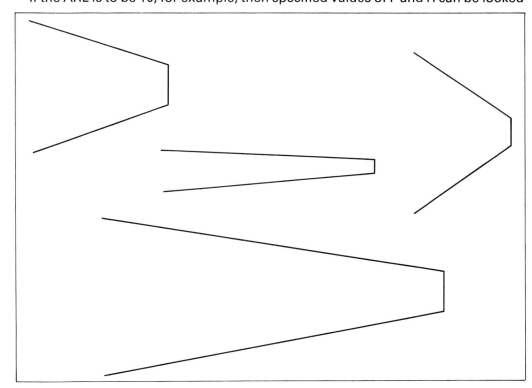

Fig. 13.24 Various V masks

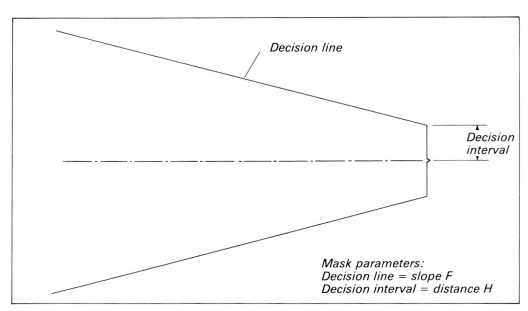

*Fig. 13.25
V mask parameters*

*Mask parameters:
Decision line = slope F
Decision interval = distance H*

up from which an appropriate mask can be designed. This mask will then detect a change in process level of a stipulated magnitude in 10 points, on average. Note that the ARL is the *average* run length: sometimes the change will be detected in less than the 10 sample points, sometimes more.

Appropriate detail on the design of the V mask is available in BS5703(2).

13.6.4   From V mask to computer

The V mask provides an effective visual technique for detecting process changes, but it does have some limitations. Masks tend to be cumbersome and their use does not always produce consistent decisions because of the variability arising from personal interpretation of results. The alternative is to use a tabular equivalent of the V mask, where the parameters F and H are represented by numerical equivalents which determine the basis for action. An example will assist in explaining the method and Fig. 13.26 introduces the notation. The method can be adapted for variables or attributes, but in view of the considerable potential of the Cusum method for monitoring and reducing the variability about the nominal, it is its use with variables that will be discussed here.

The relationship between the V mask and a tabular method is now seen in that the slope of the mask, F, is used to fix reference lines set outwards from the nominal T. The upper reference value K_1 and the lower reference value K_2 are then used to

*Fig. 13.26   Nominal
and reference values*

generate Cusum tables as before. Two sets of Cusum values result, depending on whether K_1 or K_2 is used as the target. The following example makes use of K_1, but a similar procedure would hold for K_2.

To illustrate the use of the reference value K_1 a nominal of 50 will be used. From existing knowledge of the process and the desired requirements in detecting a change, K_1 has been determined as 51. In practice F and ARL for the process would be required, but they are not necessary in order to understand the principle of using K_1 to generate a Cusum approach which can be adapted for shop floor use.

Readings (X_1, X_2, X_3, etc.) are taken from the process in the usual way and a Cusum table generated as previously. The only difference is that instead of using the nominal as the basis of providing Cusum results, the reference value K_1 is used. The reason for this can be seen by comparing the data shown in Fig. 13.27(a) and (b). With a target of 50, the nominal for the process, the positive and negative Cusum runs should balance out if the process is running on target. With a target of 51, however, positive Cusums will only materialise when values of x occur which are greater than the reference value of 51. In other words, using the reference value as the target for the Cusum allows a significant run of points above the reference value to be detected. This signals that the process mean has moved sufficiently to result in unacceptable product if action is not taken. Fig. 13.28 shows the extended Cusum table based on a target value of 51.

The Cusum results are plotted in Fig. 13.29, which also shows the conventional plot and its relation to K_1 and the nominal. As expected, it can be seen that every time the process level runs above the value of K_1 then the Cusum plot starts to move upwards. This can be used as a guide to detecting that the process is running at too high a level. An upward change in Cusum slope thus corresponds to the Cusum values becoming positive, as opposed to remaining zero or negative.

Fig. 13.30 shows the portions of the Shewhart chart and the corresponding Cusum graph corresponding to process levels which could signal the need for readjustment.

A shortened procedure is now suggested, based on the notion that every time the Cusum value reaches zero or more then a new Cusum tabulation is started, using only the positive Cusums. Fig. 13.31 shows the appropriate amendment to Fig. 13.28. These positive Cusums are plotted in Fig. 13.32. The graph illustrates the fact

Fig. 13.27 Cusum table with different targets

(a) Target, T = 50

Sample no.	X	(X − 50)	Cusum
1	49.4	−0.6	−0.6
2	50.4	0.4	−0.2
3	49.6	−0.4	−0.6
4	50.6	0.6	0
5	50.3	0.3	0.3
6	51.1	1.1	1.4
7	51.5	1.5	2.9
8	51.3	1.3	4.2

(b) Target, T = 51

Sample no	X	(X − 51)	Cusum
1	49.4	−1.6	−1.6
2	50.4	−0.6	−2.2
3	49.6	−1.4	−3.6
4	50.6	−0.4	−4.0
5	50.3	−0.7	−4.7
6	51.1	0.1	−4.6
7	51.5	0.5	−4.1
8	51.3	0.3	−3.8

Sample no.	X	X − 51	Σ (X − 51)
1	49.4	− 1.6	− 1.6
2	50.4	− 0.6	− 2.2
3	49.6	− 1.4	− 3.6
4	50.6	− 0.4	− 4.0
5	50.3	− 0.7	− 4.7
6	51.1	0.1	− 4.6
7	51.5	0.5	− 4.1
8	51.3	0.3	− 3.8
9	50.6	− 0.4	− 4.2
10	49.5	− 1.5	− 5.7
11	50.2	− 0.8	− 6.5
12	50.5	− 0.5	− 7.0
13	50.9	− 0.1	− 7.1
14	51.3	0.3	− 6.8
15	51.3	0.3	− 6.5
16	51.1	0.1	− 6.4
17	50.6	− 0.4	− 6.8
18	50.1	− 0.9	− 7.7
19	50.9	− 0.1	− 7.8
20	50.6	− 0.4	− 8.2
21	51.5	0.5	− 7.7
22	49.9	− 1.1	− 8.8
23	50.8	− 0.2	− 9.0
24	51.0	0	− 9.0
25	51.3	0.3	− 8.7
26	51.1	0.1	− 8.6
27	51.4	0.4	− 8.2
28	51.6	0.6	− 7.6
29	51.3	0.3	− 7.3
30	51.4	0.4	− 6.9

Fig. 13.28
Cusum table

that the only Cusum values of interest are those that contribute towards indicating a deteriorating situation requiring action to be taken.

What limit is there on the magnitude of the new Cusum before action is called for? To answer this use is made of the decision interval, H. A line is drawn at a value of H units, where H, in V mask terms, is half the width of the nose. If the Cusum value rises above this, then it means that the change in process level is sufficient to justify resetting. Again, values for H are determined depending on the requirements of the process. The references to BS5003 cited in Appendix I provide all the necessary detail for those who wish to pursue the topic. Suppose H is 1.8, as shown in Fig. 13.33. Then sample 30 indicates a need to take action because the process is now out of statistical control.

If a lower reference value, K_2, is used then a similar analysis follows except that it is now only negative Cusums, corresponding to a build-up of points below K_2, that are of interest.

If both K_1 and K_2 are used then a two-sided Cusum table can be generated.

This numerical equivalent of the V mask introduces an extra layer of training and understanding for those who wish to gain full benefit of the power of Cusum in an analytical sense. But the statistical rationale behind the technique is not of concern to those interested in applying as effective a monitoring method as possible at operational level. The fact that the technique is being used successfully at shop floor level, as the following case study shows, should dispel fears that it is too complicated.

Case study 12: British Steel

The Tinplate Division of   British Steel   undertook a top-down   approach to SPC based on a cascade training programme

Senior staff were trained in SPC over a period of 3 days, but in addition they were

Fig. 13.29 Shewhart chart and Cusum graph

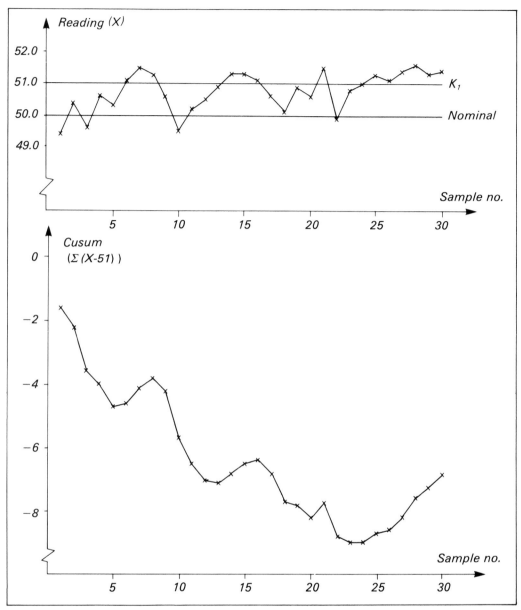

trained in presentation and other related skills so that they could then assist in the training of staff at the intermediate level. Outside trainers were also used and a blitz programme was undertaken so that, in selected pilot areas, SPC could be introduced at shop floor level as soon as possible. The earlier programme had ensured that all staff in a vertical line of responsibility corresponding to the pilot areas had been exposed to SPC training.

The Cusum method was applied to the measurement of strip shape in cold rolling operations. The use of the conventional $(\bar{X}, R)$ chart was not really viable because the period during which the same work rolls were in the mill was variable and relatively short. A new approach was required, and hence a simple application of Cusum was developed at the second cold reduction operation at the Ebbw Vale works of BS. Cusum was used to control one of the 'Shape' parameters.

'Shape' is defined as the deviation from absolute flatness in a sheet product. The particular type of deviation to be monitored is termed 'cross-bow', and is illustrated in Fig. 13.34. The condition results from differing residual strain states at the top and bottom surfaces, i.e. one surface is in tension and the other in compression.

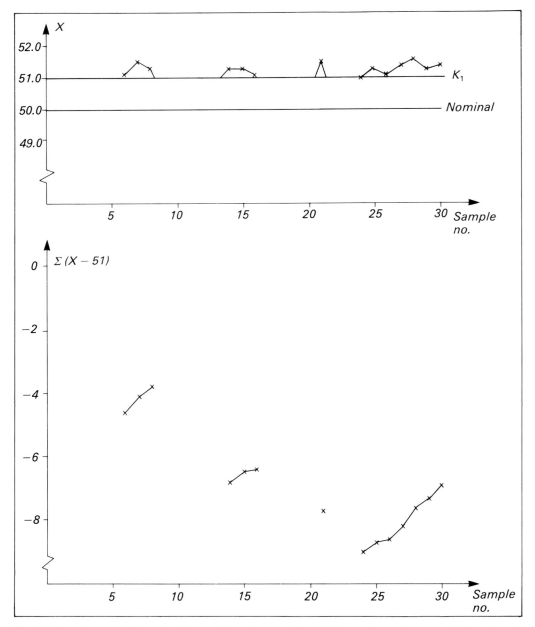

Fig. 13.30   Shewhart chart and Cusum graph showing positive Cusums only

Basically, the condition is dictated by the quality of the roll surface used. Each pair of work rolls is a singular event in its own right. As roll wear occurs during the process, the rate of wear varies sufficiently between the two roll surfaces to create the 'cross-bow' condition described.

The post-rolling processes provide facilities for strip flattening, such as stretch rollers and roller levellers. But experience has shown that the as-rolled degree of bow must be contained within limits if a commercially flat product is to result. In any case, SPC is synonymous with prevention and hence the emphasis on avoiding correcting procedures.

Process histories established a decision interval for the as-rolled degree of cross-bow. When this level is reached, the work rolls are changed. The control procedure is designed to signal the need for a roll change before a critical point is reached and to minimise the reaction to false alarms in the sequence.

The procedure is relatively simple. Each coil is tested, a standard coil length taken, and the bow measured and recorded. The bow reading, X, is then used to

Fig. 13.31 Cusum table showing positive Cusums

Sample no.	X	X − 51	Σ (X − 51)	Positive cusum
1	49.4	−1.6	−1.6	
2	50.4	−0.6	−2.2	
3	49.6	−1.4	−3.6	
4	50.6	−0.4	−4.0	
5	50.3	−0.7	−4.7	
6	51.1	0.1	−4.6	0.1
7	51.5	0.5	−4.1	0.6
8	51.3	0.3	−3.8	0.9
9	50.6	−0.4	−4.2	
10	49.5	−1.5	−5.7	
11	50.2	−0.8	−6.5	
12	50.5	−0.5	−7.0	
13	50.9	−0.1	−7.1	
14	51.3	0.3	−6.8	0.3
15	51.3	0.3	−6.5	0.6
16	51.1	0.1	−6.4	0.7
17	50.6	−0.4	−6.8	
18	50.1	−0.9	−7.7	
19	50.9	−0.1	−7.8	
20	50.6	−0.4	−8.2	
21	51.5	0.5	−7.7	0.5
22	49.9	−1.1	−8.8	
23	50.8	−0.2	−9.0	
24	51.0	0	−9.0	0
25	51.3	0.3	−8.7	0.3
26	51.1	0.1	−8.6	0.4
27	51.4	0.4	−8.2	0.8
28	51.6	0.6	−7.6	1.4
29	51.3	0.3	−7.3	1.7
30	51.4	0.4	−6.9	2.1

build up the Cusum chart. The reference value, K, is 30mm. Only upward (positive) slopes of the corresponding Cusum chart require action to be taken. At this stage the Cusum is reset to zero. Horizontal or downward slopes of the Cusum chart allow the operation to proceed.

Experience has also shown that the decision interval criterion is equivalent to the

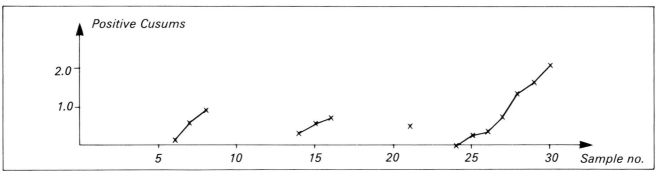

Fig. 13.32 Plot of positive Cusums

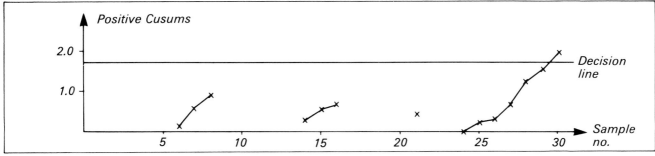

Fig. 13.33 Plot of positive Cusums plus decision line

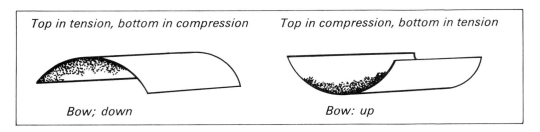

Fig. 13.34 Cross-bow effect

occurrence of two points in succession producing a positive Cusum. This fail-safe condition means that the work rolls are changed automatically.

Fig. 13.35 shows a typical Cusum chart.

It is worth noting that the Cusum values are calculated, and the chart maintained, by mill operators. Precise limiting conditions, such as run length, or control slopes, are not involved.

Cash savings on processed coils are considerable. An important feature of this case study is the use of the Cusum method at shop floor level.   British Steel   have shown that a technique traditionally couched in analytical terms and typically reserved for management activity can readily be utilised by operators provided they are given proper training and management support.

13.6.5 **Cusum versus Shewhart**

It is possible to quantify the relative efficiencies of the Shewhart chart and an equivalent Cusum chart in detecting a change in process level. Fig. 13.36 provides some basic detail. As a rough guide the Cusum method is approximately three times as effective as the Shewhart chart in detecting small changes in process level.

For example, let a process nominal be 6 units and the standard deviation for the process 0.1 units. The requirement is to detect as quickly as possible when the process level is 6.1 units, i.e. when it has moved 1 standard deviation away from the target. Reference to Fig. 13.36 shows that the Shewhart chart will detect the change, on average, in 70 sample points whereas the Cusum chart will require only 11 sample points. In other words the Cusum chart detects the change 6 times as quickly. The smaller the change in level from nominal, the more effective is the Cusum method. As the change becomes larger, then in fact the conventional control chart is better than the Cusum chart.

This has implications when designing SPC programmes. Some companies use both types of chart for some processes: the Shewhart chart for detecting larger changes and the Cusum chart for noting small changes.

The Cusum method was first put on an established footing in the UK in the 1950s. Since then it has seen various applications, from monitoring the reject levels in telephone assembly lines to checking the quality level of chocolate bars, and from assessing the performance of plant in the dairy industry to monitoring the strength of concrete. For a technique that is relatively new, the applications have been impressive. Even so, the full potential of the technique has not been realised. There is little doubt that as SPC programmes develop, the Cusum method will play a major part in reducing the variability about the nominal, or cutting down on reject levels. The current emphasis on continuous improvement will surely allow the Cusum technique to come into its own in monitoring small changes and fine-tuning processes.

Summary

- The Cusum technique applies to both variables and attributes.
- The method requires a target and a sequence of readings.
- A Cusum graph detects process changes by a change in slope.
- A slope guide, or protractor, enables process levels to be read off the chart.

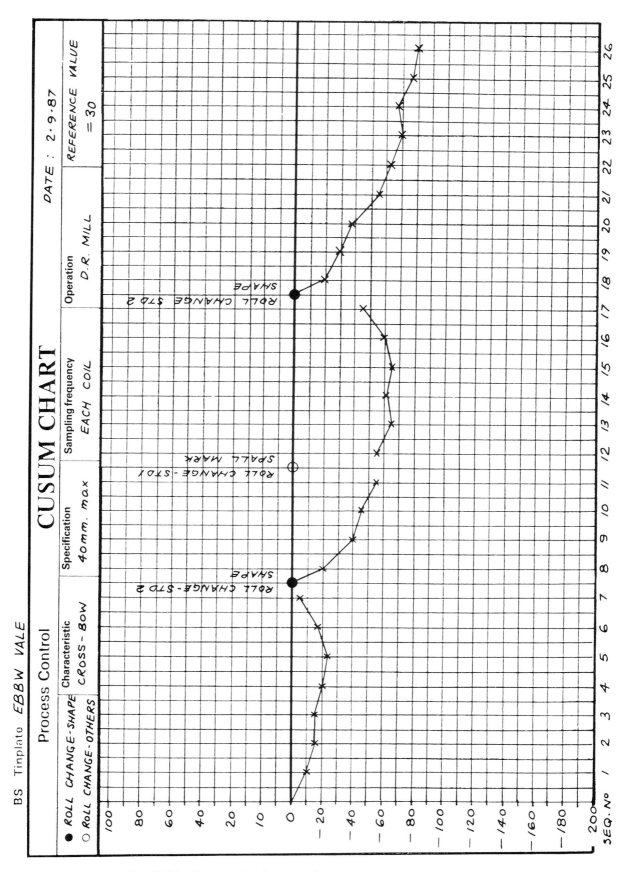

Fig. 13.35   Cusum plot for cross-bow

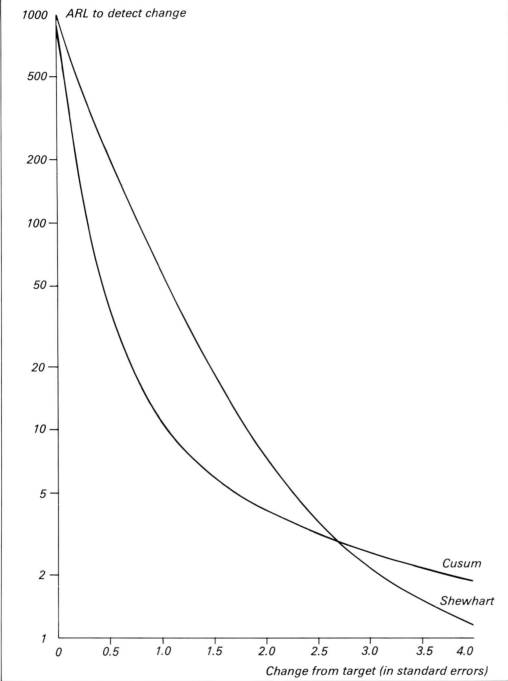

Fig. 13.36
Comparison of the
sensitivity of Shewhart
and Cusum graphs to
change

Fig. 13–36 is part of an extract from BS5703. It is reproduced by permission of BSI.
Complete copies of the standard can be obtained from them at Linford Wood, Milton
Keynes, Bucks MK14 6LE.

- The vertical scale on a Cusum graph does not indicate performance.
- The target is typically the nominal for the process.
- An appropriate Cusum scale must be chosen to emphasise real changes while
  not prompting action on common causes.
- The Cusum chart is useful retrospectively in analysing previous records.
- The V mask corresponds to control limits on a Shewhart chart.
- Average run length (ARL) is the inverse of probability.
- V mask parameters are tabulated in BS5703(ii).

- Decision interval schemes provide a tabulation equivalent of the mask.
- Cusum methods detect small changes more quickly than Shewhart methods.
- Wide-ranging applications of the Cusum method already exist but it is still under-utilised.
- Continuous improvement will provide for more extensive use of the Cusum technique.

Process control is the theme of this book and hence there has been a concentration on the process. Various charts have been described to improve these processes, but little reference has been made to analysing the performance of the machine which nearly always forms part of the process. The following chapter, the final one on the techniques side of SPC, covers machine capability analysis. It has much in common with charting techniques in that data is collected, plotted and analysed, but it also introduces a neat form of graph plotting based on the properties of the normal distribution.

Chapter 14 Machine capability analysis

14.1 Introduction

In the preceding chapters all activities have been thought of in terms of processes, which has opened up avenues of applications of SPC techniques that have little to do with the traditional environment of the control chart – the shop floor in the machine shop. For the final technique the discussion will revert to machining areas. There is every reason to do so because a machine is part of the process and the process cannot be capable unless the machine itself is inherently capable. Not only that, machine capability is another way of analysing repeatability, and this raises many issues associated with calibration and the appropriateness of any measuring equipment used by an organisation.

   The topic of machine capability needs to be discussed, therefore, but it must be put in perspective. It is now a relatively minor aspect of the complete SPC programme. It would appear that in the early days of introducing SPC, organisations somehow picked up a message that machine capability analysis was the answer. As a result, capability sheets appeared all over the place almost overnight. The message was wrong. Too much emphasis had been mistakenly placed on machine capability and not enough on monitoring the process by control charts, and machine capability studies now play a lesser role in SPC programmes.

14.2 Process or machine?

A process, as defined earlier, is a combination of manpower, method, material, machine and environment. A machine is one element of a process and analysing a machine will be different from analysing a process.

   A machine capability study helps to provide a picture of the performance of the process as a whole. Typically, it requires a maximum of 50 items, although there are ways of analysing samples of 25, or even 10. A critical requirement is that the items are taken in sequence and over a relatively short time period. If this is not so then the effect of time means that it is a process that is being analysed, not a machine, and the method which follows no longer applies.

14.3 The machine capability chart

The chart for recording and analysing the results of a machine capability study takes a slightly different form from the control chart. Fig. 14.1 shows the essential sections. The administrative section is familiar enough. Details relating to the capability study are entered here. The data block allows the 50 results to be recorded. These are then transferred in a particular format onto the tally chart.

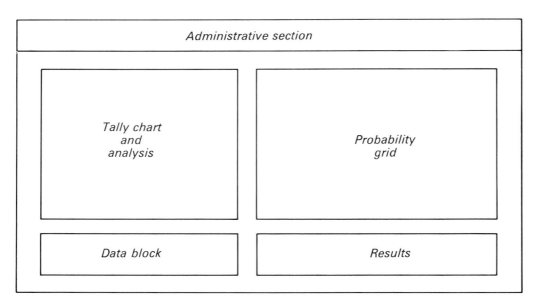

Fig. 14.1 Machine capability chart:

Further analysis enables readings to be plotted on a unique type of graph paper known as probability paper. Finally, depending on the nature of the plot, results can be entered in the final section.

The best way to describe the use of the capability chart is to work through an example. In doing so it will be seen how each of the sections comes into play.

14.4 Carrying out a capability study

The capability study to be discussed here relates to a reaming operation on a bearing bush. The specification for the hole diameter is 7.51 ± 0.03mm.

Fig. 14.2 shows the completed administrative section on the capability chart. The hole diameters are to be recorded in actual units, i.e. millimetres. If a coding were required then some amendment would be necessary. This will be covered in Section 14.4.2

14.4.1 Collecting and recording the data

Before collecting the data some initial requirements must be satisfied. The first precondition is that the measuring equipment must be accurate (a typical guideline is that it must be properly calibrated and have a precision of at least one-tenth of the tolerance of the item to be measured). Also, the parts to be used must already be acceptable prior to the operation, there must be an uninterrupted run under standard conditions, the machine must be preset at nominal, and so forth. These requirements are rather detailed but if they are not met then the study is worthless. Company-specific SPC manuals provide appropriate source material.

The initial requirements having been satisfied, 50 bushes are taken in sequence, hole diameters measured and the results recorded, in the appropriate numbered box, in the data block section of the chart. The 50 readings are shown in Fig. 14.3. It is good practice to identify each bush by its appropriate number. Once the 50 readings are available, they are used to generate a pattern.

Machine Capability Chart						Division	Department	Ref no
						AUTO	C	B/63
Part no Description BEARING BUSH	Characteristic REAMED HOLE	Specification 7·51 ± 0·03 mm	Operation REAMING	Machine no./type 78J/3	Measuring equipment AC1042	Unit of measurement mm Zero O		

Fig. 14.2 Administrative data

Fig. 14.3 Data block

READINGS									
1 7.52	6 7.49	11 7.55	16 7.52	21 7.53	26 7.53	31 7.52	36 7.51	41 7.53	46 7.52
2 7.51	7 7.53	12 7.52	17 7.53	22 7.50	27 7.52	32 7.54	37 7.51	42 7.52	47 7.54
3 7.53	8 7.52	13 7.54	18 7.52	23 7.53	28 7.55	33 7.54	38 7.52	43 7.51	48 7.53
4 7.52	9 7.51	14 7.52	19 7.53	24 7.54	29 7.50	34 7.52	39 7.51	44 7.51	49 7.53
5 7.49	10 7.52	15 7.55	20 7.53	25 7.52	30 7.51	35 7.50	40 7.52	45 7.50	50 7.50

14.4.2 Generating the tally chart

The readings are now transferred to the section of the capability chart which allows tally marks to be recorded corresponding to each reading. A visual scan of the 50 readings shows the least value to be 7.49 and the greatest to be 7.55. With the usual proviso regarding the presence of wild values, the 'Value' column needs to be scaled to cover this span of readings. The chosen scale appears in Fig. 14.4 together with the 50 tally marks.

In the example the figures represent actual readings. Often, though, the readings are coded from a zero set on nominal. For example, 7.51 would become zero and units of measurement would then be in multiples of 0.01mm measured from the nominal. The value column and administrative section would be amended accordingly, as shown in Figure 14.5. This makes no difference to the further analysis except that at the final stage the results need to be reconverted to millimetres.

VALUE	TALLY MARKS
7.58	
7.57	
7.56	
7.55	///
7.54	ЖН
7.53	ЖН ЖН /
7.52	ЖН ЖН ЖН /
7.51	ЖН ///
7.50	ЖН
7.49	//
7.48	
7.47	
7.46	
7.45	
7.44	

Fig. 14.4 Tally marks: actual readings

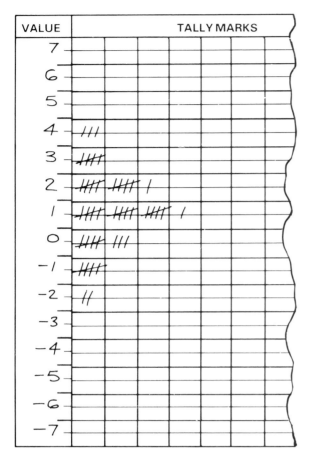

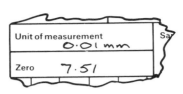

*Fig. 14.5   Tally marks:
coded readings*

It could be argued that a wider scale could have been chosen, as only half of the allocated space is being utilised. There is a reason for this, which will become clear when the probability grid zone of the chart is used, as described in Section 14.5.

The tally chart shows that the pattern of results is symmetrical and that a normal distribution would prove a good fit. A purely visual decision is not good enough, however, and a more objective method is needed to establish whether or not a normal distribution is appropriate. The first step is to complete the f, Σf and $\Sigma f\%$ columns.

14.4.3   Calculating the cumulative frequencies

The total number of tally marks in each interval is first recorded in the column headed 'f'. Each of the readings in the 'Value' column is really the central value of a group. For example, 7.49 is the centre of an interval defined by limits of 7.485 and 7.495. Fig. 14.6 shows that the limits for each interval correspond to the thicker lines on the scale.

For the first interval, there are 2 readings corresponding to a central value of 7.49. These 2 readings by definition will be numerically less than the upper limit of the interval, i.e. 7.495. For the second interval, there are 5 readings which are less than the new upper limit of 7.505. The 2 readings from the previous group must also be less than 7.505 and therefore altogether there are 2 + 5, i.e. 7 readings, less than 7.505. In the same way 2 + 5 + 8, i.e. 15 readings, must be less than 7.515 and so on. These cumulative readings are recorded at each stage in the Σf (cumulative frequency) column. Fig. 14.7 explains the procedure being adopted.

These calculations have been carried out using an upward cumulation, as this is the usual convention. If a downward cumulation is adopted the intermediate results

Fig. 14.6 Classes and limits

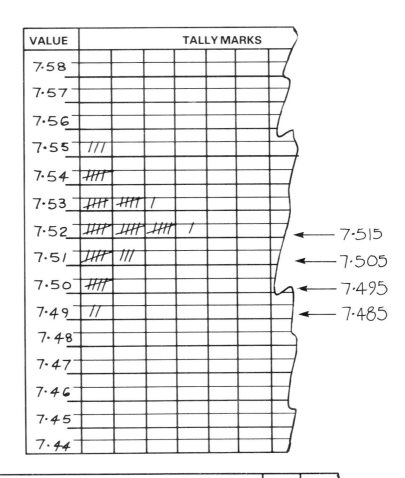

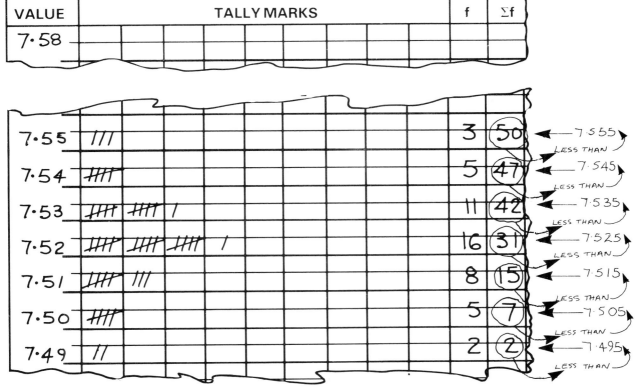

Fig. 14.7 Generation of cumulative readings

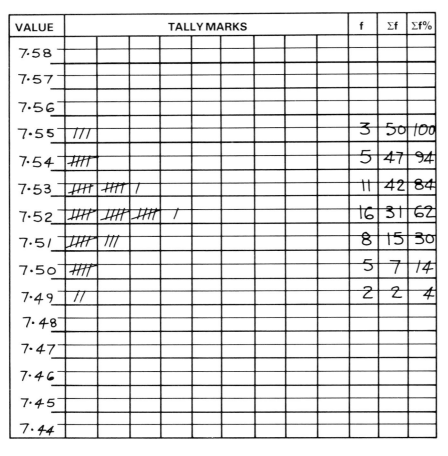

Fig. 14.8 Completed tally chart and analysis

VALUE	TALLY MARKS	f	Σf	Σf%
7·58				
7·57				
7·56				
7·55	///	3	50	100
7·54	₮₭₮	5	47	94
7·53	₮₭₮ ₮₭₮ /	11	42	84
7·52	₮₭₮ ₮₭₮ ₮₭₮ /	16	31	62
7·51	₮₭₮ ///	8	15	30
7·50	₮₭₮	5	7	14
7·49	//	2	2	4
7·48				
7·47				
7·46				
7·45				
7·44				

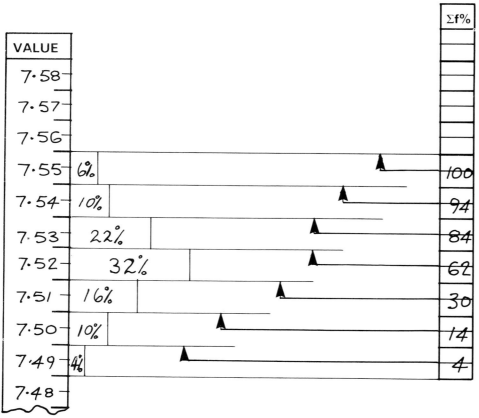

Fig. 14.9 Alternative explanation of cumulative values

will be different, as will the probability plot, but the final results will be the same.

The final column, Σf%, converts the cumulative frequencies to percentages. Since there are 50 bushes in total, each figure need only be doubled to give the results shown in Fig. 14.8.

Fig. 14.9 illustrates the reasoning in a different format. Here the tally marks have been replaced by the corresponding histogram. The areas in each interval are recorded and the accumulation of these areas then gives rise to the figures in the Σf% column.

Before moving on to the probability grid, these figures need to be related to corresponding areas under the normal curve. As visually, at least, a normal distribution provides a good fit to the pattern of tally marks, a normal curve can be superimposed on the readings as shown in Fig. 14.10. (Drawing in the curve would not be necessary in practice. The step has only been introduced here to clarify the basis of normal probability plotting.) The properties of the normal curve can now be used to provide a reference standard against which the actual areas generated in Fig. 14.9 can be compared. Fig. 14.11 shows the areas under the normal curve below fixed multiples of the standard deviation s.

If the cumulative frequencies are plotted horizontally against units of standard deviation vertically, then a curve is obtained as in Fig. 14.12. This graph is not a standard, familiar one, primarily because the horizontal scale being used is defined in equal increments. However, using a non-linear scale as shown in Fig. 14.13 converts the curve into a straight line when the distribution is normal. This particular non-linear scale forms the basis of normal probability paper, and the next section will show how it can be used.

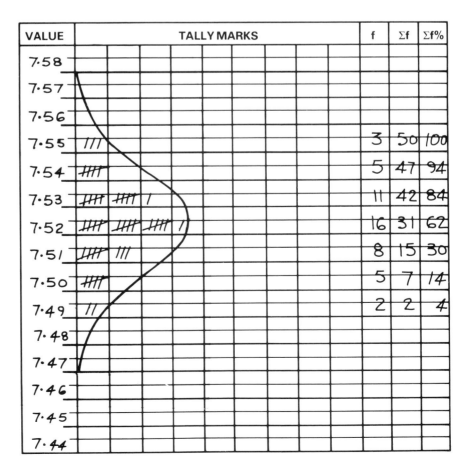

Fig. 14.10 Tally marks plus normal curve

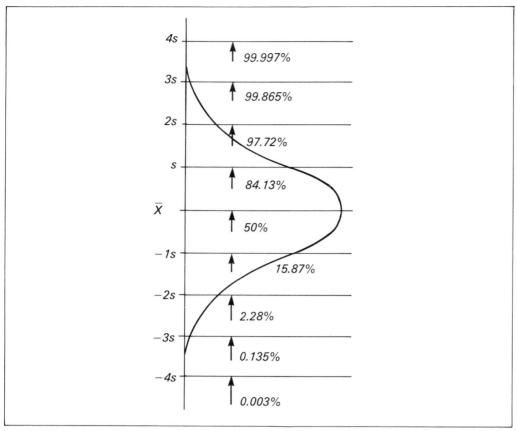

Fig. 14.11   Areas under normal curve

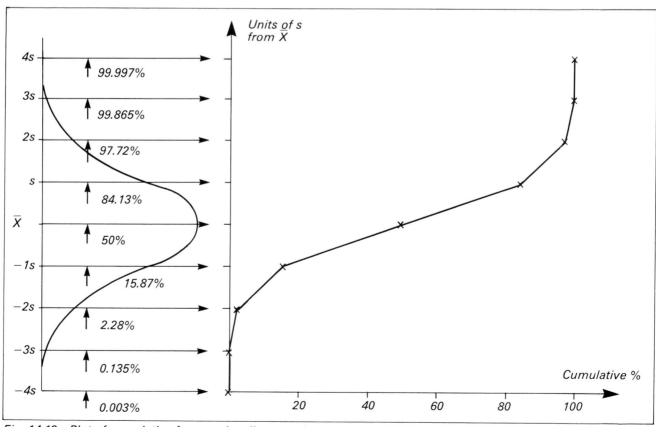

Fig. 14.12   Plot of cumulative frequencies: linear scale

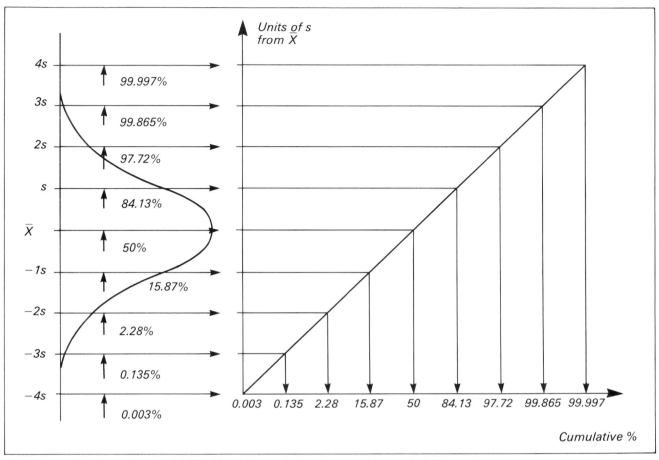

Fig. 14.13   Plot of cumulative frequencies: probability scale

14.4.4   **Plotting the readings**

The probability grid section of the sheet has been designed so that areas under the normal curve at given intervals can be transformed into a straight line.

Percentages are plotted on the horizontal (non-linear) axis of the grid. It is seen that neither 0 or 100% appears on the scale, which is to be expected from the properties of the normal curve. The vertical scale (which is linear) corresponds to the distance along the base of the normal curve.

Plotting proceeds as follows, and uses the arrows which link the tally chart section and the probability grid. With $\Sigma f\% = 4$, the line through the centre of the 4 is followed, which corresponds to the upper limit, i.e. 7.495. Using the arrow to move from the tally chart to the probability grid, the intersection of the horizontal line being followed and the vertical line corresponding to a cumulative percentage of 4% is then marked (Fig. 14.14).

The procedure is repeated for all other points. It is not possible to plot at 100% the value corresponding to 7.555. Instead the mean of 100% and 94% is taken, i.e. 97%, and a line which is exactly halfway between the arrow corresponding to 7.545 and 7.555 is followed. The intersection of this line and the vertical line corresponding to 97% is then marked.

The best straight line is then drawn through the points (Fig. 14.15). In doing so, more weighting is given to the evidence provided by points in the centre of the plot rather than at the extremes.

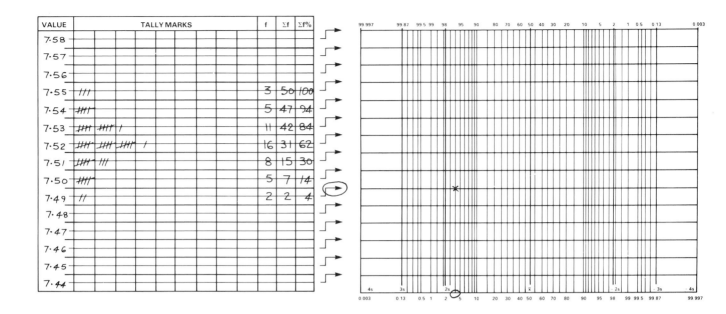

Fig. 14.14 Use of probability paper

The results provide a good straight line, verifying that the distribution is normal. If a straight line cannot be drawn through the points then a normal curve cannot be used. There are methods of analysing such data in a different way, and Section 14.6 provides a brief introduction to these. It is good practice to refer back to the pattern provided by the tally marks. If it is not symmetrical, or if there are two peaks, then it obvious that a straight line will not result.

The use of arrows to link the tally chart and the normal probability plot, and hence to determine the location of points on the grid, is a unique approach. The format adopted here, which is used extensively by the Ford Motor Company and, with variations, by other organisations, reflects well on those concerned with transforming statistical theory into a very practical technique.

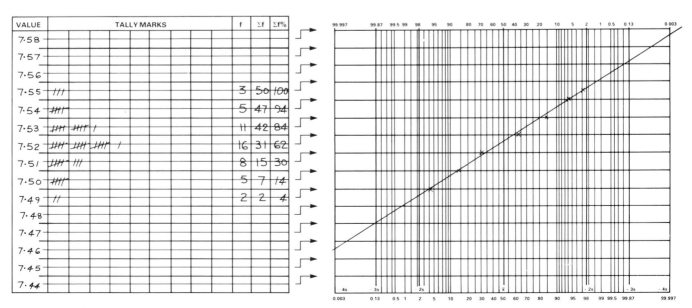

Fig. 14.15 Completed plot

Now that a straight line has been obtained it can be used to determine various properties relating to the machine performance.

14.5 Calculating results

14.5.1 The mean

It is known that for a normal distribution the mean value corresponds to the 50% cumulative frequency value. Therefore, to obtain the mean value the point of intersection of the vertical line corresponding to the 50% value and the straight line through the points is sought and read off on the 'Value' scale (Fig. 14.16). Note that in doing this there is no 'step down' in moving from the grid to the 'Value' scale. The step-up process was necessary, and allowed for, in generating the straight line; the reverse is not necessary when reading off results on the 'Value' axis.

In this case the mean is 7.521. The mean based on the plot can now be recorded in the 'estimated mean' section of the data box, as can the specified target of 7.51. (For comparison, the actual mean value of the 50 readings, obtained by calculator, is 7.521.) It is clear that the machine setting is incorrect and that a readjustment is necessary. A measure of the adequacy of the setting is provided by the index C_{mk}, to be covered in Section 14.7.

Adjusting the machine so that the mean corresponds to the nominal of 7.51 is equivalent on the normal probability plot to moving the straight line parallel to itself until it cuts the 50% value at 7.51mm (Fig. 14.17).

14.5.2 The standard deviation

Machine capability is typically defined as 8s, where s is the standard deviation – in this case of the hole diameters based on the sample of 50. Once 8s is known, therefore, s can be obtained.

There is often some misunderstanding regarding the use of 8s for the capability of a machine as opposed to 6s for a process. The reason is that if 8s falls within the

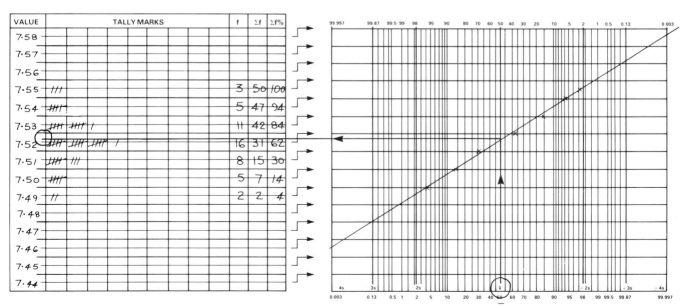

Fig. 14.16   Determining the mean

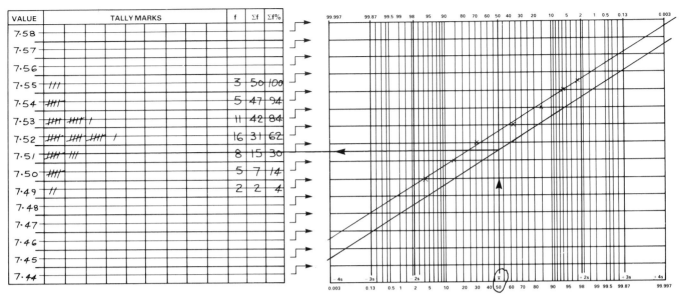

Fig. 14.17 Adjusting the setting

specification limits the variability in the machine must be lower than if only 6s fell within those same limits. There is then a better chance that by the time the machined item is combined with other items on other machines with different operators/ materials and different environments, then 6s for the process will fall within the process specifications. This feature will be emphasised when machine capability indices are defined in Section 14.7.

The value of 8s can be obtained from the probability plot as follows. First the two vertical lines corresponding to − 4s and + 4s are located. In fact the probability grid has been designed so that the vertical sides of the grid correspond to these values. (The areas in the tail of the normal distribution corresponding to each of these values is 0.003%. This figure should be familiar from the P_z tables used in Chapter 5 when discussing the properties of the normal curve.) The points where the straight line intersects the − 4s and + 4s lines are noted and read across, as before, to the 'Value' scale. The distance between these two readings is then 8s (Fig. 14.18).

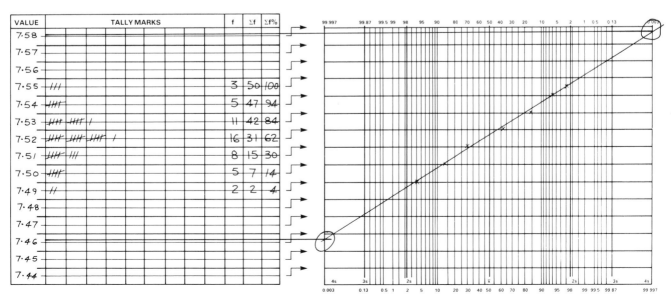

Fig. 14.18 Determining the standard deviation

Hence, in this case,

$$8s = 7.581 - 7.461 = 0.12$$

and therefore

$$s = 0.12/8 = 0.015.$$

(This compares favourably with the actual value of 0.0145 obtained by calculator.) These results can be entered in the appropriate boxes, as also can the specified tolerance of 0.06.

In some instances, reading off the points of intersection of the straight line with the $\pm$ 4s lines may not be easy. For example, if the straight line runs at a steep angle, corresponding to wider variability, then it will cut the 4s line beyond the span of the vertical scale, and some estimation will be necessary. In these cases it is preferable to use the $\pm$ 3s lines to determine 6s, from which both 8s and s can then be found. This is the reason for not choosing too large a scale for the 'Value' axis: it would mean that the straight line would slope steeply making it less easy to read off the $\pm$ 4s values directly.

Fig. 14.19 shows the effect of variability. In line with the philosophy of continuous improvement, first the machine must be correctly set at 7.51. Then the task is to reduce the variability about the nominal. This on the graph corresponds to the straight line becoming closer to horizontal.

The relationship between the variability in the hole diameters and the specification provided for the item has not yet been considered. Again, the probability plot is useful.

14.5.3 **Capability**

The specification quoted for the diameter is 7.51 $\pm$ 0.03mm. Therefore the upper specification limit (USL) is 7.54 and the lower specification limit (LSL) 7.48, and lines can be drawn across the sheet corresponding to these values (Fig. 14.20). If a normal distribution is superimposed as before, then the areas in the tails beyond these specification limits will correspond to the percentage of product that is unacceptable. Rather than try to estimate the result from the tally chart or histogram, the probability grid is used.

First the points of intersection of the plotted line and the specification lines, marked A and B in Fig.14.21, are found. Using the bottom scale , A corresponds to a value of 0.3, which is the percentage of product that is below the LSL. Similarly, B corresponds to a value of 11, which is the percentage of product above the USL. These values are entered in the results section.

These figures are far too high for the machine to be considered capable. The critical total area outside the specification limits is 2 × 0.003, i.e. 0.006%. This will occur when 8s is exactly equal to the tolerance of 0.06. For this to happen the straight line must intersect the specification limits on the +4s and −4s lines respectively, as shown in Fig. 14.22.

Moving the straight line whilst keeping its angle the same is equivalent to adjusting the setting of the machine. Hence, for the example here, it is known that in the short term the optimum position of the mean will correspond to the target of 7.51, and therefore the line will move to a position indicated in Fig. 14.23. The percentage outside specification will now change to 2.5% below LSL and 2% above USL, i.e. 4.5% in total. This is certainly a reduction in the percentage of unacceptable product, but the overall position regarding capability has not changed: the natural variability represented by the readings is excessive and no amount of machine adjustment will make the machine capable. It is no longer acceptable to think in terms of optimum machine setting so that, for example, the number of bushes with oversized holes is minimised at the expense of producing more with undersized holes which are then reworked. Continuous improvement requires working on the

VALUE	TALLY MARKS	f	Σf	Σf%
7.58				
7.57				
7.56				
7.55				
7.54	///	3	50	100
7.53	HHT	5	47	94
7.52	HHT HHT /	11	42	84
7.51	HHT HHT HHT /	16	31	62
7.50	HHT ///	8	15	30
7.49	HHT	5	7	14
7.48	//	2	2	4
7.47				
7.46				
7.45				
7.44				

VALUE	TALLY MARKS	f	Σf	Σf%
7.58				
7.57				
7.56				
7.55				
7.54				
7.53	//	2	50	100
7.52	HHT HHT	10	48	96
7.51	HHT HHT HHT HHT ///	23	38	76
7.50	HHT HHT ///	13	15	30
7.49	/	1	2	4
7.48	/	1	1	2
7.47				
7.46				
7.45				
7.44				

VALUE	TALLY MARKS	f	Σf	Σf%
7.58				
7.57				
7.56				
7.55				
7.54				
7.53				
7.52	HHT /	6	50	100
7.51	HHT HHT HHT HHT HHT HHT HHT ////	39	44	88
7.50	HHT	5	5	10
7.49				
7.48				
7.47				
7.46				
7.45				
7.44				

Fig. 14.19 Effect of reduced variability

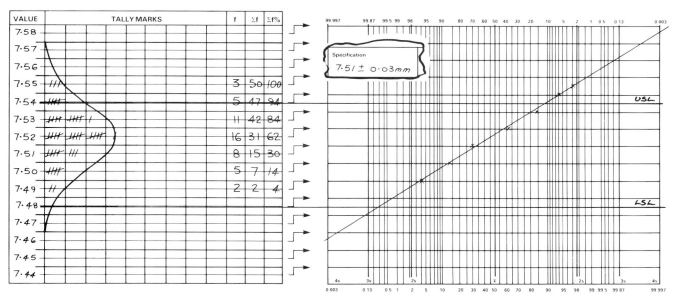

Fig. 14.20   Probability plot and specification limits

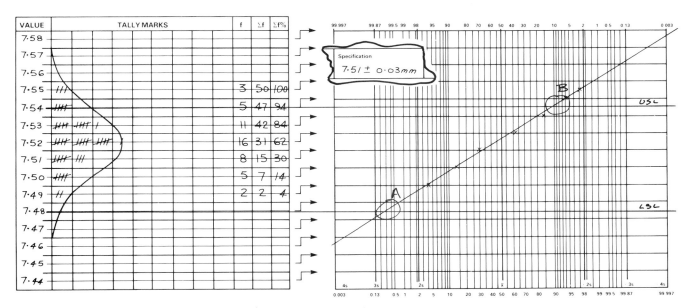

Fig. 14.21   Determining the percentage of unacceptable product

variability, i.e. changing the system under which the items were produced: the condition of the machine itself, the maintenance schedule, the material used, the type of reamer used. Permanent improvements must be sought that reduce cost and improve customer satisfaction.

14.6   Other patterns

The basis of probability plotting is that if the items under analysis are normally distributed, then a straight line is obtained. Conversely, if a straight line is not obtained, then the distribution is not normal. Two examples will serve to illustrate this.

Fig. 14.24 shows the form of the probability plot when the items generate a skew distribution. In neither case is a straight line fit possible with the usual probability paper, but a form of probability paper is available which converts a skew distribution into a straight line. Details are given in the appropriate technical references.

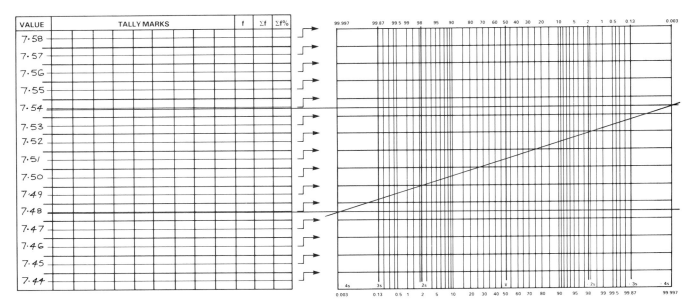

Fig. 14.22   Critical position of probability plot

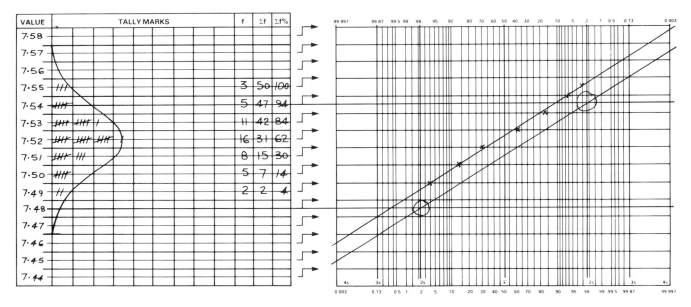

Fig. 14.23   Effect of adjusting the setting

Fig. 14.25 shows how two mixed distributions separate out on a tally chart and as a result produce a plot on the probability paper which can be treated as two distinct lines. Each line can then be analysed independently in the usual manner.

The standard form of probability paper can also be amended to allow analysis based on just 10 readings. However, too much emphasis should not be placed on the extended use of the simple graphical approach applied to 50 readings. It is better to restrict the use of these techniques to situations where they are known to be of real value and provide a return.

14.7 Measuring machine capability

In the same way as C_p and C_{pk} indices are used to assess process capability, C_m and C_{mk} provide a measure of machine capability.

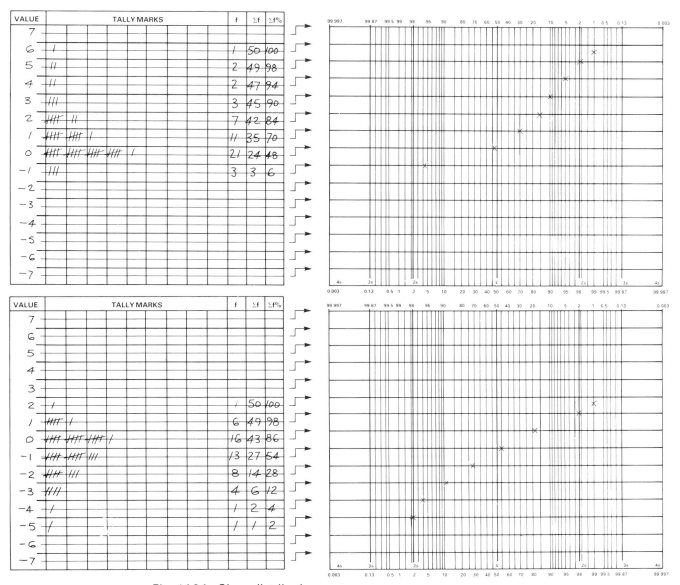

Fig. 14.24   Skew distributions

14.7.1 C_m

C_m measures the relationship between the tolerance of the item and the natural variability within the sample of 50 items. It is defined in a similar way to C_p, i.e.

$$C_m = \text{Tolerance}/6s$$

The value of s is known from the capability, 8s.
   In the example being used, tolerance is 0.06 and s is 0.015. Therefore

$$C_m = \frac{0.06}{6 \times 0.015}$$

$$= 0.67$$

Confusion often arises here as to what is the least possible value of C_m in order for the machine to be capable. The capability of the machine has already been defined

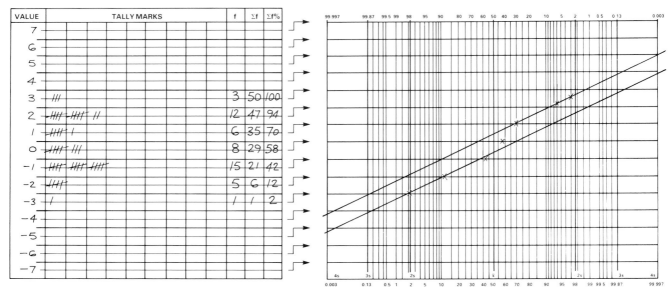

Fig. 14.25   Mixed distributions

as 8s, i.e. the variability needs to be sufficiently small that a span of 8s is equal to, or less than, the tolerance. Yet the C_m value uses 6s in the definition. How are these statements compatible?

Two alternatives present themselves. C_m could be defined as Tolerance/8s, in which case the minimum value for the machine to be capable is 1. Although there may be some advantage in having 1 as the minimum value for both C_p and C_m, the drawback is that different multiples of s are used in the expressions for C_p and C_m. There is therefore a preference for keeping 6s in the expression for both C_p and C_m and changing the minimum value of C_m. C_m, therefore, must be at least 1.33 (8 divided by 6) for the machine to be capable. In the example being used, with a value of 0.67 for C_m, the machine is therefore incapable.

The setting has to be on nominal for optimum performance. C_m alone will not provide information on the suitability of the setting and therefore C_{mk} is required.

14.7.2 C_{mk}

C_{mk} corresponds to C_{pk} and the definitions of the two are almost identical:

$$C_{mk} = \frac{USL - \overline{X}}{3s} \text{ or } \frac{\overline{X} - LSL}{3s}$$

As with C_{pk}, the smaller of the two values is chosen, which will always correspond to the form of the expression using the specification limit nearest to $\overline{X}$. In this case $\overline{X}$ is 7.521 and therefore the appropriate limit to use is 7.54. Thus

$$C_{mk} = \frac{7.54 - 7.521}{3 \times 0.015}$$

$$= 0.422$$

C_m and C_{mk} are identical when the machine setting is on nominal. The value of 0.422 for C_{mk} here, being different to the C_m value of 0.67, is a numerical measure of the inadequacy of the machine setting. The analysis of the relationship between C_m and C_{mk} is identical to that between C_p and C_{pk}.

Finally, when all the results are available, the completed results box appears as in Fig. 14.26.

SPECIFIED TOLERANCE $0 \cdot 06$	SPECIFIED TARGET $7 \cdot 51$	ESTIMATED OUT OF TOLERANCE
ESTIMATED CAPABILITY (8s) $0 \cdot 12$	ESTIMATED MEAN $7 \cdot 521$	TOP ___13___ % BOTTOM ___$0 \cdot 3$___ %
ESTIMATED STANDARD DEVIATION (s) $0 \cdot 015$	$C_m =$ ___$0 \cdot 67$___	$C_{mk} =$ ___$0 \cdot 422$___

Fig. 14.26 Completed results

Machine capability studies have naturally tended to be associated with machines as generally understood, i.e. a manufacturing unit in a traditional engineering shop floor environment. As SPC programmes develop, though, it is becoming evident that other applications of machine capability studies are possible. In particular, the capability of measuring equipment lends itself to probability plotting. As the following case study shows, inherent variability in the monitoring equipment can be a real problem.

Case study 13: PPG Industries (UK) Ltd

Case study 10 in Chapter 12 provided a background to the SPC programme introduced at PPG Industries. A key part of the SPC training programme was the use of projects. Without these the benefits of the 3-day training course would not have been realised. By becoming involved in a personal project, the trainee becomes familiar with the various techniques. More than that, previously unrecognised applications of a particular statistical tool may be found – as the following case study illustrates.

Control of colour is paramount in the paint industry. A standard procedure used in PPG, common to the paint industry as a whole, is to use a standard white tile to calibrate the spectrophotometer. Once a day the standard tile is inserted into the unit and the necessary adjustments made to provide a reference value against which other colours can be compared. In the past no allowance was made for any uncertainty present in the spectrophotometer. However, the results of a study on the instrument showed that this would no longer be acceptable.

Fifty consecutive readings based on the same standard tile showed considerable variability to be present, as indicated in the tally chart in Fig. 14.27. Not only was the variability far greater than had been assumed, there was also an indication that the readings were showing a skew distribution rather than a normal one.

The target reading was 0, i.e. if the unit was correctly calibrated with zero variability in the readings, then all values would consistently be recorded as 0. There was therefore clearly a setting problem also. There was no specification limit as such quoted, and hence calculations of C_m and C_{mk} were not possible.

A PPG representative discussed the problem of inconsistency with the instrument service engineer. He explained that rapid consecutive measurements on the same spot (four exposures to the artificial daylight source per reading) would transmit heat to the standard tile. As a result the reflectance values produced would alter and may well lead to a misrepresented capability result. A partial repeat of the exercise, allowing sufficient time for the tile to cool between measurements, indicated a more capable process, but with a consistent downward trend apparent. Further investigation was being undertaken following the information collected.

The project highlighted two issues. Firstly there was a need to allow for the inherent variability in the spectrophotometer. In the long term it was decided to upgrade the unit and replace it with a more consistent machine. Secondly there were implications regarding other items of measuring equipment: no longer could

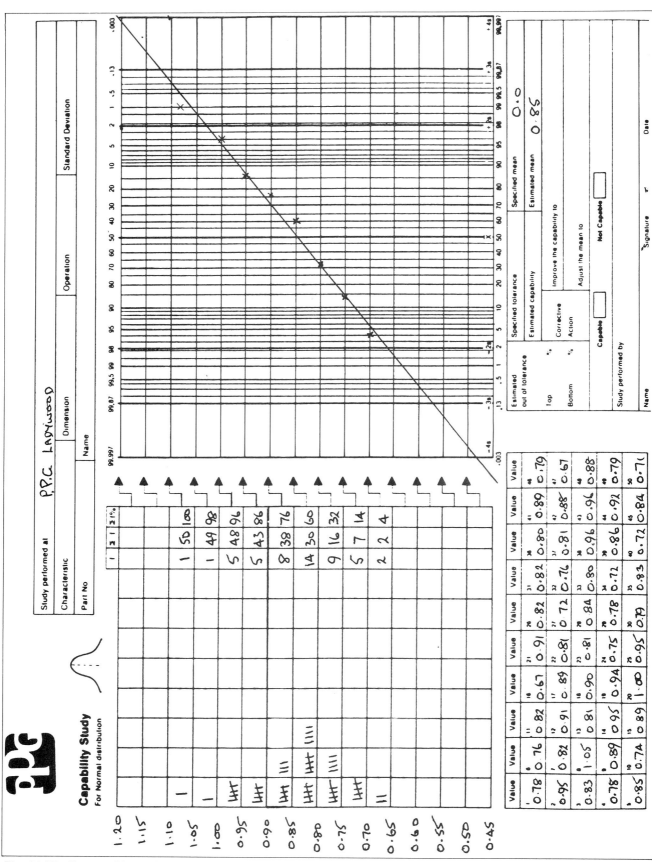

Fig. 14.27   Repeatability of results

the company take for granted the claims from manufacturers regarding the accuracy and precision of their instruments. This type of calibration problem is being repeated across Western industry as the full implications of SPC programmes begin to be recognised.

Summary

- Important as they are, machine capability studies play a minor role in SPC programmes in comparison with control charts.
- Items must be taken in sequence over a short time period.
- Probability grids enable straight lines to be generated by using specially constructed non-linear scales.
- With normal probability paper, a normal distribution converts to a straight line.
- The mean corresponds to the 50% value.
- Capability is 8s.
- The $\pm 4s$ lines enable the standard deviation to be determined.
- Reducing the variability corresponds to producing a plotted line closer and closer to horizontal.
- Moving the plotted line parallel to itself corresponds to changing the setting.
- Another form of probability paper is available to convert a skew distribution into a straight line.
- C_m measures variability in machine output.
- C_{mk} measures machine setting.
- Machine capability indices are interpreted in the same way as process capability indices, apart from the difference in the minimum values.
- The least value of C_m and C_{mk} (for capability) is 1.33.
- Machine capability analysis can be applied to measurement repeatability.
- Calibration procedures will be affected by the increasing use of SPC techniques.

A machine capability study can be very useful. For example, an out-of-control situation on a control chart may suggest that a machine capability study be carried out. Supplier Quality Assurance (SQA) engineers may require a capability study as part of their audit. More fundamentally, if the machine is not capable then the process cannot be capable. However, plastering the walls with capability plots does not indicate an effective SPC programme. In fact the reverse is true. Repeated and unnecessary use of machine capability analyses indicates a lack of understanding regarding the true meaning of SPC.

This chapter completes the techniques section of the book. Other techniques are available and are gradually being incorporated in SPC related programmes as experience with them builds up. For example, Taguchi methods based on the design of experiments are having a major impact particularly at the design stage of a product/process.

Whatever the technique, it will not work if the organisation's approach to quality improvement is not right. Without an effective policy for introducing SPC, the techniques will be no more effective than they were as part of SQC activities. They will be restricted to a few areas, and utilised in isolation from each other. They will be introduced by enthusiasts but fail because of the lack of support across the organisation. Implementing SPC requires a sequence of operations. For the larger organisations it will take years rather than months. Attitudes need to change and that will not happen as a result of short-term activities. Even with the relatively limited experience of SPC implementation in the West one message is clear: rushing the programme will almost certainly lead to failure, in part at least. There is no way that massive training can be effectively undertaken in a relatively short time. If the programmes are hurried the West will fall further behind the Japanese rather than catch up.

The next chapter provides a typical sequence for implementing SPC, with amplification of some of the major issues raised.

Chapter 15 Implementing the SPC programme

15.1 Introduction

Understandably, the bulk of this book has been taken up with describing the various SPC techniques. The need for a change in attitude has been emphasised and the point has been repeatedly made that SPC is not just a programme of chart filling. It will only operate successfully in a climate where barriers are being broken down, operators fully involved, supervisors trained to lead, managers committed, customers (internal or external) recognised as being important, and so forth. These factors and others are part of a much wider programme which requires a series of steps to be brought into effective operation.

This chapter covers the various issues which need addressing when an organisation decides to introduce SPC. There are certain guidelines which need to be followed if the programme is to have a chance of succeeding. There may be slight variations between organisations and the programme must be flexible enough to allow for adjustment to suit particular needs. However, there is now sufficient evidence to indicate that failure to follow the guidelines will, at best, delay the programme and, at worst, sabotage it .

15.2 The implementation programme

Fig. 15.1 shows a suggested sequence for introducing SPC into an organisation. Whilst this sequence is not immutable, it does provide a perfectly adequate sequence for introducing the programme, and the various stages will be used as the basis for the discussion here.

15.3 Obtaining commitment

SPC programmes are no different from any other quality improvement programmes in that unless the senior management is committed then no progress can be made. In fact it is best not to introduce the programme at all unless there is a guarantee of support from the top.

How much commitment is there to SPC programmes in industry in general? Not as much as is required for success, according to surveys carried out in a cross-section of British industry. Senior executives are not attending open 3-day SPC courses for a variety of reasons . As a result , the majority of those who do attend training courses, such as those provided for suppliers to the Ford Motor Company, are not in a position to influence the company when they return. Keen and motivated as they are, they return from the course only to be faced with resolving the problems which have accumulated on their desks in their absence.

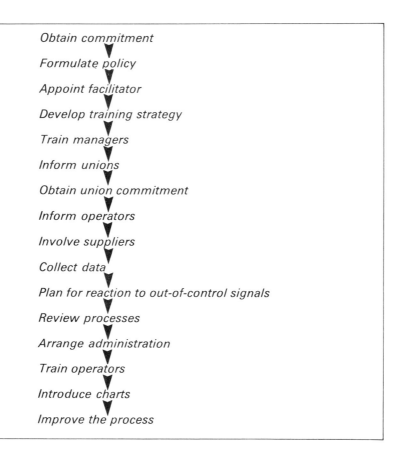

Fig. 15.1   SPC implementation

Senior executives typically believe that they understand what is meant by SPC (after all many of them studied statistics at university). They are therefore committed to the programme in theory, yet in practice avoid direct involvement by allocating the responsibility to a subordinate. As Deming states, it is always somebody else's problem.

So how can the commitment of senior management be won? Here are some suggestions:

- Senior executives need to be made aware of the true performance of the organisation by determining the quality costs.
- Executives need to understand what is meant by SPC. This requires a presentation, usually by an external consultant. This presentation should bring out the need for management commitment as well as providing experience in the use of the statistical tools and an understanding of the management philosophy associated with SPC. Typically, a day is required to provide the detail, but again much depends on the organisation. For example, it would be appropriate to give a shortened version to Main Board Directors. In a smaller organisation, the senior executives would also benefit by attending a standard 3-day course, as provided for the middle managers.
- The organisation concerned should be exposed to the SPC experiences of other organisations. Success stories, visits and (perhaps as effective as any) a presentation by a senior executive from another company who can talk at first hand of his SPC experiences, all help to get over the need for commitment from the top.

Following the initial commitment the support must be sustained. There are three ways of providing this:

- Projects are needed to provide experience and to set a leadership example.

- Senior staff should support the courses for middle managers and operators by attending at appropriate times.
- Regular visits should be paid by senior management to operational departments. There should be a genuine interest in the charts on display and a demonstrable commitment to improvement (within the necessary limitations of time and money). Actions speak louder than words.

In one organisation the operators in the packing area complained about the lack of a suitable truck, which meant that for months they could not do their job as effectively as they would have liked. With the advent of SPC under a committed General Manager a change in management attitude was apparent. SPC charts in the packing area backed up the need for a truck, and following a management visit to the area and discussion with the operators concerned, an order for not one, but two trucks was put into effect. The operators were also asked to provide details regarding the size and design of the trucks they would like.

Immediate positive response from senior management in this way is unusual. When it happens it provides a tangible example of management support for SPC. Not only that, the operators now know that the onus is on them to fulfil their part of the bargain.

This senior management commitment, therefore, is crucial. It must be seen to operate in practical ways – mere words are not enough. There are sufficient quality circle programmes littering the roadway of quality improvement to serve as reminders of the dangers of false promises, unsupported slogans and misguided management. If a programme has failed once it makes it that much more difficult to succeed with something similar in the future. For far too many companies SPC programmes may be their last chance. If they get it wrong again this time then the opportunity for change and improvement may not present itself again.

Once management commitment to SPC has been gained, a corporate plan must be formulated.

15.4 **Formulating a policy**

A plan is required to provide the correct framework for the introduction of SPC. A first step is usually the appointment of a steering committee and a facilitator. The role of the facilitator is anlaysed in Section 15.5.

Amongst other items, the plan should detail:

- The aims and objectives of the programme and the implications regarding corporate policy.
- The managerial responsibilities.
- The training strategy.
- The resources required for the programme.
- Financial requirements.
- Time scales.

The steering committee at this stage will typically consist of five or six senior managers and the facilitator. In some companies it may comprise the entire management team initially. The representation on the committee may change depending on the thrust of the programme at a particular time.

The steering committee should take into account the various quality initiatives which may be in progress or under consideration. For example, if quality circles are operating, then the SPC programme should allow for the integration of quality circle activities. Similarly, there may be a demand from the customer that BS5750/ISO9000 is implemented. Again, as far as possible, the development of the SPC programme should mesh in with any demands regarding a systems approach. Alternatively the company may already be committed to adopting a quality improvement programme based on the views of Crosby or Juran rather than Deming. Whatever the circumstances in the company at the time, the SPC policy must take into account these various factors, perhaps picking the best of each philosophy, adapting it to the advantage of the organisation, integrating it with existing programmes in other disciplines, and fine-tuning it over time in line with the philosophy of continuous improvement.

Each organisation will devise its own policy within the general guidelines proposed above. The approach of one company in setting up a policy statement may be of interest here, and is described in the following case study.

Case Study 14: Century Oils

Century Oils, which has its headquarters in Stoke-on-Trent, is a medium-sized British company with offices all over the world. The company has an excellent reputation as a producer of specialist lubricants. Over the last four or five years it has invested heavily in new plant and buildings and made extensive use of the latest technology in enhancing its image as a professional organisation. There is a top-down commitment to provide customer satisfaction and to increase its share of the market, and SPC is a part of this programme.

At a very early stage in the SPC programme the company produced a policy document on SPC. It had always adopted the approach of trying to make programmes unique to itself and hence SPC was defined as a major activity within the wider concept of Century Process Control (CPC). The policy document included:

- A general policy statement.
- General objectives.
- The benefits of CPC.
- Operating principles.
- Specific CPC responsibilities.

Within these sections were definitions and statements concerning the spectrum of proposed SPC activity. Of particular interest here is the detail relating to the specific CPC responsibilities of senior management, the CPC Centre Manager (the facilitator), activity team leaders, activity team members, steering committee members, departmental managers, research and development, quality assurance, and staff/operators in general.

Throughout the document there is a theme of group working, problem-solving, communication, performance indices (C_p, C_{pk}), control charts, presentations, training, improvement and control plans – all of which reflect the new way of operating based on prevention rather than detection.

With the policy document prepared, training followed on a top-down basis. Projects were undertaken, a 1-day seminar arranged for suppliers, and a professionally produced marketing video on SPC made available to suppliers and other interested parties. Not surprisingly the predicted benefits of SPC soon began to materialise. The decision to set up a policy document had played a major part in the successful first step down the road of continuous improvement.

15.5 Appointing a facilitator

The facilitator plays a key role in establishing, monitoring and developing the SPC programme. Ideally, the post should be full-time, but this will depend on the size of the organisation. In the smaller companies it may mean a shared responsibility with another activity. In larger organisations it could result in a facilitator in each plant, reporting to a senior facilitator based centrally.

The responsibilities associated with the position of facilitator include a core list as follows:

- To offer practical advice to managers and others who are beginning to use SPC techniques.
- To assist in the development of a training strategy for the organisation.
- To provide information on all aspects of the SPC programme on a company-wide basis.
- To monitor progress in the various areas and to provide assistance, directly or indirectly, as necessary.
- To ensure that action taken is prompt and effective.
- To provide advice on the various statistical methods. Ideally the facilitator should be able to provide this directly. If not, in the short term external support is necessary.
- To set up a monitoring system on the progress of the implementation. A major reason for this is to be able to assess the commitment of the organisation to continuous improvement.

The facilitator should report directly to the Site Director.

Without a facilitator an SPC programme will not succeed, yet two surveys (Appendix I) have shown that far too many organisations undertaking such a programme have not appointed one. It is not surprising that these organisations find it difficult to make the progress they are looking for.

Where does the facilitator come from? It is interesting that whenever a new programme is being implemented in an organisation, an evangelist comes forward. It is the same with SPC. The facilitator often stands out prior to appointment. He or she is a good communicator, is interested in statistical analysis in a practical way, can work with senior management, has a good rapport with the shop floor, has the trust and confidence of all levels of the organisation and generally enthuses about SPC and all its implications. The background of the person would tend to be technical/quality.

The post of facilitator is not a permanent one. As the programme develops and SPC runs under the control of operational and managerial employees, then the facilitator is no longer necessary. It would seem, however, that time-scales have been misjudged in the early attempts at introducing SPC. Programmes are long-term, and the position of facilitator is therefore likely to be more permanent than was first thought.

One of the responsibilities of the facilitator is to develop a training strategy.

15.6 **Developing a training strategy**

Experience with major organisations and their suppliers is showing that training strategies are conspicuous by their absence. Senior management faced with a need to institute SPC react, in typical crisis fashion, by running a course or two. Faced with an impending Ford assessment, or an insistent customer who needs to be impressed, another course may be added on. This *ad hoc* response to what is a very long-term programme involving changes of attitude as well as familiarity with technical skills means that SPC gets off to a bad start. In some cases the position is even worse. It has been very much the norm for training in SPC to be set at the middle management/technical level with no senior management representation. The training has been in the traditional SQC mode rather than being based on the philosophies of SPC and continuous improvement. It means that perhaps two to three years later the requirements of a major customer will necessitate the planning and implementing of a new training programme. Hence, not only has time and money been wasted but the previous training has not really been appropriate for the task.

The training strategy must take into account a series of issues, such as: How many people need training? What different types of courses need to be provided? Who will do the training? When will people be trained? Will the training take place off-site or in-house? What back-up is there to the training? What other training requirements is SPC exposing? There is a real need for a training plan to take into account the requirements of the organisation. The position is not helped by the savage pruning of training departments, certainly in the UK if not in the West as a whole. With no training staff to call on, organisations are forced to use outside agencies or to use other staff from, for example, technical/quality departments as trainers. Having said that, it does seems to be the case that technical staff are often preferable to those in training. It has much to do with the fact that trainers are often instructors in craft skills and not able to step quickly into SPC training roles. Whatever the scale of the programme it falls far short of a typical training programme in Japan (Fig. 15.2). Training programmes on this scale, which have now been operating for many years, indicate the task that faces the West.

With a training strategy in operation it is possible to see the sequence of training requirements. The programme starts with the managers.

15.7 Training managers and supervisors

Unless managers are familiar with SPC techniques, and recognise that it is their responsibility to take the lead in changing the system and the attitudes that go with it, then the operators will not respond.

Typically, managers would be trained in SPC over 3 days. This would be a minimum period. The course would include videos, presentations and, particularly important, hands-on exercises. There is no substitute for learning by practice. Measuring rods, throwing dice or counting beads all provide a visual, practical introduction to the statistical ideas on which SPC is based. At the same time, the various exercises relating to the charts themselves provide an opportunity to reinforce the philosophies associated with the programme. Deming's 14 points, for example (Appendix A), are relevant in one way or another to all the patterns produced on the various charts.

Typically, supervisors are trained at the same time as the managers. For both, however, allowance must be made for age and educational background. Some staff will be long in experience and short in numerical confidence. Handling numbers, determining control limits and using calculators may not come easily. Some preparatory training is therefore beneficial. However, the implications of management attitudes probably register earlier with this age group than with the younger staff who have degrees or diplomas. The technical aspects of SPC may not be such a problem with the younger age group, although even here the educational experience of three or four years at a university or polytechnic will have left its mark. Major industrially based SPC initiatives have now shown up the relative failure of the UK educational system to turn out technicians, managers and supervisors who have knowledge of the tools, techniques and management philosophies that they need to do their job in the industrial and commercial world. This issue opens up a debate regarding the role of higher education which can only take place elsewhere than in this book. It is sufficient to suggest here that the current heavy demand for training in practical understanding of simple statistical ideas would not have come about had educational establishments put emphasis on practical training needs rather than educational ideals.

1 Top management	5 days
2 Middle management	10 days
3 Basic course	23 days
4 Elementary course	8 days

Fig. 15.2 Statistical training in Japan

Supervisors play a key role in SPC programmes and therefore further skills are required of them. They need expertise in presentation methods, leadership, communication, problem-solving, etc. Those organisations with a commitment to training in the widest sense will have provided these skills as part of standard company policy. For too many organisations, SPC training is exposing a failure to fulfil these other training requirements. The use of project teams, inter-departmental groups, and problem-solving teams to identify the reasons for the special causes shown by the control charts demands a level of expertise in the supervisor which is unfortunately lacking in many companies.

A major difficulty associated with the standard 3-day SPC training courses is that after the course many still lack confidence in the use of the charts. Is it an $(\bar{X}, R)$ chart or a moving mean chart? Do I use a c chart or an np chart? How do I calculate C_{pk}? These are typical questions raised following the standard course and they emphasise the need for follow-on training.

It is at this point that projects are so important. At the end of each course a project is chosen by the course member. The project should relate to a process directly relevant to that person's activity and will give experience in determining the type of chart to use, collecting the data, calculating the control limits, determining the special causes and, if possible at this stage, pinning them down to factors which have been operating on the system. The project should not be used as a weapon against another person, section or department. A first step in breaking down barriers is to get one's own activities under control. Only then is there justification in approaching the internal customer to work on problems which may be more his responsibility.

The facilitator plays an important part in assisting with projects. With his or her guidance the projects are developed to a stage where some 6 weeks after the course a further 1-day course is held at which all the trainees give a brief presentation on their project to the other course members.

The projects may or may not be developed further. A management committed to SPC will have accepted that training courses at management level will involve projects which could be mainly training exercises only. Administrative issues regarding the projects need resolving at this stage (see Section 15.15). As a result of this some will be shelved, some developed, and some held over until operator training is in progress.

The importance of projects cannot be over-emphasised. Most managers recognise where their problems lie, but the frequency of occurrence of a problem does not reflect its importance. Projects provide a true picture of events and often belie the unsubstantiated views of operations which were previously held.

15.8 **Informing the unions**

The trade unions in the organisation need a presentation and/or training on SPC. This may mean the Managing Director talking informally with the Joint Negotiating Committee, or a brief formal presentation, or a formal training course. Whatever the format, the unions need to be told about the company's intention to introduce SPC. This does not mean that union agreement is being sought. Rather, the unions are being informed, with a commitment from management to respond positively to the issues arising from the programme which conflict with traditional union responsibilities such as payment schemes, bonus rates and job descriptions.

If the presentation to the unions can be coupled with training, then so much the better. A typical half-day course which covers the statistical training programme in Japan, quality costs, the emphasis on prevention, an exercise involving measuring and plotting, the limitations of tolerances, and the philosophy of continuous improvement is more than sufficient to indicate the basic ideas involved.

In some cases the union personnel may attend a 3-day training course but in their capacity as supervisors not union representatives. This is helpful because increasing the exposure of union staff to the techniques and philosophies of SPC programmes can only benefit the cause.

Experience to date indicates that trade unions will co-operate in SPC programmes as long as there is consultation and their views are taken into account. The initial presentation, whatever its form, should lead on to union commitment. If it does not, then other factors are operating. It may be that management and workforce view the factory floor as a battleground on which each tries to score a victory. With long-standing traditional attitudes such as these in operation, it will take more than SPC to soften the situation. Thankfully, with more enlightened management and more professional unions, these situations are becoming far less common. If there is still a poor industrial relations record, however, then co-operation between union and management to implement SPC may not be possible.

15.9 Obtaining union commitment

There is little evidence of union opposition to SPC. The unions can see that SPC programmes will help them to put right some of the wrongs which have pervaded Western industry over the past few decades. Unions know more than anyone that scrap parts, reject material and incorrect paperwork have been used to keep production lines in continuous operation and administrative processes in constant activity. Unions recognise the barriers that exist – the fear engendered in the workers by autocratic managers. They know only too well the injustice whereby operators are criticised or even penalised for poor performance when it is the system that is at fault. Hence unions will support SPC programmes because the philosophy they are based on is right. However, they do have reservations, the extent of which depends on the nature of the organisation, the way that management and union relationships have developed and the results of these relationships. Two concerns in particular are evident: piecework and collective bargaining.

15.9.1 Piecework

By its very nature, SPC requires a break in production so that charts can be plotted. Hence piecework earnings will be reduced. Operators cannot be expected to take a loss in earnings as a result of the introduction of SPC and therefore in the short term there is a requirement for management to compensate them financially. It is understandable, for example, that traditional unions will seek that their members receive money for plotting crosses on a chart: the role of the operators is changing, and as such demands a payment for carrying out a different job. These first skirmishes should gradually be replaced by more constructive discussions as SPC becomes more operational and as the unions see that the management commitment to change is genuine.

In the long term, SPC will mean that piecework is phased out and replaced by group or plant bonus schemes. Piecework and good quality do not go together in any case, and any organisation committing itself to SPC and the Deming philosophy should regard the elimination of the piecework system as a high priority. Again, full consultation with the workforce is the key. Once the termination of piecework has been successfully accomplished the operators will be afforded a pride in their job and more decision-making responsibility than they have ever had before.

15.9.2 Collective bargaining

The introduction of quality circles in Western industry in the late 1970s first alerted the trade unions to the fact that their traditional role in collective bargaining was being undermined. In the 1980s SPC could be seen as imposing another threat. This should not be so. If unions are consulted at a very early stage, their views listened to, information on company performance provided, and so forth, then mistrust and conflict should be avoided.

With the right approach, union responses can exceed expectations. One shop steward in a traditional industry recognised the widespread company benefits of SPC and recommended to management that SPC principles be incorporated in the company's induction programme for all new employees. This approach to SPC typifies all that is best in the trade union movement and points the way to far closer positive collaboration between management and unions in making the organisation competitive.

Once the unions are committed, the operators need to be informed.

15.10 **Informing the operators**

The operators will already be aware of possible changes in operating conditions, either directly through the unions or indirectly through sight of control charts being plotted by management or supervisors. It is imperative that as soon as possible after union commitment is forthcoming, the operators are informed about the SPC programme.

How this is handled depends on the size of the organisation. For smaller companies a talk from the Managing Director may be sufficient. In a larger organisation the communication exercise needs to be carefully thought through. In one company an effective information brochure was designed and sent out with the pay packets. In another case a half-hour video was specially produced. The first sequence on the video featured a senior company executive who explained the need for SPC in terms of the company's performance in the marketplace. The remainder of the video covered SPC in a simplistic way and explained how the role of the operator would change. Some 8000 employees were shown the video to minimise communication problems as the management training was undertaken.

It must be emphasised that this awareness programme does not involve training. It is a case of letting the operational staff on the shop floor or in the office know that a comprehensive programme is being instigated and that they will play a key part in the drive for improved company performance.

Once the organisation is fully informed of the programme, it is appropriate to look outwards to the supply base.

15.11 **Involving suppliers**

Major manufacturers face a dilemma regarding their suppliers. Once a company has accepted the partnership philosophy of working with its suppliers in a long-term contractual relationship based on quality not on cost, at what stage does it go out to the supply base with details of its requirements regarding quality? Does it lead by example and put its own house in order first, accepting that in doing so it could be delaying overall improvement because a large percentage of the company's activities depend on the supply base, or does it approach its suppliers immediately?

The answer lies somewhere between the two extremes and is unique to each organisation. Whatever the decision, the result will depend to a large extent on the company's relationships with suppliers in the past. Has there been frustration rather than co-operation? Has one supplier been bought off against another? Has it been the case of the cheapest quote winning? Has the supplier been dictated to and/or cajoled rather than being treated as an equal partner? If an organisation is to compete effectively, it can only do so if long-term relationships are set up with suppliers chosen on the basis of quality, in all respects, not cost. Single-sourcing now becomes the order of the day. The supplier must be seen as a partner, to advise and co-operate with the customer in a continuing programme based on trust, respect and mutual collaboration.

Companies have approached this task in different ways. Suppliers have been invited to sit in on SPC training courses. Presentations to the customer's senior executives have been made by the supplier's representative. Suppliers have been informed as to where help and advice on SPC matters can be obtained. Alternatively, 1-day presentations to the main suppliers have been organised which seek the co-operation of the supply base in producing a highly competitive product.

Any creditable organisation must seek the support and co-operation of the supplier without a 'big stick' operation. After all, an organisation is only as good as its supply base. If the prime manufacturer has some 60% of its final product coming from outside the organisation, then it follows that the success of the customer is dependent on the success of the supplier involved.

Various company guidelines for introducing SPC to the supply base are available. For example, the Ford Q101 document sets out detailed supplier requirements with reference to the statistical implications in the Ford SPC manual. Hewlett-Packard and others also provide schedules that make reference to the need for statistically based operations.

Depending on the size of the organisation, the sequence of SPC implementation described in this chapter may well be repeated in the supply base, and again in the sub-supply base, and so forth.

Many references have been made to the influence of a major customer in this programme of continuous improvement. Fig. 15.3 illustrates the point. A vast, continuing programme of supplier involvement, requiring training and experience in SPC, is being set up. Some companies caught up in the Ford supply chain, for example, are ones which at first appear to have little connection with the automotive industry.

The major customers would prefer their suppliers to introduce SPC of their own violition , and not because it is being demanded of them.It is disappointing that so few companies have seen the possible advantages of introducing SPC without external pressure. Those which chose to implement SPC programmes four or five years ago will have rightly gained an edge on their competitors.

15.12 **Collecting of data**

All the previous stages in the implementation sequence have been leading up to the

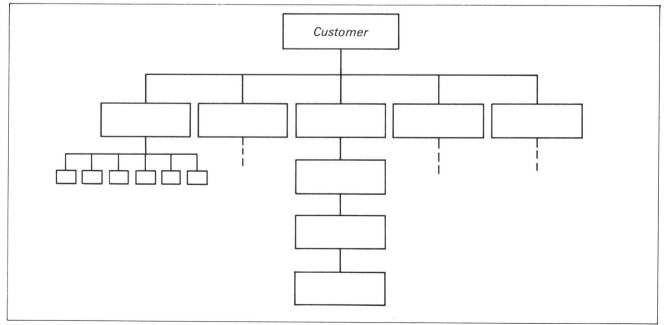

Fig. 15.3 SPC chain

crucial step whereby data is collected. There is rarely a shortage of data in a company. However, SPC requires correct data, taken at the right time and in the right place by the right people. Hence it is necessary to set up the SPC framework before data is collected.

The data collection stage can be looked at in two ways. Firstly, it may mean collecting, and probably reformatting, data which is already available so that priorities for SPC application are highlighted. There are various methods of doing this but they all involve a systematic approach and not a gut feeling based on experience. The task will involve referring to computer-generated lists of information relating to the processes, assessing quality control records, and looking up charts and tables which could be of value. The data required is unlikely to exist already in exactly the right form for analysing and plotting. SPC programmes are already exposing the shortcomings of many computer systems: not only do they provide far too much data in tabular form, which makes realistic decisions impossible, but the information required for control chart purposes is lacking.

Secondly, fresh data is required to build up control charts relating to the processes concerned. At this stage the charts will refer to management processes, i.e. they will not involve activities directly relating to operators. For example, a personnel manager could analyse absentee records and injuries, while an accountant could monitor the percentage of incorrect invoices. It is unwise to involve the operators until the stage is reached when plotting charts will shortly be a major activity in their area and on the machines and processes they are working on.

A vital requirement, which could form the basis of a project, is the collection of quality cost figures. Someone in the organisation must be made responsible for this activity so that a figure can be obtained which indicates the true problem regarding the cost of poor quality.

It will become apparent that collecting data is a bigger task than it at first appeared. As the programme develops other data will be required, such as machine capability figures.

One of the major reasons for collecting data is to provide guidance on the location of the pilot areas. Management will already have some ideas on where the main problems lie, but must accept that until the data is collected the exact nature and degree of the problems are not known. A Pareto analysis on the data will identify the 'blitz' areas. As a result decisions can be taken on the nature of the charts required. For example, if visual defects in the product are a major problem, then attribute charting will take top priority. However, it could be that a secondary analysis indicates the need for an $(\bar{X}, R)$ chart earlier in the system, in which case a second type of chart is needed.

Life being what it is, processes will go out of control, for worse or for better, irrespective of the type of chart, and this must be allowed for in setting up the SPC programme.

15.13 Planning for reaction to out-of-control signals

A strategy must be developed for dealing with special causes when they occur. The SPC programme becomes a meaningless exercise if there is no positive reaction to what the control chart is saying.

The presence of a special cause requires action to find the reason. The first step is to use the skill and experience of the operator. If that is not sufficient – and as discussed earlier, special causes on the R chart, for example, are not ones that operators can influence – then a problem-solving group is required. The operator, supervisor and technical personnel will form the nucleus of this group but its exact nature will depend on the particular problem. Groups may be inter-departmental, for example, or involve a mixture of day-shift and night-shift staff. However they are organised, they must have a strategy for action.

Before such groups meet, the organisation must have a policy on the implementation of the four rules for determining special causes. Is the process stopped as a matter of policy when a point is outside the upper or lower control limit? Is there firm management commitment to assess the situation, realising all the implications, when the process is out of control yet still capable and there is pressure to keep the line running in order to satisfy an urgent need for the product? Are operators who have been trained in SPC charting techniques and promised management support in practice told not to bother when they inform management of issues raised by the charts? Are operators reacting to hard-line supervision by plotting false values because they fear that the charts are being used to penalise them?

Clear guidelines are necessary, just as much when a special cause indicating improvement is noted as when a signal showing deterioration is recorded. As has been repeatedly stated, problem-solving teams working on improvement signals provide the key to breakthrough and company survival.

15.14 Reviewing processes

Once the decision has been taken on the location of the major thrust of SPC at operator level, the particular processes are reviewed. What is to be measured, where, and how? When are readings to be taken, and how often?

The first step is to eliminate traditional, but now outdated, methods of quality assessment. Go/no-go gauges must be disposed of and replaced by monitoring equipment which gives measurements of the process, or product. Comments in earlier chapters have warned against premature use of such equipment with computerised visual displays. Similarly, equipment should not be acquired which is excessively accurate. A careful analysis of the requirements of the company regarding monitoring equipment is required which will allow for the gradual introduction of SPC outwards across the organisation from the pilot area.

15.15 Arranging administration

Introducing SPC in an effective manner will, naturally, require some administration, but this must be kept to a minimum and introduced only as and when necessary. If SPC is seen to be associated with bureaucracy, empire building and excessive paperwork it will receive the same adverse comments as all other programmes which are overweight in systems and procedure and underweight in effectiveness and involvement. Any systems required to back up the SPC programme will be sufficiently provided for by ISO 9000. In any case, the SPC programme does require a formal quality system behind it in order to operate successfully.

The administration associated with SPC is concerned instead with different issues: preparing the appropriate charts, handling the projects and storing the project reports, collating the various projects, updating the charts, monitoring the problem-solving groups, assessing C_p/C_{pk} values, and so forth. If the necessary documentation is not kept up-to-date, then true progress cannot be assessed. Even worse, the operators will see management slipping back to their old ways of ignoring signals and making up the rules as they go along.

Other aspects of the administrative side need consideration, such as the publication of an SPC newsletter, SPC notice boards, the development of a company-specific SPC manual, and introductory brochures. A common message from those organisations making successful inroads into the programme is that there is much to be gained by making the programme as unique to the company as possible. However, the support material must not appear to be mere decoration lacking in management substance.

Now that the programme has been set up the stage of training the operators is at last reached.

15.16 **Training the operators**

Operator training in an area should not be undertaken until everyone in a vertical line of responsibility in that area has already been trained in SPC. It is only natural to try to rush things – get some charts going, train a few operators, impress the customers that the company is serious about SPC. However, the customers are not naive. They can tell whether there is a serious commitment to the programme and one failing which is only too easy to detect is operators using charts with no support for action on the special causes. Hence there is a need for a logical structural approach to operator training which has been laid out in the overall training strategy.

This strategy must allow for the fact that with a large organisation of, say, a thousand or more employees, it could take an unacceptably long time for the operators to come into the training programme. Therefore a vertical 'slice' of employees in one area should be trained initially. The area chosen may relate to a product for a certain customer, or to a certain process. By restricting senior management training initially to those directly involved with the pilot area it is possible to reach operator-level training in a few months.

Operators cannot be effectively trained in less than 3 days. Allowance also has to be made for additional training in handling numbers and in the use of a calculator. It is important that the training is made directly relevant to the work area involved. It should therefore involve the components, packages and paperwork which are familiar to the workforce concerned.

Various forms of training are available, including several SPC videos, computer-based systems, interactive packages and trainer-led resources. Each has its advantages and an organisation must choose the most appropriate for its particular needs. Two comments can be offered. First, it is advisable not to assume that one package will cover everything. 'Instant expert' training programmes in SPC appear with as much regularity as instant computerised charting systems. Secondly, there is a requirement for personnel in the company who are familiar with the philosophy and techniques of SPC and are also trained in passing on this knowledge in a way appropriate to the level of employees they are teaching. Such people are not a common commodity. There is no substitute at some stage for tutor-led groups, or even one-to-one training, if that is appropriate for the company.

Training should involve experience of filling in the actual charts that operators will be using. The training should also be timed so that the techniques will be introduced onto the shop floor very soon after the training course.

It is open to question how much should be covered in the course. Some organisations have been training their operators to calculate the control limits. This has a lot to recommend it. If the operators can be trained to do as much as possible on the charting and calculation side, it not only gives them greater involvement but also releases quality and technical staff to do other tasks.

15.17 **Introducing the charts**

There are a number of practical issues involved with actually introducing the charts. Are the charts to be in special workstations, on stands, or on the machine itself? Are pens or pencils provided? Are there rulers for drawing the lines? Are calculators necessary and if so are they available?

It is vital that the charts are filled in correctly and a useful procedure is to provide an operating schedule for each chart that makes it quite clear what is expected of the operator. The operator should be encouraged to take on as much responsibility as possible regarding the chart and the process, and should also be provided with tangible evidence that there is strong management support for the programme.

15.18 **Improving the process**

With the charts operating, the cycle of continuous improvement begins. Data is collected, limits calculated, special causes eliminated if they make the process worse or retained if they improve the process, limits recalculated, and so on indefinitely. The critical role of the control charts in this cycle, and the importance of problem-solving groups in determining the special causes, should both be famililar by now.

The cycle cannot be hurried: it takes time and money to tackle special causes. However, there is no other way forward. There is certainly no future in returning to a system in which quality is achieved by mass inspection, action is based on opinions rather than data, and the views of operators are ignored. SPC is a vital tool in the struggle for profitability and success.

Summary

The 16-point programme for introducing SPC:
- Obtain senior executive commitment.
- Formulate an SPC policy.
- Appoint a facilitator.
- Develop a training strategy.
- Train the managers and supervisors.
- Inform the unions.
- Obtain union commitment.
- Inform the operators.
- Involve suppliers.
- Collect data.
- Plan for reaction to out-of-control signals.
- Review processes.
- Arrange administration.
- Train the operators.
- Introduce the charts.
- Improve the process.

If these guidelines are adopted, progress can be made. Continuous improvement could result in slight modifications to the sequence, building on the experiences of others in introducing SPC. Those who have already gone down the road of process improvement have encountered pitfalls on the way. The final chapter discusses these and provides guidance on how to avoid them.

Chapter 16 Avoiding the pitfalls

16.1 Introduction

The previous chapter provided guidelines for implementing an SPC programme. If followed, they should result in an organisation introducing SPC without facing avoidable obstacles. It would be insincere, however, to claim that the programme is infallible and that difficulties will not arise. The implementation programme proposed is by its very nature generic, and what may suit one organisation will not of necessity suit another. The size of an organisation and the nature of its business will affect the approach to be adopted, and therefore whatever the organisation pitfalls litter the road of continuous improvement.

This final chapter covers these pitfalls. The list of 27 items is based on typical experiences of organisations that have already implemented SPC programmes. Some of the items relate to points in the proposed implementation programme while others come as fresh topics. In either case they should assist a company embarking on SPC in avoiding the same mistakes that others have made.

16.2 Lack of understanding and commitment among top management

It is appropriate that this factor is first on the list as it is one of the most common problems observed in various SPC-related research projects. The lack of understanding takes different forms: it can operate at the technical level, it can be a failure to see the implications of SPC for the roles of managers and operators, it can relate to product quality itself, and in particular it can be a failure to understand that SPC programmes must be company-wide.

If senior executives do not understand the techniques and philosophies, then their commitment is paper-thin. It has been said before, but it needs saying again, that unless the senior executives are fully trained in SPC methodology, and are aware of the responsibility they carry to improve the system within which everyone in the company operates, then any so-called commitment to SPC is unproven. The organisation would then be well advised not to implement, at a lower functional level, any form of SPC programme because it will surely fail.

16.3 Lack of a plan

It would seem that SPC is typically introduced with little thought being given to the full consequences of the programme. It reflects short-term policies based on a quick response to try to impress a major customer. For example, SPC software is acquired before any training has been given, or any commitment forthcoming. Training

330 SPC AND CONTINUOUS IMPROVEMENT

courses are provided but almost on a random basis, with no thought being given to the nature and mix of the group to be trained or the duration and location of the course. No back-up project work is provided, no facilitator appointed, in some cases not even an overhead projector available to assist in the training. Taguchi methods suddenly become the order of the day and a rush of training is introduced in this field without recourse to analysing the relative position of SPC training, or the place of ISO 9000. In the great majority of such companies training strategies are not available. Whatever training is carried out under these conditions is of limited benefit and could even be self-defeating.

Major customers do not expect SPC miracles overnight. In fact if they see control charts appearing in rapid sequence they know that it must be pure camouflage. What they are expecting is a positive response to their questions regarding SPC. A well thought out training strategy, with clear guidelines as to the nature, extent and timing of the various programmes for SPC training is necessary. This should start with a commitment that all senior executives will attend a 1-day (as a minimum) SPC presentation, that projects will be undertaken and that a facilitator will be appointed to back up the programme. Without such a plan it is impossible to manage the SPC programme effectively, and the result will probably be more problems rather than fewer.

16.4 SPC is not company-wide

Again, the research available (see Appendix I) shows that SPC has far too often been restricted to narrow areas of company activity, typically ones which are influenced by the demands of a specific major customer. If SPC is limited to manufacturing areas, and possibly not even all of these, dual standards will be operating in the company.

It is acceptable for SPC to be applied first in those areas which will provide the maximum return, and these are likely to be in manufacturing and production. But the resulting improvement in product quality will not help the company if there are still administrative failures. There is little gained if the quality of components provided has improved out of all recognition yet they are still being delivered to the customer in inadequate packaging. Similarly customers want deliveries on time, not 2 weeks late. Delivering short numbers, sending an incorrect invoice or sending the product to the wrong plant all work against customer satisfaction.

It is not necessary to train all senior management initially. Earlier comments on training programmes have suggested that in larger organisations 'vertical slices' of personnel will be trained, enabling the operators in pilot areas to be trained as quickly as possible. However, as soon as it is felt appropriate, supporting departments must be put through the training courses and brought into the programme.

Quality activities are not the prerogative of one department. Quality is company-wide and therefore every employee needs to understand, to be trained, to appreciate his or her role, to recognise the internal customer relationship and to be made to feel part of an organisation which is genuinely open and committed to continuous improvement.

16.5 Lack of long-term commitment

SPC programmes are not easy; they require patience and perseverance They usually need a change of attitude in Western organisations and this will not take place in the short term. The programmes call for changes in attitude between department and department, manager and manager, manager and operator, and supplier and customer. Personalities do not change overnight, and for some personnel a change in attitude will be difficult to achieve at all. A long-term commitment to SPC is therefore required, which at present is all too often lacking. Management has to recognise from the outset that, as Deming says, there is no 'instant pudding'. SPC programmes require dedication and conviction over a long period.

Unfortunately, company personnel have seen it all before. Quality circles have come and gone. Quality campaigns were promoted in a blaze of publicity only to die when the first positive response was required from management. Posters have been displayed and nothing more done. Improvement programmes have been introduced, instant success ventures started, quick-fix consultants brought in. In all these cases the improvement in performance would have been small and not sustained. Unless improvement activities take place against a long-term plan for company growth and prosperity, no real progress is possible.

16.6 **Inadequate training**

Given the right format for the training programmes, understanding the SPC techniques should not prove too difficult. The tools and methods are basically simple, though they have often been made more complicated and too theoretical in the formal education courses available in the West. It is clear that a vast retraining exercise is needed to put right the previous shortcomings.

However, designing SPC training programmes does require careful planning. Here are some guidelines to follow:

- Train the senior executives thoroughly.
- Use appropriate outside assistance so that SPC experience from other organisations, not necessarily similar ones, can be utilised.
- Make the training practical, relevant and non-theoretical.
- Choose the right trainers.
- Provide follow-up assistance.
- Train all new employees.
- Provide training during working hours (essential at operator level).
- Make sure all employees are released at the appropriate time to attend training courses.
- Provide other relevant training.

16.7 **Failure to involve suppliers**

If organisations are serious in their approach to SPC then they must work with suppliers in a long-term partnership. Improving the organisation's own internal activity to near-perfection is fruitless if there is a major input from a supplier that is unsatisfactory. Hence there is a need to improve supplier performance, but through collaborative arrangements not punitive ones. If suppliers are not willing to engage in a partnership programme of continuous improvement, then it will be necessary to replace them by others who are.

16.8 **Emphasis on short-term profits**

It has been repeatedly emphasised that SPC programmes are long term. Hence if the organisation typically works on short-term profit margins and limited time scales for returns, it generates an environment in which it is impossible for an effective SPC programme to develop. Typically, companies are told that they have funds allocated for SPC with the proviso that within 18 months financial returns must materialise. Condensing everything into a short time scale leads to crash training programmes, an excess of control charts and a general feeling that buckshot has been fired at random in the hope of hitting an SPC target.

This emphasis on short-term profits is inevitable if senior executives are not made aware early enough of both the full implications of undertaking an SPC programme and their responsibilities in carrying it through.

16.9   **Commitment in only one department**

The Quality Department would appear to be the major catalyst for introducing SPC. If the programme is to run effectively then the SPC message must be directed from here to senior executive level for action. In the resulting company strategy the initial department involved, be it Quality or any other, should play a role no greater and no less than that of any other department. Unfortunately it is too often the case that only that original department remains committed to SPC. If that department happens to be the Quality Department then other difficulties result, as was discussed in an earlier chapter.

16.10   **Lack of funds**

Lack of funds is clearly a direct reflection of lack of management commitment. Examples abound: failure to send people on open training courses because of the expense, a reluctance to buy equipment or materials that the charts indicate are clearly needed, skimping on training, lack of cash incentives in general, the demand for additional work from operators without providing extra rewards, failure to release a member of staff as facilitator. The list is endless. SPC programmes do cost money, but management that holds back on releasing funds fails to appreciate that the company is probably carrying excessive quality failure costs.

16.11   **Failure to consult the workforce**

If the organisation has introduced SPC within generally agreed guidelines, then this pitfall can be avoided. However, it is not unusual for a programme to be introduced with no presentation to the union and a general disregard for the views of the workforce. It is small wonder that little progress will ensue.

Throughout the book the theme of operator involvement has been constantly stressed. No apology is offered for doing so. Given senior management support, the key to successful SPC programmes is the involvement of the operators. With consultation, training and encouragement they will play their part in helping the organisation to survive and prosper. The real doubt is whether management will fulfil its part of the bargain.

In progressive organisations there will be no implications of SPC for working practices, as problems of piecework and bonus rates will not exist. In the more traditional organisation these issues must be faced and resolved. A refusal to eliminate bonus schemes or an insistence that inspectors maintain the charts means that management has not fully understood and recognised that the solution to improved productivity depends principally on the action of management and not the operators.

16.12   **Under-estimating the workforce**

Quality circle programmes first started to question the traditional roles of management and operators in Western industry. As a result, the behavioural theories of McGregor, Taylor and others were compared and re-analysed. There appeared to be agreement on the need for more involvement of the operator in decision-making rather than just product-making activities, and for the encouragement of operator control in the widest sense.

SPC programmes are now building on the early successes of quality circle programmes. They are showing that over the years the workforce's ability to contribute to the success of the organisation has been vastly under-estimated. However, there is still a belief that operators are only concerned with the money they earn and have little real interest in the job they do whilst at work. SPC programmes are providing evidence that nothing could be further from the truth.

When the training programmes reach operator level and projects develop relating to the operators' processes, management is finding that it has unleashed an enthusiasm and a degree of involvement on the part of operators not seen before. This upsurge of progressive activity needs careful handling. It does not mean that management must vainly try to respond to the results of scores of projects. The operators are as aware as anyone of the constraints on time, money and resources in general. Positive action on solving one or two long-standing problems is all that is needed initially.

16.13 Failure to acquire adequate statistical support

There is usually no shortage in an organisation of personnel who, in the course of obtaining degrees, diplomas or certificates of one sort or another, will have taken a course on statistical techniques. However, it is becoming increasingly evident that the courses have been far too theoretical and do not match the requirements of industrial and commercial life.

This failure has been highlighted by the training programmes for suppliers to the Ford Motor Company. Such courses have revealed that instead of providing real, hands-on experience in applying simple statistical ideas to controlling and improving processes, educational courses have consistently exhorted students to satisfy requirements based on academic excellence. Pressure from industry and commerce is needed to instigate a change of approach within the education system. In the meantime, organisations implementing SPC programmes need counselling and advice, often from outside, in the practical application of statistics.

16.14 Lack of market research

Before an SPC programme is introduced, some groundwork is necessary regarding customer needs and requirements within the overall company programme for development. Is SPC being undertaken simply because of the demands of a large customer? Is the programme based on a herd instinct: everyone is doing it so we may as well follow suit? Not every organisation is introducing SPC for the right motives, and this must raise questions about whether management commitment to the programme is genuine. Far better than the customer-driven programme is a recognition that in the open market it is only by matching and then improving on the performance of a competitor that any organisation can remain in business. Single-sourcing arrangements make the commitment to SPC even more necessary.

16.15 Management by fear

Deming's 14 points for management (Appendix A) are easy to understand, but implementing some of them is far from easy. Management by fear is one of the points that is not easily resolved. Fear expresses itself in different ways: the hesitancy of a manager to express true scrap figures for the department, for example, or the reluctance of operators to record out-of-control signals on a chart because they believe that the results will be used by management as a weapon against them.

Removal of fear requires changing attitudes, and for some that may prove impossible. An autocratic manager will probably have achieved managerial status because of an ability to get the job done: a strong will, coupled with an element of fear, achieves results. It will not be possible to change that approach without changing personality. Hence a long time scale is required, in that it may require that manager to move on before open collaborative arrangements are possible.

16.16 Lack of middle management support

Even with commitment from the top, there are often difficulties at middle management and supervisory level because the staff there feel threatened by SPC. They see senior executives leading by example and recognise that the operators will now have charts to back up their assertions about the shortcomings of the organisation. Middle management may also feel threatened if operators know more about charts and problem-solving techniques than they do.

The solution is a commitment on the part of the organisation to train supervisors extensively in a range of disciplines which will enable them to provide the leadership required for the operators.

16.17   Lack of quality materials

As the SPC programme develops, goods inward inspection will gradually be replaced by assessment at the point of manufacture, using control charts as the basis of performance. The elimination of inspection-based systems needs to take place as quickly as possible. In the meantime, however, there must must be insistence that good-quality materials are required for use in production. If operators are still asked to work with material bought because it is the cheapest available, then they will be unable to do a good job. In a similar way, maintenance and housekeeping need to be upgraded as management fulfils its obligations to the quality improvement programme.

It should not be forgotten that quality materials are required by clerks, secretaries and typists just as much as by shop floor operators. If the typewriter does not work, the form is badly designed, pencils are cheap and unreliable, or documents poorly laid out for transcript, then it is not surprising if errors build up on the administrative side. These errors reflect just as badly on the organisation as would the delivery of a component which is not to specification.

16.18   Over-emphasis on computers

The tendency to over-emphasise the role of computers has been commented on in earlier chapters. Software packages abound, as do advertisements for computerised displays linked to electronic gauging. These features are an important and necessary aspect of SPC programmes, but they will not be effective without adequate prior training. Also a balance is needed between centralised systems which provide management with information and the requirement to maintain operator involvement in process control and improvement.

16.19   Moving too quickly

It is natural to be impatient for success. The desire to train a few operators in a pilot area and get charts up and running may seem overwhelming, but patience is necessary. It makes sense to involve the workforce as soon as practicable, but the dangers of doing so too quickly are illustrated by the number of SPC programmes which fail for this very reason. Within a strategy for training it is worth while using a guideline of not proceeding to the next level of training until there is confidence in the techniques and familiarity with the tools at the current level. It is better to delay the programme than to carry on knowing that staff are uncomfortable with it and feel hesitant about exposing their lack of knowledge to those next in line for training.

16.20   Lack of projects

Open SPC courses provide the opportunity of gaining basic exposure to SPC techniques and philosophies but they do fall short in not allowing personnel first-hand experience of setting up charts in their own work area. Properly constructed in-company training programmes make use of personal projects and these are essential in consolidating the training course.

16.21 **Pilot areas not chosen carefully**

Use of the Pareto technique will quickly indicate those areas where the greatest problems lie and hence the greatest benefits from SPC are to be gained. Some organisations have opted to introduce SPC in such pilot areas almost as a first step in their programme. There may be possible benefits, such as sorting out problems early on before the full programme is implemented. However, any such benefits could be put at risk by introducing charts on the shop floor without prior training . It is advisable not to involve operators until all the necessary previous stages of the implementation programme have been achieved.

The pilot areas chosen should involve operators who are keen to use control charts. Also the processes concerned should be easy to define and the charting techniques kept simple. There is always a danger of overcomplicating things at any point in the programme.

In the early stages it may be preferable to provide the charts with the control limits drawn in. The final decision on this would depend on the managerial view of how much is being asked of the operators. Whatever the chart, there should be encouragement for the recording of notes and comments.

16.22 **Monitoring products instead of processes**

SPC requires attention to the process rather than the product. The common chart is that used by operators in typical manufacturing environments such as the machine shop. The change-over from an inspection-based system to one based on prevention means that as a first step it will be, for example, component dimensions or package weights that are measured and plotted. This is still strictly product control rather than process control. If it can be shown that there is a close relationship between the process and the product, and the latter is easier to monitor, then this is acceptable in the short term. However, as the SPC programme develops and confidence builds, constructive use of the techniques demands that one concentrates on the process and monitors instead the parameters affecting the process (batch temperature, machine speed etc.). In the machine shop this would mean concentrating on the tooling involved rather than the component dimensions.

16.23 **Over-emphasis on one technique**

Just because the $(\overline{X}, R)$ chart is the most commonly applied chart, it does not follow that it is applicable to all situations. Over-emphasising one technique not only restricts the introduction of others which could be more appropriate; it also means that in a much wider sense the full potential of SPC is not realised. For example, the control chart, of whatever variety, is of no value unless action is taken to deal with the special causes. This requires other techniques such as problem-solving skills, Failure, Mode and Effect Analysis (FMEA), simple flow diagrams and performance indicator graphs. In some cases these techniques may be sufficient in themselves, without the necessity for performance-based control charts.

Of late there has been a wave of interest in the application of Taguchi methods to the design of experiments. Organisations seem to be switching their efforts from SPC to Taguchi. But Taguchi techniques, powerful as they are, do not replace SPC, just as, for example, SPC does not replace quality circles. Fig. 16.1 indicates the different phases in the development of quality improvement in Japan. It can be seen that some 10 years' experience of charting techniques was available before statistical methods were applied through the design of experiments. If it is to remain competitive with Japanese industry the West must use the Japanese experience to telescope this time scale. Some organisations are already recognising the considerable benefits which can accrue by using Taguchi methods. However, there is a failure to understand what is required if, as appears to be the case, some organisations are introducing Taguchi methods yet still operating tolerance-based systems.

All the various inputs relating to a total quality control programme need to be meshed together in a cohesive long term strategy.

16.24 Failure to respond to chart signals

Setting up the charts is one thing; responding to the charts is another. The onus is very much on supervisors to react to signals which the operators indicate will require action.

Operators, in the main, will be committed to SPC programmes because they realise that in the long term, if not in the short term, they will gain by having greater involvement in their job. Of course not all operators are honest, committed and sincere, but it is certain that a far greater proportion are than management has previously given credit for.

16.25 Failure to understand SPC

Many organisations claim to be running SPC programmes. Charts are on display in the production areas; employees have been sent on SPC courses; 1000 people have been trained in 12 months; money has been spent on control chart stands, Ford manuals, print runs of control charts, calculators, and so on. Unfortunately financial expenditure in itself does not mean that management understands SPC. In fact spending money to dishonestly impress the customer is as bad as not spending money at all. Once again, the understanding and commitment of the senior executives is the key.

16.26 Reluctance to change

Management of change now provides the basis of courses, presentations, seminars, and even after-dinner speeches. SPC requires a change in attitude as much, if not more than, any other programme related to organisational improvement. Organisations tend to feel that if the way they have done things in the past has worked, then there is no need to change. It is all too easy to rest on past achievements and fail to recognise the need to change and improve. But unless a company learns from earlier experiences, and builds them into its plan for the future, then

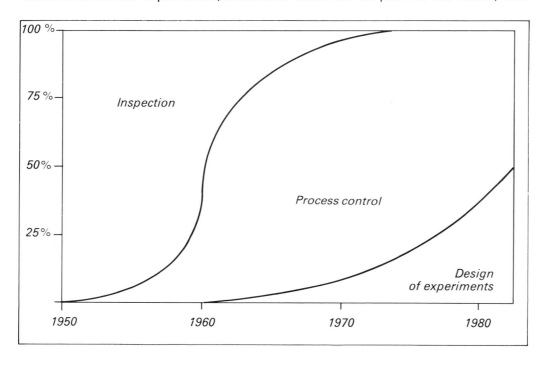

Fig. 16.1 Quality philosophies in Japan

there is little doubt that its customers will go elsewhere and that it will face cutbacks, loss of market share, and eventually failure.

16.27 General lack of knowledge and expertise in SPC

A major difficulty with SPC programmes seems to be a lack of knowledge. In contrast to the superficial impression given by many companies, surveys indicate that senior management and others generally lack knowledge and expertise.

Improvement can result from a series of actions – in particular training staff and determining quality costs. SPC is a relatively new topic and so the more senior staff will not have come across it when obtaining a formal educational qualification. Whatever experience the more junior staff and operators have had of educational establishments it is clear that further retraining is necessary, as is now taking place, to instruct as many personnel as possible in the tools and techniques which have been used extensively by the Japanese to achieve world leadership in a range of product areas.

16.28 Lack of concern for detail

Whilst there is a need not to overcomplicate matters, a consideration to detail in the planning and application stages of an SPC programme will avoid unnecessary problems. The administration of the control chart is a good example, as the following case study illustrates. The depth of detail which has been applied in the SPC administrative system of Ford New Holland Limited is impressive.

Case Study 15: Ford New Holland

To ensure that a production-driven SPC programme can expand, it is important to set up a practicable system to monitor the main parameters of data collection, statistical control and process capability. A second consideration is to minimise the amount of clerical work necessary to ensure blank charts are available to production areas when required. Since a performance-based system is dynamic, the system must also ensure that control limits and process capability (C_p, C_{pk}) status are up to date.

The system which has been implemented at Ford New Holland, and which has proved successful in addressing these considerations, is outlined in Figs. 16.2 to 16.6. Fig. 16.2 shows a block diagram of the system, which is based on a set of master charts allocated to each Superintendent. Each SPC application has a master chart (A) where all the information, including ($\overline{X}$, R) control limits, scales and C_p, C_{pk} information is in pencil and thus easily modified. A copy of this master is taken and the relevant control limits drawn on the copy. This becomes the print master (B). Copies of the print masters are stored in a file (C) to which the Production Foreman has ready access, and it is he who is responsible for ensuring that blanks are available on his line for data collection (D).

The Area Superintendent chairs a weekly review meeting which is attended by the Production Foreman, Manufacturing Engineer, a Maintenance representative, the SPC coordinator and operators as necessary. At these meetings the current charts are reviewed for the three parameters of data collection, statistical control and capability. The Superintendent then completes his status report (Fig. 16.3).

Those charts with control and capability problems are analysed and possible corrective action agreed. Charts which indicate significant change in either range or setting are recalculated by the SPC coordinator, on the completion of which the master chart and print masters are amended. All other completed charts are reduced to A4 size and stored in a binder which is retained by the Superintendent.

The completed status report is passed to the Area Manager's clerk who, on receipt of the reports from all the Superintendents, prepares an area summary (Fig.

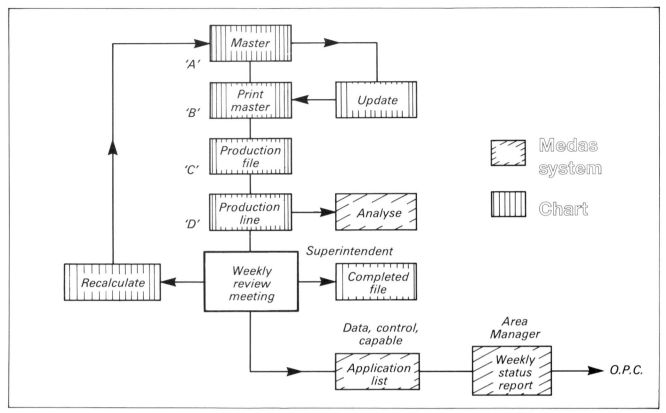

Fig. 16.2   SPC administration system

16.4) which is presented to the Plant Management Team. Examples of a typical completed Superintendent's report and Area Manager's status report are shown in Figs. 16.5 and 16.6.

The system is run by existing production personnel with statistical input from the SPC coordinator. It ensures that production staff are involved in the SPC programme and the weekly meetings, as well as ensuring that corrective action is ongoing on problem charts. In addition, it provides the coordinator with an effective forum for building confidence in the Production Department by using chart format and analysis. This provides the necessary foundation for production supervision eventually to take responsibility for the whole system. (Delegation of responsibility without first supplying this knowledge and confidence would result in failure.)

The final advantage of this system is that the Plant Management Team receive weekly overviews of the production process whilst the responsibility for the detail remains where it is most effective – at the Superintendent level.

The Medas system referred to in Fig. 16.2 is a Mechanical Manufacturing Engineering system which gives the shop floor information on processes, gauges, tools, work standard, etc. VDU monitors are available to all shop floor areas and therefore SPC analysis and administrative software has been developed and integrated into this system.

The case study highlights some of the issues covered in this chapter. It emphasises the importance of manual charting, and how computerised systems can support that and need not replace it. It stresses the need for problem-solving groups and effective communication. Above all it shows the importance of attention to detail in the SPC programme.

Summary

The pitfalls in an SPC implementation programme are:

Superintendent – Area A2

APPL	CHART	AREA		DESCRIPTION	DATA	CONTROL	CAPABLE
1	1C	CYL BLOCK	X0600	CYL BORES CENTRAL			
2	2C	CYL BLOCK	X4600	BRG LOCK DEPTH – SIDE 1			
3	3C	CYL BLOCK	X4600	BRG LOCK DEPTH – SIDE 2			
4	4C	CYL BLOCK	X4600	BRG LOCK WIDTH – FRONT			
5	4C	CYL BLOCK	X4600	BRG LOCK WIDTH – REAR			
6	6C	CYL BLOCK	X1100	POSIT HOLE 67 – BOTTOM			
7	7C	CYL BLOCK	X1100	POSIT TACHO BORE			
8	8C	CYL BLOCK	X1200	POSIT HOLE 70			
9	9C	CYL BLOCK	X1200	POSIT HOLE 71			
10	10C	CYL BLOCK	X1200	POSIT TAPPETS 1–3–5			
11	11C	CYL BLOCK	X1200	POSIT TAPPETS 7–9–11			
12	12C	CYL BLOCK	X1200	POSIT TAPPETS 2–4–6			
13	13C	CYL BLOCK	X1200	POSIT TAPPETS 8–10–12			
14	14C	CYL BLOCK	X4100	GEAR FACE DEPTH			
15	15C	CYL BLOCK	X4200	HOLE 67 – DIA			
16	16C	CYL BLOCK	X4200	HOLE 67 – POSITION			
17	17C	CYL BLOCK	X4200	HOLE 67 – SQUARENESS			
18	COMP	CYL BLOCK	X6500	CYL BORE – DIA			
19	COMP	CYL BLOCK	X6500	CYL BORE – QUALITY			
20	—	CYL BLOCK		DIRT INGRESS			

Fig. 16.3 Superintendent's status report

- Lack of understanding and commitment among top management.
- Lack of a plan.
- SPC is not company-wide.
- Lack of long-term commitment.
- Inadequate training.
- Failure to involve suppliers.
- Emphasis on short-term profits.
- Commitment in only one department.
- Lack of funds.
- Failure to consult the workforce.
- Under-estimating the workforce.
- Failure to acquire adequate statistical support.
- Lack of market research.
- Management by fear.
- Lack of middle management support.
- Lack of quality materials.
- Over-emphasis on computers.

AREA	NO. OF CHARTS	%			REMARKS
		DATA	CONTROL	CAPABLE	
A1					
A2					
A3					
A4					
TOTAL ENGINE					

Fig. 16.4   Area summary

Superintendent – Area A2   *example only*

APPL	CHART	AREA		DESCRIPTION	DATA	CONTROL	CAPABLE
1	1C	CYL BLOCK	X0600	CYL BORES CENTRAL	✓	✗	✗
2	2C	CYL BLOCK	X4600	BRG LOCK DEPTH – SIDE 1	✓	✓	✓
3	3C	CYL BLOCK	X4600	BRG LOCK DEPTH – SIDE 2	✓	✓	✓
4	4C	CYL BLOCK	X4600	BRG LOCK WIDTH – FRONT	✓	✓	✗
5	4C	CYL BLOCK	X4600	BRG LOCK WIDTH – REAR	✓	✓	✓
6	6C	CYL BLOCK	X1100	POSIT HOLE 67 – BOTTOM	✗	✗	✗
7	7C	CYL BLOCK	X1100	POSIT TACHO BORE	✓	✗	✗
8	8C	CYL BLOCK	X1200	POSIT HOLE 70	✓	✗	✗
9	9C	CYL BLOCK	X1200	POSIT HOLE 71	✓	✗	✗
10	10C	CYL BLOCK	X1200	POSIT TAPPETS 1–3–5	✓	✓	✓
11	11C	CYL BLOCK	X1200	POSIT TAPPETS 7–9–11	✓	✓	✓
12	12C	CYL BLOCK	X1200	POSIT TAPPETS 2–4–6	✓	✓	✓
13	13C	CYL BLOCK	X1200	POSIT TAPPETS 8–10–12	✓	✓	✓
14	14C	CYL BLOCK	X4100	GEAR FACE DEPTH	✗	✗	✗
15	15C	CYL BLOCK	X4200	HOLE 67 – DIA	✓	✓	✓
16	16C	CYL BLOCK	X4200	HOLE 67 – POSITION	✓	✓	✓
17	17C	CYL BLOCK	X4200	HOLE 67 – SQUARENESS	✓	✓	✗
18	COMP	CYL BLOCK	X6500	CYL BORE – DIA	✓	✓	✓
19	COMP	CYL BLOCK	X6500	CYL BORE – QUALITY	✓	✓	✓
20	—	CYL BLOCK		DIRET INGRESS	✓	✓	✓

Fig. 16.5   Completed status report

	90	70	60

AREA	NO. OF CHARTS	%			REMARKS
		DATA	CONTROL	CAPABLE	
A1	50	100	66	60	
A2	20	90	70	60	NEW OPERATOR—TRAINING ARRANGED
A3	34	100	91	85	
A4	18	89	89	78	GAUGE IN CRIB FOR REPAIR 2 DAYS—DATA COLLECTION NOW O.K
TOTAL ENGINE	122	98	77	67	

Fig. 16.6 Completed area summary   *example only*

- Moving too quickly.
- Lack of projects.
- Pilot areas not chosen carefully.
- Monitoring products instead of processes.
- Over-emphasis on one technique.
- Failure to respond to chart signals.
- Failure to understand SPC.
- Reluctance to change.
- General lack of knowledge and expertise in SPC.
- Lack of concern for detail

At the beginning of the book the point was made that control charts have been available since the 1920s. The Japanese have been using simple statistical ideas in industry since the early 1950s and few would question that their devotion to training in statistical and problem-solving techniques has been a major factor in them attaining their prime position in world markets. Yet, here in the West we are still struggling. Productivity may have gone up but an analysis of quality costs indicates a much more worrying situation. High failure costs are operating yet there seems an inability by management to do something about them. The tide is gradually turning, however, in that quality is becoming a key measure of company survival. There is a long way to go, though. There needs to be not only a much greater awareness and commitment on the part of senior management to change the systems to allow improvement to take place, but a change of attitudes in general, and a more positive response from educational establishments in providing industry with staff well trained in quality principles, problem-solving skills and a thorough understanding of the practical use of statistics.

If organisations do not change their attitudes to quality, competing effectively in world markets will be difficult, even impossible. The extent to which British industry will introduce SPC in an organised and committed way is open to question. It has to be said that the results of recent surveys do not encourage optimism. Unfortunately, there is no alternative. The West is only too aware of the progress made by Japan in the last 20 years, and organisations know what needs to be done in order for them to compete and survive. They must commit themselves to involving all their employees in programmes of quality improvement.

SPC provides one method of maintaining the competitive edge. Although it is long term, there are now sufficient examples of organisations where SPC has been introduced with effect. Not only have quality costs been reduced, operators involved and productivity genuinely increased, but the organisation is seen to operate more as a unit than a combination of diverse groups. For any organisation assessing its position in the market over the next few years, the decision to implement SPC could prove to be the turning point that leads to success rather than failure.

Appendix A. Deming's 14 points for management

1. Create constancy of purpose toward improvement of product and service with the aim to become competitive, stay in business and provide jobs.

2. Adopt the new philosophy. We are in a new economic age, created by Japan. A transformation of Western style of management is necessary to halt the continued decline of industry.

3. Cease dependence on inspection to achieve quality. Eliminate the need for mass inspection by building quality into the product in the first place.

4. End the practice of awarding business on the basis of price. Purchasing, design, manufacturing and sales departments must work with the chosen suppliers so as to minimise total cost, not initial cost.

5. Improve constantly, and for ever, every activity in the company so as to improve quality and productivity and thus constantly decrease costs.

6. Institute education and training on the job, including management.

7. Institute improved supervision. The aim of supervision should be to help people and machines do a better job.

8. Drive out fear so that everyone may work effectively for the company.

9. Break down the barriers between departments. People in research, design, sales and production must work as a team to tackle production and usage problems that may be encountered with the product or service.

10. Eliminate slogans, exhortations and targets for the workforce asking for new levels of productivity and zero defects. The bulk of the causes of low quality and low productivity belong to the system and will not be in the direct control of the workforce.

11. Eliminate work standards that prescribe numerical quotas. Instead, use resources and supportive supervision, using the methods to be described for the job.

12. Remove the barriers that rob the hourly worker of the right to pride of workmanship. The responsibility of supervision must be changed from sheer numbers to quality. Equally, remove barriers that rob people in management and engineering of their right to pride of workmanship.

13. Institute a vigorous programme of education and retraining. New skills are required for changes in techniques, materials and services.

14. Put everybody in the organisation to work in teams to accomplish the transformation.

Appendix B.   Table of P_z values

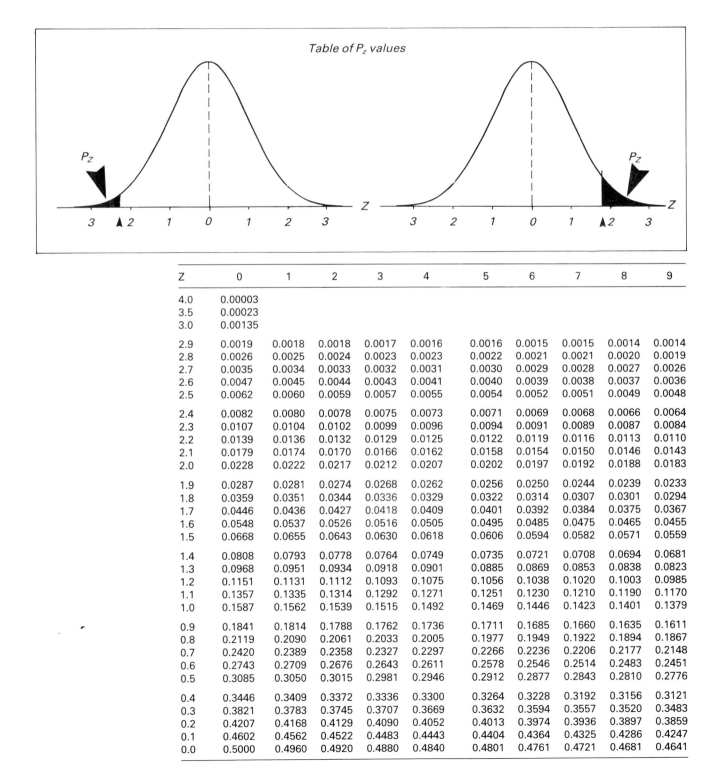

Table of P_z values

Z	0	1	2	3	4	5	6	7	8	9
4.0	0.00003									
3.5	0.00023									
3.0	0.00135									
2.9	0.0019	0.0018	0.0018	0.0017	0.0016	0.0016	0.0015	0.0015	0.0014	0.0014
2.8	0.0026	0.0025	0.0024	0.0023	0.0023	0.0022	0.0021	0.0021	0.0020	0.0019
2.7	0.0035	0.0034	0.0033	0.0032	0.0031	0.0030	0.0029	0.0028	0.0027	0.0026
2.6	0.0047	0.0045	0.0044	0.0043	0.0041	0.0040	0.0039	0.0038	0.0037	0.0036
2.5	0.0062	0.0060	0.0059	0.0057	0.0055	0.0054	0.0052	0.0051	0.0049	0.0048
2.4	0.0082	0.0080	0.0078	0.0075	0.0073	0.0071	0.0069	0.0068	0.0066	0.0064
2.3	0.0107	0.0104	0.0102	0.0099	0.0096	0.0094	0.0091	0.0089	0.0087	0.0084
2.2	0.0139	0.0136	0.0132	0.0129	0.0125	0.0122	0.0119	0.0116	0.0113	0.0110
2.1	0.0179	0.0174	0.0170	0.0166	0.0162	0.0158	0.0154	0.0150	0.0146	0.0143
2.0	0.0228	0.0222	0.0217	0.0212	0.0207	0.0202	0.0197	0.0192	0.0188	0.0183
1.9	0.0287	0.0281	0.0274	0.0268	0.0262	0.0256	0.0250	0.0244	0.0239	0.0233
1.8	0.0359	0.0351	0.0344	0.0336	0.0329	0.0322	0.0314	0.0307	0.0301	0.0294
1.7	0.0446	0.0436	0.0427	0.0418	0.0409	0.0401	0.0392	0.0384	0.0375	0.0367
1.6	0.0548	0.0537	0.0526	0.0516	0.0505	0.0495	0.0485	0.0475	0.0465	0.0455
1.5	0.0668	0.0655	0.0643	0.0630	0.0618	0.0606	0.0594	0.0582	0.0571	0.0559
1.4	0.0808	0.0793	0.0778	0.0764	0.0749	0.0735	0.0721	0.0708	0.0694	0.0681
1.3	0.0968	0.0951	0.0934	0.0918	0.0901	0.0885	0.0869	0.0853	0.0838	0.0823
1.2	0.1151	0.1131	0.1112	0.1093	0.1075	0.1056	0.1038	0.1020	0.1003	0.0985
1.1	0.1357	0.1335	0.1314	0.1292	0.1271	0.1251	0.1230	0.1210	0.1190	0.1170
1.0	0.1587	0.1562	0.1539	0.1515	0.1492	0.1469	0.1446	0.1423	0.1401	0.1379
0.9	0.1841	0.1814	0.1788	0.1762	0.1736	0.1711	0.1685	0.1660	0.1635	0.1611
0.8	0.2119	0.2090	0.2061	0.2033	0.2005	0.1977	0.1949	0.1922	0.1894	0.1867
0.7	0.2420	0.2389	0.2358	0.2327	0.2297	0.2266	0.2236	0.2206	0.2177	0.2148
0.6	0.2743	0.2709	0.2676	0.2643	0.2611	0.2578	0.2546	0.2514	0.2483	0.2451
0.5	0.3085	0.3050	0.3015	0.2981	0.2946	0.2912	0.2877	0.2843	0.2810	0.2776
0.4	0.3446	0.3409	0.3372	0.3336	0.3300	0.3264	0.3228	0.3192	0.3156	0.3121
0.3	0.3821	0.3783	0.3745	0.3707	0.3669	0.3632	0.3594	0.3557	0.3520	0.3483
0.2	0.4207	0.4168	0.4129	0.4090	0.4052	0.4013	0.3974	0.3936	0.3897	0.3859
0.1	0.4602	0.4562	0.4522	0.4483	0.4443	0.4404	0.4364	0.4325	0.4286	0.4247
0.0	0.5000	0.4960	0.4920	0.4880	0.4840	0.4801	0.4761	0.4721	0.4681	0.4641

Appendix C. Control chart constants and formulae

$\overline{X}$ and R charts and $\overline{X}$ and s charts*

Subgroup size	$\overline{X}$ and R charts				$\overline{X}$ and s charts			
	Chart for averages ($\overline{X}$)	Chart for ranges (R)			Chart for averages ($\overline{X}$)	Chart for standard deviations (s)		
	Factors for control limits	Divisors for estimate of standard deviation	Factors for control limits		Factors for control limits	Divisors for estimate of standard deviation	Factors for control limits	
n	A_2	d_2	D_3	D_4	A_3	c_4	B_3	B_4
2	1.880	1.128	—	3.267	2.659	0.7979	—	3.267
3	1.023	1.693	—	2.574	1.954	0.8862	—	2.568
4	0.729	2.059	—	2.282	1.628	0.9213	—	2.266
5	0.577	2.326	—	2.114	1.427	0.9400	—	2.089
6	0.483	2.534	—	2.004	1.287	0.9515	0.030	1.970
7	0.419	2.704	0.076	1.924	1.182	0.9594	0.118	1.882
8	0.373	2.847	0.136	1.864	1.099	0.9650	0.185	1.815
9	0.337	2.970	0.184	1.816	1.032	0.9693	0.239	1.761
10	0.308	3.078	0.223	1.777	0.975	0.9727	0.284	1.716
11	0.285	3.173	0.256	1.744	0.927	0.9754	0.321	1.679
12	0.266	3.258	0.283	1.717	0.886	0.9776	0.354	1.646
13	0.249	3.336	0.307	1.693	0.850	0.9794	0.382	1.618
14	0.235	3.407	0.328	1.672	0.817	0.9810	0.406	1.594
15	0.223	3.472	0.347	1.653	0.789	0.9823	0.428	1.572
16	0.212	3.532	0.363	1.637	0.763	0.9835	0.448	1.552
17	0.203	3.588	0.378	1.622	0.739	0.9845	0.466	1.534
18	0.194	3.640	0.391	1.608	0.718	0.9854	0.482	1.518
19	0.187	3.689	0.403	1.597	0.698	0.9862	0.497	1.503
20	0.180	3.735	0.415	1.585	0.680	0.9869	0.510	1.490
21	0.173	3.778	0.425	1.575	0.663	0.9876	0.523	1.477
22	0.167	3.819	0.434	1.566	0.647	0.9882	0.534	1.466
23	0.162	3.858	0.443	1.557	0.633	0.9887	0.545	1.455
24	0.157	3.895	0.451	1.548	0.619	0.9892	0.555	1.445
25	0.153	3.931	0.459	1.541	0.606	0.9896	0.565	1.435

$$UCL_{\overline{X}} = \overline{\overline{X}} + A_2\overline{R}$$
$$LCL_{\overline{X}} = \overline{\overline{X}} - A_2\overline{R}$$
$$UCL_R = D_4\overline{R}$$
$$LCL_R = D_3\overline{R}$$

$$\hat{\sigma} = \frac{\overline{R}}{d_2}$$

$$UCL_{\overline{X}} = \overline{\overline{X}} + A_3\overline{s}$$
$$LCL_{\overline{X}} = \overline{\overline{X}} - A_3\overline{s}$$
$$UCL_s = B_4\overline{s}$$
$$LCL_s = B_3\overline{s}$$

$$\hat{\sigma} = \frac{\overline{s}}{c_4}$$

*Reprinted, with permission, from STP 15D – *Manual on the Presentation of Data and Control Chart Analysis.* Copyright American Society for Testing and Materials.

347

Median charts*

Charts for medians (X̃)		Charts for ranges (R)		
Subgroup size	Factors for control limits	Divisors for estimate of standard deviation	Factors for control limits	
n	Ã₂	d₂	D₃	D₄
2	1.880	1.128	—	3.267
3	1.187	1.693	—	2.574
4	0.796	2.059	—	2.282
5	0.691	2.326	—	2.114
6	0.548	2.534	—	2.004
7	0.508	2.704	0.076	1.924
8	0.433	2.847	0.136	1.864
9	0.412	2.970	0.184	1.816
10	0.362	3.078	0.223	1.777

$$UCL_{\tilde{x}} = \overline{\overline{X}} + \tilde{A}_2\overline{R}$$

$$LCL_{\tilde{x}} = \overline{\overline{X}} - \tilde{A}_2\overline{R}$$

$$\hat{\sigma} = \frac{\overline{R}}{d_2}$$

$$UCL_R = D_4\overline{R}$$

$$LCL_R = D_3\overline{R}$$

* Reprinted, with permission, from STP 15D – *Manual on the Presentation of Data and Control Chart Analysis*. Copyright American Society for Testing and Materials.

Ã₂ factors derived from ASTM ATP 15D – *Data and Efficiency Tables*. Contained in Dixon and Massey (1969). *Introduction to Statistical Analysis*, 3rd edn. McGraw-Hill, New York. Reprinted with permission.

Appendix D. Control chart constants (British standards)

Control chart limits for sample mean ($\bar{X}$).

To obtain limits multiply $\bar{w}$ by the appropriate value of $A'_{0.025}$ and $A'_{0.001}$ then add to, and subtract from, $\bar{X}$.

Sample size n	For inner limits $A'_{0.025}$	For outer limits $A'_{0.001}$
2	1.23	1.94
3	0.67	1.05
4	0.48	0.75
5	0.38	0.59
6	0.32	0.50

The above extract from BS 2564:1955 is reproduced by permission of BSI. Copies of the complete standard can be obtained from them at Linford Wood, Milton Keynes, Bucks MK14 6LE.

Appendix E. Derivation of US and British control chart constants

Derivation of control chart constants based on US standards (ANSI)

For a distribution of individual readings, lines corresponding to 3 standard deviations (i.e. a probability level of 0.0013) away from the ground mean $\overline{X}$ are given by

$$\overline{\overline{X}} \pm 3\sigma$$

When considering samples of size n rather than individual readings, the standard deviation σ is replaced by the standard error $\sigma / \sqrt{n}$. But σ can be replaced by the estimate $\overline{R}/d_2$, where d_2 depends on sample size. Therefore

$$\text{Control limits are at} \quad \overline{\overline{X}} \pm \frac{3}{\sqrt{n}} \cdot \frac{\overline{R}}{d_2}$$

$$= \overline{\overline{X}} \pm \frac{3}{\sqrt{n}\ d_2} \cdot \overline{R}$$

The expression $\dfrac{3}{\sqrt{n}\ d_2}$ is denoted as A_2 and hence

$$UCL_{\overline{x}} = \overline{\overline{X}} + A_2\overline{R}$$

$$LCL_{\overline{x}} = \overline{\overline{X}} - A_2\overline{R}$$

If n = 5, for example, then

$$A_2 = \frac{3}{\sqrt{5}\ d_2}$$

$$= \frac{3}{\sqrt{5} \times 2.326}$$

$$= 0.577$$

Tables of A_2 for different n values are tabulated in Appendix C.

Derivation of British control chart constants based on British Standards (BS2564)

For a distribution of individual readings, lines corresponding to 3.09 standard deviations (i.e. a probability level of 0.001, or exactly 1 in 1000) away from the grand mean are given by

$$\overline{\overline{X}} \pm 3.09\sigma$$

351

When considering samples of size n rather than individual readings, the standard deviation σ is replaced by the standard error $\sigma / \sqrt{n}$. Therefore control limits, which are called action limits, are at

$$\overline{\overline{X}} \pm 3.09 \frac{\sigma}{\sqrt{n}}$$

But σ can be replaced by the estimate $\overline{R} / d_2$, where d_2 depends on sample size. Therefore

Action limits are at $\quad \overline{\overline{X}} \pm \frac{3.09}{\sqrt{n}} \cdot \frac{\overline{R}}{d_2}$

$$= \overline{\overline{X}} \pm \frac{3.09}{\sqrt{n} \, d_2} \cdot \overline{R}$$

The expression $\frac{3.09}{\sqrt{n} \, d_2}$ is denoted by $A'_{0.001}$ and hence

Upper action limit (UAL) $= \overline{\overline{X}} + A'_{0.001} \overline{R}$

Lower action limit (LAL) $= \overline{\overline{X}} - A'_{0.001} \overline{R}$

If n = 5, for example, then

$$A'_{0.001} = \frac{3.09}{\sqrt{5} \, d_2}$$

$$= \frac{3.09}{\sqrt{5} \times 2.326}$$

$$= 0.594$$

Tables of $A'_{0.001}$ for different n values are tabulated in Appendix D.

In addition to the action limits, the British system uses warning limits set at a 1 in 40 probability level, i.e. at

$$\overline{\overline{X}} \pm 1.96 \frac{\sigma}{\sqrt{n}}$$

With σ replaced by $\overline{R} / d_2$ this becomes

$$\overline{\overline{X}} \pm \frac{1.96}{\sqrt{n}} \cdot \frac{\overline{R}}{d_2}$$

$$= \overline{\overline{X}} \pm \frac{1.96}{\sqrt{n} \, d_2} \cdot \overline{R}.$$

The expression $\frac{1.96}{\sqrt{n} \, d_2}$ is denoted by $A'_{0.025}$ and hence

Upper warning limit (UWL) $= \overline{\overline{X}} + A'_{0.025} \overline{R}$

Lower warning limit (LWL) $= \overline{\overline{X}} - A'_{0.025} \overline{R}$

If n = 5, for example, then

$$A'_{0.025} = \frac{1.96}{\sqrt{5} \, d_2}$$

$$= \frac{1.96}{\sqrt{5} \times 2.326}$$

$$= 0.377$$

Tables of $A'_{0.025}$ for different n values are also tabulated in Appendix D.

Appendix F. Derivation of constants for individual/moving R chart

Individual chart

$$UCL_X = \bar{X} + 3s$$

$$LCL_X = \bar{X} - 3s$$

where s is obtained using the relationship $\bar{R} / d_2$.

With a moving range of 2, as in the PPG chart, d_2 is 1.128. Thus

$$3s = \frac{3\bar{R}}{1.128}$$

$$= 2.66\bar{R}$$

Therefore

$$UCL_X = \bar{X} + 2.66\bar{R}$$

$$LCL_X = \bar{X} - 2.66\bar{R}$$

Moving R chart

$$UCL_R = D_4\bar{R}$$

D_4, for n = 2, is $3.267\bar{R}$.

Therefore

$$UCL_R = 3.267\bar{R}$$

Appendix G. Symbols and definitions (in order of appearance)

f	Frequency: the number of observed values in a class interval of a frequency distribution
X	Observed value of variable or attribute
$X_1, X_2, \ldots$	Specific values of X
$\bar{X}$	X bar: arithmetic mean of a sample
n	Number of observed values in a sample
Σ	Sigma: sum of
$\bar{\bar{X}}$	X double bar: mean of a series of $\bar{X}$ values. Sometimes called the grand mean
$\tilde{X}$	Curly X: median
R (or w)	Range of a number of readings
s	Standard deviation of a sample
μ	Arithmetic mean of a population
σ	Standard deviation of a population
$\wedge$	Best estimate; e.g. $\hat{\sigma} = s = \bar{R}/d_2$
$\bar{R}$	Mean of a series of R values
C_p	Process capability index. A measure of the variability of the process in relation to the specification
Z	Units of standard deviation away from the mean of a normal distribution
P_z	Area in the tail of a normal curve
k	Number of samples
UCL	Upper control limit
LCL	Lower control limit
A_2	Constant, dependent on sample size, used in determining control limits for $\bar{X}$ using $\bar{R}$
D_3/D_4	Constants, dependent on sample size, used in determining control limits for R
ϕ	Nominal (of a process)
d_2	Constant, dependent on sample size, used to determine $\hat{\sigma}$ from $\bar{R}/d_2$
C_{pk}	Process capability index. A measure of the variability of the process, and its setting, in relation to the specification
$\bar{\tilde{X}}$	Curly X bar: mean of a series of $\tilde{X}$ readings
$\tilde{A}_2$	Constant, dependent on sample size, used in determining control limits for $\tilde{X}$
$\bar{s}$	Mean of a series of s values
B_3/B_4	Constants, dependent on sample size, used in determining control limits for s
A_3	Constant, dependent on sample size, used in determining control limits for $\bar{X}$, using $\bar{s}$
c_4	Constant, varying with sample size, used to determine $\hat{\sigma}$ from $\bar{s}/c_4$
p	Proportion of non-conforming units in a sample
np	Number of non-conforming units in a sample of size n
c	Number of nonconformities in a sample

u	Number of nonconformities in a sample of size n
$\bar{p}$	Mean proportion of non-conforming units in a series of samples of varying size
$\overline{np}$	Mean number of non-conforming units in a series of samples of constant size n
$\bar{u}$	Mean number of nonconformities in a series of samples of varying size
$\bar{c}$	Mean number of nonconformities in a series of samples of constant size n
T	Target value, used in Cusum analysis
ARL	Average run length: the number of points required, on average, to detect a signal that the process has changed
K_1/K_2	Reference values used in numerically based Cusum procedures
C_m	Machine capability index. A measure of the variability of the machine in relation to the specification
C_{mk}	Machine capability index. A measure of the variability of the machine, and its setting, in relation to the specification

Appendix H. Various charts

$(\bar{X}, R)$ chart

$(\bar{X}$ moving R) chart

Attribute chart

Multiple characteristics chart

Cusum chart

Machine capability studies chart

Process log sheet

Century Oils chart

PPG chart

Process Control Chart – Variables ($\overline{X}$/R)

MEAN $\overline{X}$

RANGE R

| | | | | | | Shift | Time | Date | X_1 | X_2 | X_3 | X_4 | X_5 | $\sum X$ | $\overline{X}$ | R |

Week no.

Sampling frequency

Department

Unit of measurement

Division

Zero

C_p

C_{pk}

Measuring equipment

Machine no./type

Operation

Operator

Specification

LCL

Characteristic

UCL

Part no./Description

$\overline{\overline{X}} =$

UCL $=$

$\overline{R} =$

MEAN $\overline{X}$

RANGE R

Shift																
Time																
Date																
X_1																
X_2																
X_3																
X_4																
X_5																
$\sum X$																
$\overline{X}$																
R																

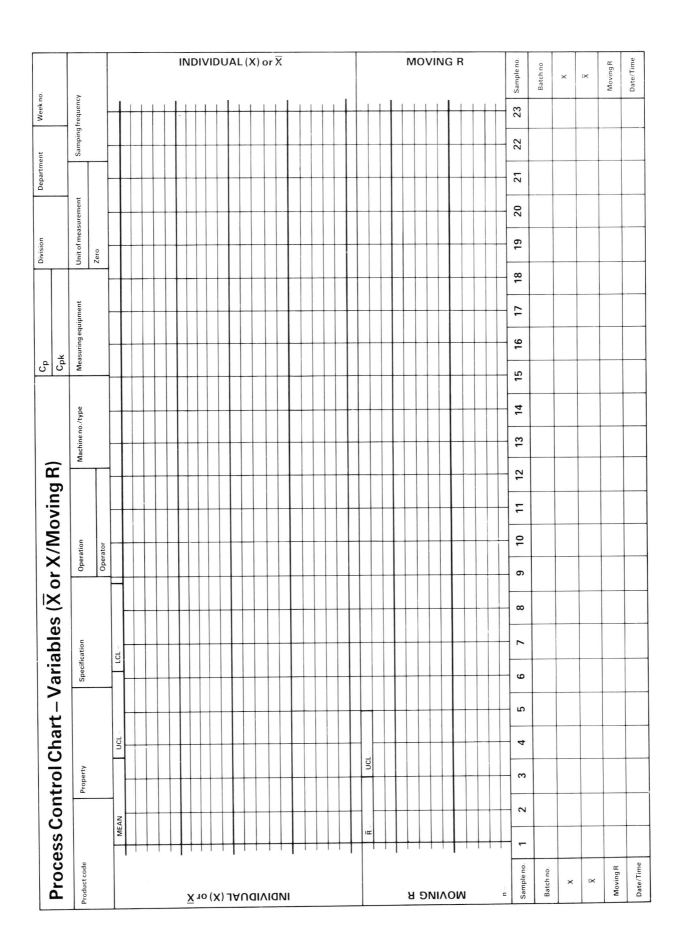

Process Control Chart – Attributes

Part no./Description	Operation	Machine no./Type	Sampling frequency	Target sample size	Mean =	UCL =	LCL =	Week no.	Division	Department

p ☐ c ☐
np ☐ u ☐

Number (np, c)/Proportion (p, u)

Number (np, c)/Proportion (p, u)

Sample size	Number	Proportion	Date/Time

Sample size	Number	Proportion	Date/Time

Process Control Chart – Multiple Characteristics

Part no./Description	Operation	Machine no./Type	Sampling frequency	Target sample size	Division	Department	Week no.

☐ p ☐ c
☐ np ☐ u

Mean =   UCL =   LCL =   %

f   %

Defective/Defect frequency

Characteristic

No.

Total Defective/Defect

Sample size

Proportion

Date/Time

Process Control Chart – Cusum

Division	Department	

Characteristic	Target T =	

CUSUM

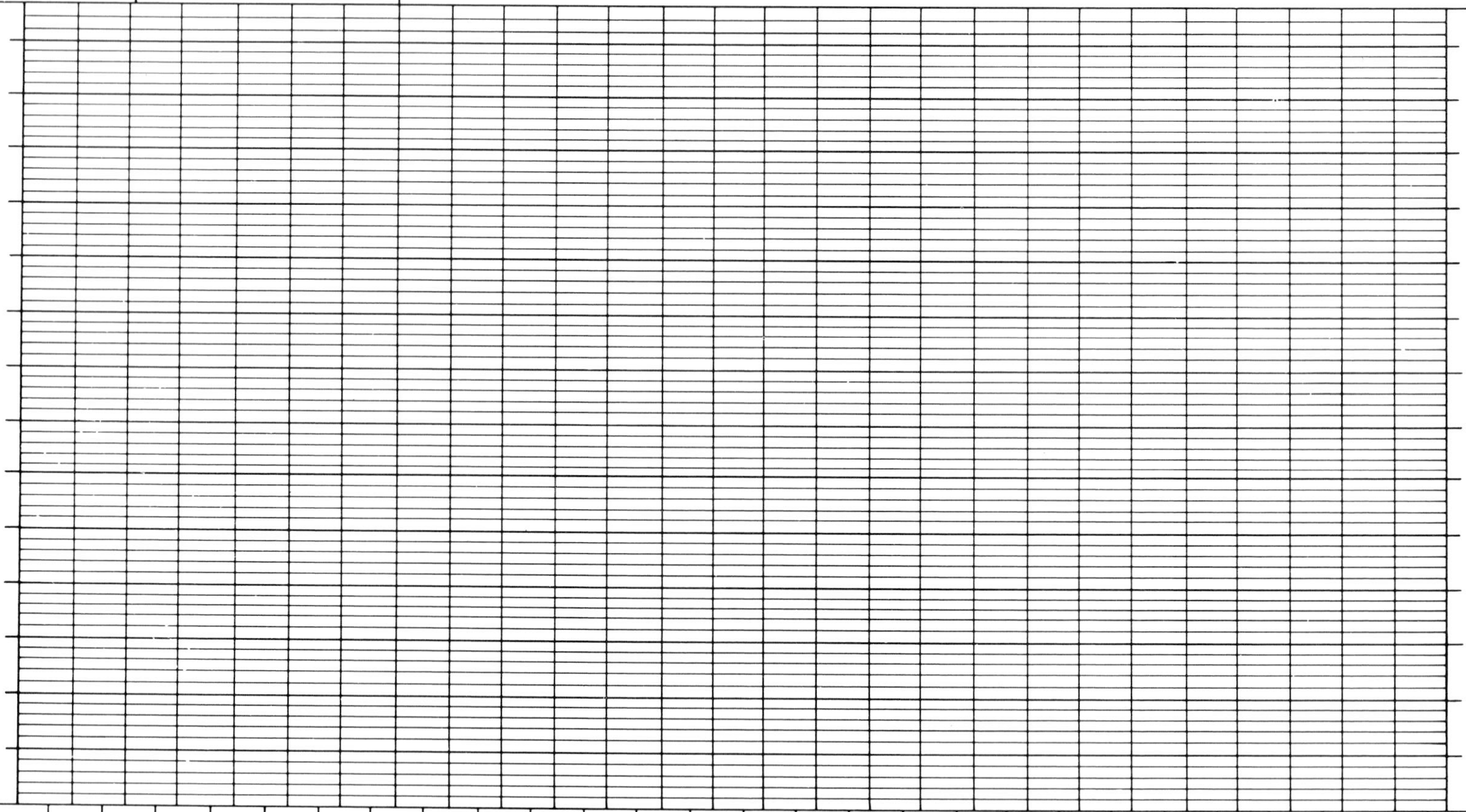

Sample no.	1	2	3	4	5	6	7	8	9	10	11	12	13	14	15	16	17	18	19	20	21	22	23	24	25	26	27	28	Sample no.
Ref. no.																													Ref. no.
Reading X																													Reading X
$X - T$																													$X - T$
$\Sigma(X-T)$																													$\Sigma(X-T)$

Machine Capability Chart

Part no./Description	Characteristic	Specification	Operation	Machine no./type	Measuring equipment	Division	Department	Ref.no.

Unit of measurement

Zero

VALUE

TALLY MARKS

f Σf $\Sigma f\%$

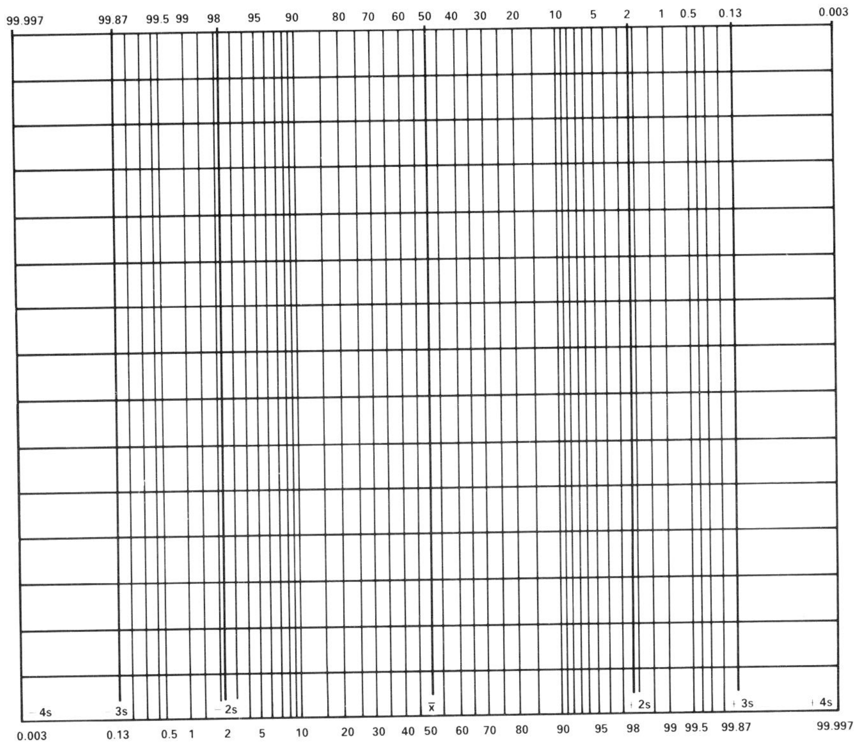

0.003 0.13 0.5 1 2 5 10 20 30 40 50 60 70 80 90 95 98 99 99.5 99.87 99.997

4s 3s 2s $\bar{x}$ 2s 3s 4s

READINGS

1	6	11	16	21	26	31	36	41	46
2	7	12	17	22	27	32	37	42	47
3	8	13	18	23	28	33	38	43	48
4	9	14	19	24	29	34	39	44	49
5	10	15	20	25	30	35	40	45	50

SPECIFIED TOLERANCE

SPECIFIED TARGET

ESTIMATED OUT OF TOLERANCE

TOP ____%

BOTTOM ____%

ESTIMATED CAPABILITY (8s)

ESTIMATED MEAN

ESTIMATED STANDARD DEVIATION (s)

$C_m =$ ____

$C_{mk} =$ ____

Name

Signature

Date

PROCESS LOG SHEET

DATE	TIME	COMMENTS

DATE	TIME	COMMENTS

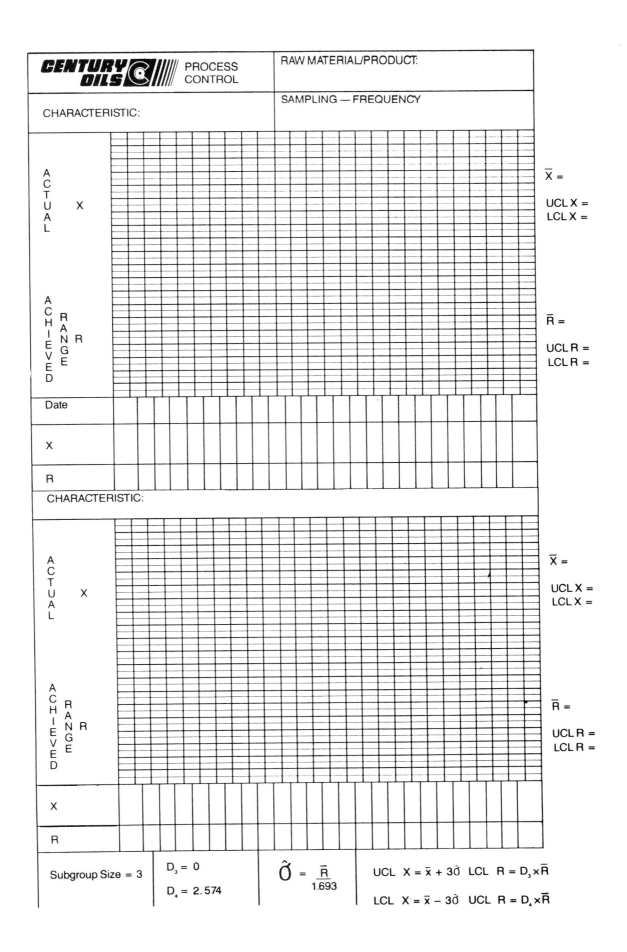

DEPARTMENT:

PRODUCT CODE:

PROPERTY:

SPECIFICATION:

STATISTICAL PROCESS – QUALITY CONTROL CHART
Ladywood (UK)

FOR VARIABLES
INDIVIDUAL / MOVING RANGE

UCL $_{MR}$ = = 3·268 x $\overline{MR}$

LCL x = = $\overline{x}$ – (2·66 x $\overline{MR}$)

UCL x = = $\overline{x}$ + (2·66 x $\overline{MR}$)

RECALCULATION OF CONTROL LIMITS:

$\overline{MR} = \dfrac{TOTAL\ (MR)}{No.\ RESULTS} =$ _____

$\overline{\overline{x}} = \dfrac{TOTAL\ (X)}{No.\ RESULTS} =$ _____

INDIVIDUAL (X)

NOTE

MOVING RANGE (MR)

	1	2	3	4	5	6	7	8	9	10	11	12	13	14	15	16	17	18	19	20
DATE / TIME																				
BATCH No.																				
X																				
1																				
2																				
MR																				

SPCC/1/10/87

Appendix I.   Further information

Books

Anon. (1976). *QC Circles: Applications, Tools and Theory.* American Society for Quality Control.

Annon. (1980). *Code of practical guidance for packers and importers. Weights and Measures Act 1979. HMSO, London.*

Annon. (1980). *Manual of practical guidance for Inspectors. Weights and Measures Act 1979. HMSO, London.*

Anon. (1982). *Japan Quality Control Circles.* Asian Productivity Association.

Anon. (1984). *The Pursuit of Product Excellence: A Guide to SPC.* GKN Automotive.

Anon (1985). *Statistical Process Control: Course Notes.* Ford Motor Company, Brentwood, Essex.

Anon. (1985). *Statistical Process Control Experiments.* Ford Motor Company, Brentwood, Essex.

Anon. (1985). *Statistical Process Control Methods.* Chrysler Corporation, USA.

Anon. (1985). *Continuing Process Control and Process Capability Improvement.* Statistical Methods Office, Ford Motor Company, USA.

Anon. (1986). *Statistical Process Control: Instruction Guide.* Ford Motor Company, Brentwood, Essex.

Anon. (1986). *Guidelines to Statistical Process Control.* The Society of Motor Manufacturers and Traders, London.

Amsden, R.T., Butler, H.E. and Amsden, D.M. (1986). *SPC Simplified: Practical Steps to Quality.* Unipub/Kraus International Publications, White Plains, New York.

Crosby, P.B. (1979). *Quality is Free: The Art of Making Quality Certain.* McGraw-Hill.

Crosby, P.B. (1985). *Quality Without Tears: The Art of Hassle-Free Management.* McGraw-Hill.

Cullen, J. and Hollingum, J. (1988). *Implementing Total Quality.* IFS Publications, Kempston, Bedford.

Deming, W. Edwards (1989). *Out of the crisis*, MIT, Centre for Advanced Engineering Study, Cambridge, Mass.

Grant, E.L. and Leavenworth, R.S. (1980). *Statistical Quality Control.* McGraw-Hill.

Feigenbaum, A.V. (1983). *Total Quality Control.* McGraw-Hill.

Guaspari, J. (1985). *I Know It When I See It.* American Management Association.

Harrington, H.H. (1987). *The Improvement Process: How America's Leading Companies Improve Quality.* McGraw-Hill.

Ishikawa, K. (1982). *Guide to Quality Control.* Asian Productivity Association.

Juran, J.M. and Gryna, F. (1980). *Quality Planning and Analysis.* McGraw-Hill.

Juran, J.M., Gryna, F. and Bingham, F. (1979). *Quality Control Handbook*, 3rd edn. McGraw-Hill.

Kilian, Cecilia S. (1988). *The World of W. Edwards Deming*, Ceepress Books, The George Washington University, Washington DC 20152.

Mann, N.R. (1985). *The Keys to Excellence: The Story of the Deming Philosophy.* Prestwick Books, Santa Monica, California.

Nixon, F. (1971). *Managing to Achieve Quality and Reliability.* McGraw-Hill.

Oakland, J.S. (1986). *Statistical Process Control.* Heinemann.

Ott, E.R. (1975). *Process Quality Control: Troubleshooting and Interpretation of Data.* McGraw-Hill.

Ouchi, W.G. (1981). *Theory Z.* Addison Wesley.

Peters, T.J. and Austin, N.K. (1985). *A Passion for Excellence: The Leadership Difference.* Random House, New York.

Price, F. (1984). *Right First Time.* Gower Press, Aldershot, Hants.

Scherkenbach, W.W. (1986). *The Deming Route to Quality and Productivity: Roadmaps and Roadblocks.* Cee Press Books, Washington, DC.
Shewhart, W.A. (1931). *Economic Control of Quality of Manufactured Product.* Van Nostrand.
Walton, Mary (1989). *The Deming Management Method*, Mercury Books, London.

Articles and others

Anon. (1988). *SPC Applications to Short Production Runs.* Ford Motor Company, Brentwood, Essex.
Clark, I. (1988). *Implementing Statistical Process Control: Guidelines and Pitfalls.* Training for Quality Unit, Bristol Polytechnic.
Dale, B.G. and Shaw, P. (1988). A study of the use of Statistical Process Control in the automotive-related supplier community. Department of Management Sciences, University of Manchester Institute of Science and Technology. Occasional Paper No. 8801.
Grubbs, F.E. (1983). An optimum procedure for setting machines or adjusting processes. *Journal of Quality Technology*, Vol. 15(4), October.
Lascelles, D. and Dale, B. (1985). Quality training and education in the UK: a state of the art survey. *Quality Assurance*, vol. 11/12, pp. 39–42.
Lockyer, K.G., Oakland, J.S., Duprey, C.H. and Followell, R.F. (1984). Research into the use of statistical methods of Quality Control in selected sectors of UK manufacturing industry. University of Bradford Management Centre.

Various editions of *Quality Progress, Journal of the American Society for Quality Control* and *Annual Transactions of the American Society for Quality Control.*

British Standards relating to SPC

BS 2564   Control chart technique when manufacturing to a specification, with special reference to articles machined to dimensional tolerances

BS 2846   Guide to statistical interpretation of data Part 4/Part 7

BS 5532   Statistical terminology

BS 5700   Guide to process control using quality control chart methods and Cusum techniques

BS 5701   Guide to number-defective charts for quality control

BS 5702   Guide to quality control charts for measured variables (In preparation)

BS 5703   Guide to data analysis and quality control using Cusum techniques

　　　　　Part 1:   Introduction to Cusum charting

　　　　　Part 2:   Decision rules and statistical tests for Cusum charts and tabulations

　　　　　Part 3:   Cusum methods for process/quality control by measurement

　　　　　Part 4:   Cusum for counted/attribute data

BS 6143   Guide to the determination and use of quality-related costs